Abel-Gontcharoff Pseudopolynomials and Stochastic Applications

Series Editor
Nikolaos Limnios

Abel-Gontcharoff Pseudopolynomials and Stochastic Applications

Philippe Picard
Claude Lefèvre

iSTE

WILEY

First published 2025 in Great Britain and the United States by ISTE Ltd and John Wiley & Sons, Inc.

ISTE Ltd
27-37 St George's Road
London SW19 4EU
UK

www.iste.co.uk

John Wiley & Sons, Inc.
111 River Street
Hoboken, NJ 07030
USA

www.wiley.com

Library of Congress Control Number: 2025941859

British Library Cataloguing-in-Publication Data
A CIP record for this book is available from the British Library
ISBN 978-1-83669-072-6

The manufacturer's authorized representative according to the EU General Product Safety Regulation is Wiley-VCH GmbH, Boschstr. 12, 69469 Weinheim, Germany, e-mail: Product_Safety@wiley.com.

Contents

Preface

Our objective in this brief preface is twofold. On the one hand, we explain how, first in epidemic theory and then for first-passage problems, we came to use the polynomials initially introduced by Gontcharoff, here called Abel–Gontcharoff (A–G) polynomials. On the other hand, we explain how we were then led to generalize this mathematical tool in the form of so-called A–G pseudopolynomials in order to be able to deal with similar but more complex problems in applied probability.

Abel–Gontcharoff polynomials: historically, Markovian stochastic theory of epidemics, cluttered with partial differential equations, struggled to obtain explicit results. But, starting in the 1980s, things changed with the use of martingale theory, a theory that until then had remained fairly confined to areas considered theoretical and was therefore largely under-exploited in the field of applications. Its application to epidemic models of the SIR (susceptible, infected, removed) type has indeed led, unexpectedly but very naturally, to perfectly explicit results, provided that A–G polynomials are used to express them. These constitute a relatively little-known generalization of Abel polynomials and, like the latter, they lend themselves to the writing of Abelian expansions, which are remarkable extensions of Taylor expansions and are of great theoretical and practical interest. Initially, the A–G polynomials appeared in the form of multiple integrals, the Abel polynomials corresponding to the very simple special case where this multiple integral highlights all the zeros of the polynomials. Expressions of these A–G polynomials were also known in the cumbersome form of complicated determinants. The extension to the complex field was immediate and, in fact, already implicit in Abel's work.

Moreover, it was of course recognized in epidemic theory that multi-population models are useful for taking into account the heterogeneity of individuals. As the martingale approach proved to be perfectly suited to their study, it immediately became clear (still in the 1980s) that a multidimensional extension of the A–G polynomials was both possible and necessary. Necessary if we wanted to have explicit results for multi-population models, but also certainly possible since the martingale tool accepted without difficulty the passage from dimension 1 to dimensions $2, 3, \ldots$. To our knowledge, no extension of this type had yet been carried out, either because its usefulness had not been seen (no foreseeable application), or because the current symbolism (multiple integrals or determinants) was an obstacle to this extension, or even more fundamentally because the algebraic study of the univariate A–G polynomials remained little developed. As is often the case, it is the potential applications that will be the source of unexpected but welcome theoretical developments.

The conjunction of the possible and the necessary suggested an easy task of generalization, and this is what happened since before the end of the 1980s, the multidimensional generalization of A–G polynomials had become operational, the very useful concept of Abelian expansion being naturally preserved and extended. To achieve this, it was sufficient to sacrifice the (unnecessarily cumbersome) integral or determinant representations and instead focus on the basic algebraic structures involved. At the same time, having now awakened our minds, it was easy for us to flush out A–G polynomials hidden in many other chapters in the field of applied probability, which we only had to recognize and bring to light. The most notable success in this treasure hunt was provided by the class of first-crossing probabilities for the trajectories of counting processes (a Poisson or birth process, for example) with an increasing lower boundary. Apart from a few simple factors or changes of variables, these probabilities correspond exactly to the A–G polynomials, which implies that these polynomials can be included in the concert of tools usable in applied probability. We can go further with this question of crossing for a counting process trajectory by replacing the increasing lower boundary with the trajectory of another counting process, independent of the first. This led us to introduce polynomials A–G whose parameters are no longer fixed but are stochastic in nature. All this, of course, opens up a rich mine of new problems to be posed and solved. Some of these will be addressed here in a broader framework, but there is naturally still much to be done in that direction.

Extension to A–G pseudopolynomials: only the first step is difficult! Once the possibility and usefulness of generalizing the A–G polynomials were recognized, the paths to explore presented themselves, with new problems, certainly, but also each time clues indicating how to overcome these obstacles. For example, regarding the first cross-question mentioned above, it is natural to substitute a compound Poisson trajectory for that of a standard Poisson process, or a trajectory of a multiple birth process for that of a standard birth process excluding, in particular, twins. Such generalizations, even if they appear minimal, no longer allow us to express the probabilities sought in terms of simple A–G polynomials.

The solution is, however, within reach. In fact, if the family of fundamental monomials $x^n/n!$ (Appell's family generated by the usual differential operator D) are replaced by a suitable family of polynomials $e_n(x)$ (Appell's family generated by a different suitable operator Δ), a careful examination shows that we can replace the A–G polynomials (related to the $x^n/n!$) by other polynomials (this time related to the $e_n(x)$) enjoying similar algebraic properties and mainly allowing us to obtain Abelian expansions again. These new polynomials can therefore be considered as generalized polynomials A–G.

However, we can go further. Taking $e_n(x)$ that are polynomials is a completely useless constraint because what really matters is the linear independence of these $e_n(x)$ and not their polynomial character. We thus arrive at the somewhat enigmatic expression of A–G pseudopolynomials, which has been adopted in the present monograph and at the mathematical methodology that it covers. By this, we mean first giving ourselves a family of linearly independent functions $e_n(x)$, then defining on it an operator Δ satisfying the conditions $\Delta e_{n+1} = e_n$, finally passing from the $e_n(x)$ to the A–G pseudopolynomials in a similar way to the passage from $x^n/n!$ to the A–G polynomials. It will be shown that these A–G pseudopolynomials satisfy many elegant and efficient properties, and can be easily evaluated numerically by recursion. Thanks to this mathematically powerful tool, we will be able to master a wide variety of problems encountered in applied probability. This is the subject of the current monograph. Note that our interest in this application area is based on our own research, but readers with different expertise will certainly develop other applications.

Acknowledgements

We warmly thank our families for their encouragement (and patience) during our research over all these years. We are also grateful to ISTE Group and Professor Nikolaos Limnios, editor of the "Mathematics and Statistics" series, for their welcome and support.

August 2025

1

Historical Abel-Gontcharoff Polynomials

1.1. Abel identity

In 1826, Niels Henrik Abel (1802–1829)[1] published, in the first volume of Crelle's Journal, the following identity:

$$(x+\alpha)^k = \sum_{n=0}^{k} \binom{k}{n} \alpha(\alpha-\beta n)^{n-1}(x+\beta n)^{k-n}, \quad k \in \mathbb{N} \ (\equiv \{1,2,3,\ldots\}),$$

[1.1]

where x, α, β are real numbers. A similar formula, less compact but analytically equivalent, is also found, through many other formulas, in a memoir by Augustin-Louis Cauchy (1789–1857)[2].

Note that there is a strange asymmetry between x and α in the right part (r.h.s.) of [1.1], whereas on the contrary its left part (l.h.s.) is perfectly symmetrical. It is obvious that β corresponds only to a parameter but there remains an ambiguity between x and α. Indeed, which one, chosen as a variable, will play the main role, when the other will be reduced to being only a simple parameter? Abel made no comment on this issue and simply presented [1.1] as a natural generalization of the binomial formula.

1 Abel, N.H. (1826). Beweis eines ausdruckes, von welchem die Binomial-formel ein einzelner fall ist. *Journal für die Reine und Angewandte Mathematik*, 1, 159–160.

2 Cauchy, A.-L. (1826). *Exercices de mathématiques*. De Bure Frères, Paris.

The recursive proof given by Abel for his formula does not shed light on its hidden algebraic structure. Nevertheless, some information about [1.1] can be found in a posthumous document which gives the following formula for an infinitely differentiable function ϕ:

$$\phi(x+\alpha) = \sum_{n=0}^{\infty} \frac{\alpha(\alpha-\beta n)^{n-1}}{n!} \phi^{(n)}(x+\beta n). \qquad [1.2]$$

Here, $\phi^{(n)}$ denotes the nth derivative of ϕ. Clearly, [1.2] gives [1.1] when $\phi(u) = u^k/k!$. Not having obtained a correct proof for this remarkable formula, Abel did not publish it and only made public the particular case of the binomial for which he possessed a proof. Dying at a young age, he did not have time to complete the general case [1.2].

Today, it is the algebraic structure of the r.h.s. in [1.2], which holds our attention. In the 19th century – where it was called the "Abel series" – mathematicians were only interested in its possible convergence (or not) and by the equality (or not) between the two members of the formula. These topics were discussed by Halphen (1881)[3] and Jensen (1902)[4]. Even Pólya and Szegő (1972)[5] spent a few pages on these problems in their famous book. All proofs were based on Lagrange expansions that are effective in attacking convergence problems but are not very illuminating for our purpose of understanding.

1.2. Abel polynomials and expansions

The formula [1.2] has a certain similarity with the classical Taylor expansion. The relation is even more meaningful if we first perform an exchange between x and α, which gives:

$$\phi(x+\alpha) = \sum_{n=0}^{\infty} \phi^{(n)}(u_n) \frac{x(x-\beta n)^{n-1}}{n!},$$

3 Halphen, G. (1881). Sur une série d'Abel. *Bulletin de la S.M.F.*, 10, 67–87.

4 Jensen, J. (1902). Sur une identité d'Abel et sur d'autres formules analogues. *Acta Mathematica*, 26, 307–318.

5 Pólya, G. and Szegő, G. (1972). *Problems and Theorems in Analysis I.* Classics in Mathematics, Springer, Berlin, Heidelberg, New York.

after putting

$$u_i = \alpha + \beta i, \quad i \in \mathbb{N}_0 \ (\equiv \{0, 1, 2, \ldots\}). \tag{1.3}$$

Since the r.h.s. is also

$$\sum_{n=0}^{\infty} \phi^{(n)}(u_n) \frac{(x + \alpha - u_0)(x + \alpha - u_n)^{n-1}}{n!},$$

$\phi(x + \alpha)$ can be expanded as:

$$\phi(x + \alpha) = \sum_{n=0}^{\infty} \phi^{(n)}(u_n) \, A_n(x + \alpha | u_n), \tag{1.4}$$

where we used the definition recalled below of the Abel polynomials.

DEFINITION 1.1.– *For $\{u_i, i \in \mathbb{N}_0\}$ of the affine form [1.3], the Abel polynomials are given by:*

$$A_n(x|u_n) = \frac{(x - u_0)(x - u_n)^{n-1}}{n!}, \quad n \in \mathbb{N}_0 \tag{1.5}$$

when $\beta = 0$, they reduce to the monomials $A_n(x|\alpha) \equiv (x - \alpha)^n/n!$.

The expansion [1.4] is written more compactly:

$$\phi(.) = \sum_{n=0}^{\infty} \phi^{(n)}(u_n) \, A_n(. \, | u_n),$$

hence the following accompanying standard definition.

DEFINITION 1.2.– *The Abel expansion of the function $\phi(x)$ is given by:*

$$\sum_{n=0}^{\infty} \phi^{(n)}(u_n) \, A_n(x|u_n). \tag{1.6}$$

1.3. Gontcharoff contribution

For a long time, analysts only paid attention to the Abel series [1.6] and ignored the Abel polynomials that could be found inside these series.

The situation changed when Gontcharoff (1930)[6], (1937)[7], sometimes written Gontcharov or Gončarov, introduced the family of polynomials of degree n in x, denoted $P_n(x|u_0,\dots,u_{n-1})$, which depend on n real parameters $u_0,\dots,u_{n-1}$ and are characterized by the functional property:

$$P_n^{(m)}(u_m|u_0,u_1,\dots,u_{n-1}) = \delta_{m,n}, \quad 0 \leq m \leq n, \tag{1.7}$$

where $P_n^{(m)}$ denotes the mth derivative of P_n and $\delta_{m,n}$, the classic Kronecker symbol. Note that our notation differs slightly from that of Gontcharoff in which the u_i in particular are considered complex.

From [1.7], we see that these polynomials can be expressed in the multiple integral form:

$$P_n(x|u_0,\dots,u_{n-1}) = \int_{u_0}^{x} dx_1 \int_{u_1}^{x_1} dx_2 \dots \int_{u_{n-1}}^{x_{n-1}} dx_n, \tag{1.8}$$

which implies:

$$P_n^{(m)}(x|u_0,\dots,u_{n-1}) = P_{n-m}(x|u_m,\dots,u_{n-1}). \tag{1.9}$$

Alternative equivalent expressions of P_n exist, notably in terms of determinants, but these are not very convenient in practice.

As noticed by Gontcharoff, in the special case $\{u_i = \alpha+\beta i, i \in \mathbb{N}_0\}$, each P_n given by [1.8] has the simple explicit expression:

$$P_n(x|u_0,\dots,u_{n-1}) = \frac{(x-\alpha)(x-\alpha-\beta n)^{n-1}}{n!}, \tag{1.10}$$

6 Gontcharoff, W. (1930). Recherches sur les dérivées successives des fonctions analytiques. Généralisation de la série d'Abel. *Annales Scientifiques de l'E.N.S.*, 47, 1–78.

7 Gontcharoff, W. (1937). *Détermination des fonctions entières par interpolation*. Hermann et Cie, Paris.

that is, referring to [1.5]:

$$P_n(x|u_0,\ldots,u_{n-1}) = A_n(x|\alpha+\beta n).$$

REMARK.– The identity between [1.10] and [1.7] can be checked as follows. Let us write:

$$A_n(x|\alpha+\beta n) = \frac{(x-\alpha-\beta n)^n}{n!} + \beta\frac{(x-\alpha-\beta n)^{n-1}}{(n-1)!},$$

so that by differentiating m ($\leq n$) times:

$$A_n^{(m)}(x|\alpha+\beta n) = \frac{(x-\alpha-\beta n)^{n-m}}{(n-m)!} + \beta\frac{(x-\alpha-\beta n)^{n-m-1}}{(n-m-1)!}.$$

For $x = \alpha + \beta m$, this gives:

$$A_n^{(m)}(\alpha+\beta m|\alpha+\beta n) = \frac{[(m-n)\beta]^{n-m}}{(n-m)!} + \beta(n-m)\frac{[(m-n)\beta]^{n-m-1}}{(n-m)!},$$

which reduces to $\delta_{m,n}$ as desired.

Because of this particular case, we choose to call Abel–Gontcharoff (A–G) polynomials, the polynomials $P_n(x|u_0,\ldots,u_{n-1})$ defined by [1.7] with arbitrary parameters.

1.3.1. *A digression to the Appell polynomials*

At first glance, there seems to be a great similarity between the Gontcharoff polynomials in multiple integral form [1.8] and the well-known Appell polynomials[8]. Denoting them $Q_n(x)$, these Appell polynomials indeed have the multiple integral representation:

$$Q_n(x) = \int_{v_{n-1}}^{x} dx_1 \int_{v_{n-2}}^{x_1} dx_2 \ldots \int_{v_0}^{x_{n-1}} dx_n. \qquad [1.11]$$

Observe that [1.11] gives by differentiation the remarkable relation $Q_n^{(1)} = Q_{n-1}$, $n \geq 1$.

8 Appell, P.E. (1880). Sur une classe de polynômes. *Annales de l'E.N.S.*, 9, 119–144.

In fact, it is possible to go from the formulas [1.8]–[1.11] and vice versa, simply by taking:

$$u_i = v_{n-1-i}, \quad 0 \le i \le n-1.$$

However, such identification is more confusing than effective, for at least four reasons:

i) This possibility of identification is only local, that is, for any finite polynomial families, but it does not work for infinite families.

ii) The family of A–G polynomials [1.8] existentially depends on all u_n: without explicit reference to these parameters, it is quite impossible to work with the polynomials. On the contrary, the family of Appell polynomials [1.11] is essentially based on the functional relation $Q_n^{(1)} = Q_{n-1}$: thus, the v_n have here only an anecdotal role and are generally ignored in the notation.

iii) In an A–G family, all the polynomials vanish at the same value $x = u_0$. We will see that many interesting results rely on this simple collective property. The situation differs with an Appell family where each polynomial has its own roots, these often being undetermined.

iv) By using a family of A–G polynomials, it is possible to write Abel-like expansions (i.e. similar to [1.6]) that are extremely robust, even more so than Taylor expansions. This is not true with an Appell family.

In summary, the families of Appell and A–G polynomials, even if they locally admit a common integral representation, are far from being twin copies of each other.

1.3.2. *Back to the A–G polynomials*

Let us summarize the key points recalled. From now on, it will be convenient to rewrite the A–G polynomials $P_n(x|u_0, \ldots, u_{n-1})$ as:

$$G_n(x|U), \quad n \in \mathbb{N}_0,$$

where U is the family of reals:

$$U = \{u_i, \, i \in \mathbb{N}_0\},$$

which is an infinite sequence because it serves to build the whole family of polynomials.

DEFINITION 1.3.– *By virtue of [1.7]:*

$$G_n^{(m)}(u_m|U) = \delta_{m,n}, \quad 0 \leq m \leq n. \tag{1.12}$$

In particular, $G_n(x|U)$, $n \geq 1$, *all have a common root at the point* $x = u_0$. *Moreover, from [1.9]:*

$$G_n^{(m)}(x|U) = G_{n-m}(x|E^m U), \quad 0 \leq m \leq n, \tag{1.13}$$

where $E^m U$ *represents the shifted family*

$$E^m U \equiv \{u_{m+i}, \, i \in \mathbb{N}_0\}. \tag{1.14}$$

In the affine case [1.3], they reduce to the Abel polynomials, that is,

$$G_n(x|U) = A_n(x|u_n), \quad n \in \mathbb{N}_0.$$

For a function $\phi(x)$, *its Abel–Gontcharoff expansion is given by:*

$$\sum_{n=0}^{\infty} \phi^{(n)}(u_n)\, G_n(x|U) \tag{1.15}$$

The convergence of the sum [1.15] and its ability to reconstruct ϕ have been a major concern in the work of Abel and Gontcharoff. This subject is not the main focus of this book. Nevertheless, a basic result that will be useful to us is when ϕ is a polynomial.

PROPOSITION 1.1.– *Any polynomial is the sum of its A–G expansion, whatever family* U *is chosen.*

Proof. A polynomial $R_l(x)$ of degree l can always be expressed as $R_l(x) = \sum_{n=0}^{l} c_n\, G_n(x|U)$ with appropriate coefficients c_n. So, by differentiating m ($\leq l$) times and then choosing $x = u_m$, we get, thanks to [1.12]:

$$R_l^{(m)}(u_m) = \sum_{n=0}^{l} c_n\, G_n^{(m)}(u_m|U) = \sum_{n=0}^{l} c_n\, \delta_{m,n} = c_m,$$

so that after substitution:

$$R_l(x) = \sum_{n=0}^{l} R_l^{(n)}(u_n)\, G_n(x|U).$$

This is precisely the expansion [1.15] of $R_l(x)$. □

Finally, let us mention that an affine transform of U, of the form $\alpha + \beta U$, has a simple effect on the A–G polynomials since by [1.8]:

$$G_n(\alpha + \beta x|\alpha + \beta U) = \beta^n\, G_n(x|U), \quad n \geq 0. \qquad [1.16]$$

1.4. Increased recognition

Reference to Gontcharoff remained confined to the mathematical theory of interpolation until H.E. Daniels, working in statistics and probability, identified the A–G polynomials also in applied mathematics, to his astonishment. This identification, however, was not too clear as the analysis was sometimes obscured by the previously mentioned confusion between the A–G and Appell polynomials. Moreover, the working conditions were made difficult because the polynomials were used via their explicit expressions with determinants.

What is most unexpected is that these encounters occurred three times and in different areas. The first (Daniels (1945)[9]) was during the war when, serving in a science laboratory of the Royal Air Force, he was studying the strength of parachute cords. The second (Daniels (1963)[10]) was in the study of the first meeting between the trajectory of a standard Poisson process and a nonlinear fixed boundary. The third (Daniels (1967)[11]) was for the determination of the total size in the so-called general epidemic model. The diversity of these three fields of application brought to light that the A–G

9 Daniels, H.E. (1945). The statistical theory of the strength of bundles of threads. *Proceedings of the Royal Society of London*, A183, 405–435.

10 Daniels, H.E. (1963). The Poisson process with a curved absorbing boundary. *Bulletin of the I.S.I.*, 40, 994–1008.

11 Daniels, H.E. (1967). The distribution of the total size of an epidemic. *Proceedings of the Fifth Berkeley Symposium on Mathematical Statistics and Probability*, 4, 281–293.

polynomials had to leave the strict domain of the theory of interpolation and had to play their own role in the vast field of applied mathematics. When pure and applied mathematics dialogue together, there is certainly a benefit for everyone. In particular, new applications appear and simultaneously new questions arise which lead to further developments.

It was first the case when Picard (1980)[12], (1984)[13] studied epidemic processes using martingales and met again the A–G polynomials (independently of the work of Daniels). This methodology has been continued and developed in a number of papers, starting with Lefèvre and Picard (1990) and Picard and Lefèvre (1990) (see the list of our papers on A–G polynomials at the end). In particular, we have underlined apparent links with the Appell polynomials, and we have constructed the multivariate extension of the A–G (and Appell) polynomials that we have used in the study of epidemic risk (and insurance) models.

This research has attracted some interest among biomathematicians (and actuaries) (see, e.g., Ball et al. 1997)[14]. At this point, it is important to highlight a nice surprising result obtained by Ball and O'Neill (1999)[15], which deals with randomness in a multiplicative form within the parameter family U. Finally, we mention that the main properties of A–G polynomials have been derived more recently in algebra and combinatorics by Kung and Yan (2003)[16] and Khare et al. (2014)[17].

We will now present three simple examples in which the intervention of A–G polynomials is quite natural but where the need for their generalization clearly appears.

12 Picard, P. (1980). Applications of martingale theory to some epidemic models. *Journal of Applied Probability*, 17, 583–599.

13 Picard, P. (1984). Applications of martingale theory to some epidemic models (II). *Journal of Applied Probability*, 21, 677–84.

14 Ball, F.G., Mollison, D., Scalia-Tomba, G. (1997). Epidemics with two levels of mixing. *The Annals of Applied Probability*, 7, 46–89.

15 Ball, F.G. and O'Neill, P.D. (1999). The distribution of general end-state random variables for stochastic epidemic models. *Journal of Applied Probability*, 36, 473–491.

16 Kung, J. and Yan, C. (2003). Gončarov polynomials and parking functions. *Journal of Combinatorial Theory*, A102, 16–37.

17 Khare, N., Lorentz, R., Yan, C. (2014). Bivariate Gončarov polynomials and integer sequences. *Science China, Mathematics*, 57, 1561–1578.

1.5. A first meeting problem

1.5.1. *Poisson trajectory with lower boundary*

Consider the right-continuous trajectory of a standard homogeneous Poisson process that begins at level 0 at some time x, possibly negative; it is written as $\{N_{t-x}, t \geq x\}$. We are interested in the first crossing of this trajectory with a continuous increasing boundary that starts below the initial level 0 of the trajectory; it is denoted as $t \to f(t)$, $t \geq x$.

Note that the crossing of the Poisson trajectory with the lower bound necessarily corresponds to a meeting. In fact, the useful part of the lower bound is made up of the only points having a natural number as the ordinate. Therefore, we will focus on the useful part of the boundary:

$$F \equiv \{(v_i, i),\ i \in \mathbb{N}_0\}.$$

in which v_i represents the first instant when the boundary reaches the level i (i.e. $f(v_i) = i$). The set of these instants is denoted $V = \{v_i, i \in \mathbb{N}_0\}$; of course, v_i increases with i.

An operation that will often be necessary is to move the boundary vertically to a higher positive value, or the trajectory to a lower negative value, with integer length steps. Note that these two movements are effectively equivalent for a meeting problem. Given a boundary F which is characterized by the set V, its rise of length 1 (or $r \in \mathbb{N}_0$) will be characterized by the shifted set EV (or $E^r V = \{v_{r+i}, i \in \mathbb{N}_0\}$) as precisely defined by [1.14].

Let λ be the parameter of the Poisson process initiated at time x. Recall that:

$$\mathbb{P}(N_{t-x} = n) = e^{-\lambda(t-x)} \frac{[\lambda(t-x)]^n}{n!}, \quad t \geq x,\ n \geq 0.$$

Thus, the Poisson trajectory starts at the point $(x, 0)$ where x is $\leq v_0$ (remember that the boundary considered is a lower boundary). Later we also consider the starting points (x, j) with $j \in \mathbb{N}_0$. The first meeting between the trajectory and the boundary will take place at an instant T (possibly infinite) and at the level K such that $T = v_K$.

1.5.2. *First meeting level distribution*

Let us insert x and V in the notation below to indicate the initial conditions, namely, the starting time of the Poisson trajectory and the critical points of the lower boundary, respectively.

PROPOSITION 1.2.–

$$\mathbb{P}(K = n|x, V) = e^{-\lambda(v_n - x)}(-\lambda)^n \, G_n(x|V), \quad n \geq 0, \qquad [1.17]$$

where $V = \{v_i, \ i \in \mathbb{N}_0\}$.

Proof. Denote by T_0 the time interval between the start time and the time of the first jump of the trajectory. For a meeting at level 0, we have:

$$\mathbb{P}(K = 0|x, V) = \mathbb{P}(T_0 > v_0|x, V) = e^{-\lambda(v_0 - x)}. \qquad [1.18]$$

For a level $n \geq 1$, T_0 takes a value $t \in (x, v_0)$ with the density function $\lambda \exp(-\lambda(t - x))$. By a recursive argument, we get:

$$\mathbb{P}(K = n|x, V) = \int_x^{v_0} \mathbb{P}(K = n-1|t, EV) \, \lambda e^{-\lambda(t-x)} dt, \quad n \geq 1, \ [1.19]$$

where EV is defined above. Indeed, before time $T_0 = t$, the trajectory remains at level 0, and after $T_0 = t$, it begins a new life starting at level 1 but the relative positions of the trajectory and the boundary have changed this time. That is why in the r.h.s. t and EV appear instead of x and V. Also, remember that a positive jump in the trajectory is equivalent to a translation of the boundary in the opposite direction. In the present case, we chose the second strategy in order to keep the same symbolism under the integral in [1.19].

To simplify [1.19], let us introduce an unknown function R_n defined by:

$$\mathbb{P}(K = n|x, V) = e^{-\lambda(v_n - x)} \, R_n(x|V), \quad n \geq 0. \qquad [1.20]$$

Then, [1.18] gives $R_0(x|V) = 1$, and [1.19] becomes:

$$R_n(x|V) = \int_x^{v_0} \lambda \, R_{n-1}(t|EV) dt, \quad n \geq 1, \qquad [1.21]$$

after noting that:

$$e^{-\lambda(v_n - x)} = e^{-\lambda(v_{1+n-1} - t)} e^{-\lambda(t-x)}.$$

The relations [1.21] recursively provide the R_n in the form of polynomials in x expressed as multiple integrals.

To obtain a formula in terms of A–G polynomials, we change the notation by:

$$\begin{cases} v_i = -u_i \text{ (thus, } V = -U), \quad x = -y, \\ \text{and } R_n(x|V) = \lambda^n \tilde{R}_n(y|U). \end{cases} \qquad [1.22]$$

Making this substitution allows us to write that:

$$\tilde{R}_n(y|U) = \int_{-y}^{-u_0} \tilde{R}_{n-1}(-t|EU)dt = \int_{u_0}^{y} \tilde{R}_{n-1}(\tau|EU)d\tau,$$

by successive iterations, we then have:

$$\tilde{R}_n(y|U) = \int_{u_0}^{y} dx_1 \int_{u_1}^{x_1} dx_2 \ldots \int_{u_{n-1}}^{x_{n-1}} dx_n,$$

and comparing with [1.8],

$$\tilde{R}_n(y|U) = G_n(y|U) = G_n(-x| - V) = (-1)^n G_n(x|V), \qquad [1.23]$$

where we used [1.16].

Inserting [1.22] and [1.23] into [1.20] leads to formula [1.17]. □

Note that when the trajectory starts at point $(0, 0)$, [1.17] becomes:

$$\mathbb{P}(K = n|0, V) = e^{-\lambda v_n} (-\lambda)^n G_n(0|V), \quad n \geq 0.$$

In particular, with a linear boundary $f(t) = (t - \alpha)/\beta$, $\beta > 0$, then $v_i = \alpha + \beta i$ for all i and using [1.10], we have:

$$\begin{aligned} \mathbb{P}(K = n|0, V) &= e^{-\lambda(\alpha+\beta n)} (-\lambda)^n G_n(0|\{\alpha + \beta i, i \in \mathbb{N}_0\}) \\ &= e^{-\lambda(\alpha+\beta n)} \lambda^n \alpha(\alpha + \beta n)^{n-1}/n!, \end{aligned}$$

which is the probability mass function (p.m.f.) of a generalized Poisson distribution.

Moreover, a direct corollary of [1.17] is:

$$\mathbb{P}(T < \infty | x, V) = \sum_{n=0}^{\infty} \mathbb{P}(K = n | x, V) = \sum_{n=0}^{\infty} e^{-\lambda(v_n - x)} (-\lambda)^n \, G_n(x|V),$$

which implies the inequality of interest:

$$\sum_{n=0}^{\infty} e^{-\lambda v_n} (-\lambda)^n \, G_n(x|V) \leq e^{-\lambda x}.$$

Finally, it is easy to see that for a Poisson process now initiated in (x, j):

$$P(K = n|(x, j), V) = e^{-\lambda(v_n - x)} (-\lambda)^{n-j} \, G_{n-j}(x|E^j V), \quad n \geq j.$$

1.6. A final epidemic outcome

1.6.1. *Reed–Frost epidemic process*

An SIR-type epidemic model describes the spread, in a closed population, of an infectious disease that follows the pattern: susceptible → infected → removed (immune). The standard Markovian process for such an infection is the so-called general epidemic process for the continuous-time version and the so-called Reed–Frost process for the discrete time version. We focus here on the latter.

More specifically, the Reed–Frost model is based on the following assumptions. Every susceptible can become infected after sufficiently close contact with an infective. If ever infected, this susceptible first passes through a latency period of a fixed duration equal to 1, then through an infectious period contracted at a single instant. At this instant, the infected susceptible transmits the disease agents to any susceptible present with a constant probability p; all possible infectious contacts are independent of each other. After that, this individual is removed by isolation or loss of infectivity and becomes immune to further infection.

So, consider a discrete time scale $t \in \mathbb{N}_0$, and let S_t and I_t denote numbers of susceptibles and infectives at time t, just after the tth latency period. The initial values (S_0, I_0) are given by (n, m), the total population size being $n + m$. At time t, each susceptible present escapes infectious contact with a given infective with probability $q = 1 - p$. Thus, given (S_t, I_t), the number of susceptibles remaining at time $t + 1$, that is, S_{t+1}, follows a binomial distribution with exponent S_t and parameter q^{I_t}, which is denoted by:

$$S_{t+1} =_d Bin(S_t, q^{I_t}), \quad t \in \mathbb{N}_0. \qquad [1.24]$$

After its contagious instant, each infective is removed permanently, so that:

$$I_{t+1} = S_t - S_{t+1}, \quad t \in \mathbb{N}_0. \qquad [1.25]$$

The epidemic process continues as long as there are infectives in the population. Thus, the end of the epidemic is the random instant $T = \inf\{t : I_t = 0\}$. At this time, S_T represents the number of susceptibles who have finally escaped the disease. In parallel, the final size of the epidemic is defined as the total number of infections $n - S_T$ which, by [1.25], is equal to $I_1 + \ldots + I_{T-1}$. Our main objective is to determine the exact distribution of S_T (or $n - S_T$).

1.6.2. *Final state distribution*

We start by constructing a family of n martingales for the epidemic process. Let $\{\mathcal{F}_t, t \in \mathbb{N}_0\}$ be the σ–field generated by $\{(S_u, I_u), 0 \leq u \leq t\}$. For $k, s \in \mathbb{N}_0$, we denote $s_{[k]} = s(s-1)\ldots(s-k+1)$ with the convention $s_{[0]} \equiv 1$.

LEMMA 1.1.– *For each $k = 1, \ldots, n$, the process:*

$$\{(S_{t,[k]}\, q^{k(S_t+I_t)}, \mathcal{F}_t), t \in \mathbb{N}_0\} \qquad [1.26]$$

constitutes a martingale.

Proof. Let us introduce the probability generating function (p.g.f.) of S_{t+1} with argument $x \in \mathbb{R}$, conditionally on $\mathcal{F}_t$. From [1.24]:

$$\mathbb{E}(x^{S_{t+1}}|\mathcal{F}_t) = [1 + (x-1)q^{I_t}]^{S_t}, \quad t \in \mathbb{N}_0. \qquad [1.27]$$

By differentiating [1.27] k ($\leq n$) times, then putting $x = 1$, we get:

$$\mathbb{E}\big(S_{t+1,[k]}|\mathcal{F}_t\big) = S_{t,[k]}\, q^{kI_t}, \quad t \in \mathbb{N}_0. \qquad [1.28]$$

Multiply both sides of [1.28] by q^{kS_t}. Using [1.25], we see that:

$$\mathbb{E}\big(S_{t+1,[k]}\, q^{k(S_{t+1}+I_{t+1})}|\mathcal{F}_t\big) = S_{t,[k]}\, q^{k(S_t+I_t)}. \qquad [1.29]$$

Since $S_{t,[k]}\, q^{k(S_t+I_t)}$ has a finite mean for each k, [1.29] implies the announced assertion. □

Now, the end T of the epidemic is a stopping time with respect to $\{\mathcal{F}_t, t \in \mathbb{N}_0\}$, finite a.s. and such that $I_T = 0$. Therefore, applying the optional stopping theorem for martingales directly gives the following n identities.

COROLLARY 1.1.– *For each* $k = 1, \ldots, n$*:*

$$\mathbb{E}\big(S_{T,[k]}\, q^{kS_T}\big) = n_{[k]}\, q^{k(n+m)}. \qquad [1.30]$$

Thanks to the relations [1.30], we can obtain the p.g.f. of S_T, $\phi(x) = \mathbb{E}(x^{S_T})$, in terms of a particular family of A–G polynomials.

PROPOSITION 1.3.– *For* $x \in \mathbb{R}$*:*

$$\phi(x) = \sum_{k=0}^{n} n_{[k]} q^{k(n+m-k)}\, G_k(x|U), \qquad [1.31]$$

where $U = \{q^i, \, i \in \mathbb{N}_0\}$.

Proof. The p.g.f. $\phi(x)$ of S_T is, by definition, a polynomial of degree n in x. By proposition 1.1, it admits an A–G expansion:

$$\phi(x) = \sum_{k=0}^{n} \phi^{(k)}(u_k)\, G_k(x|U), \qquad [1.32]$$

whatever the family $U = \{u_i, i \in \mathbb{N}_0\}$.

Choose $u_i = q^i$ for all i in [1.32]. Using [1.30], we then obtain:

$$\begin{aligned}\phi^{(k)}(u_k) &= \mathbb{E}\big[S_{T,[k]}\,(q^k)^{S_T-k}\big]\\ &= n_{[k]}\,q^{k(n+m-k)},\end{aligned}$$

hence the formula [1.31]. □

From this expression for $\phi(x)$, the properties of $G_k(x|U)$ allow us to obtain different explicit results including the m.p.f. and the factorial moments of S_T (or $n - S_T$) and various asymptotics when n is large (as often).

1.7. A goodness-of-fit test

1.7.1. *Kolmogorov–Smirnov statistic*

Consider the null hypothesis $H_0 : F(x) = L(x)$ for all $x \in \mathbb{R}$ where $F(x)$ is the unknown distribution function (d.f.) and $L(x)$ is a hypothetical d.f. completely specified. As an alternative, we examine the one-sided case $H_1 : F(x) < L(x)$ for some x.

To make a decision, we draw a sample $X_1, \ldots, X_n$ of n independent identically distributed (i.i.d.) random variables. It is known that the empirical d.f. $F_n(x) = (1/n)\sum_{j=1}^{n} \mathbf{1}(X_j \leq x)$ is a strongly consistent estimator of F, uniformly in x. When testing H_1, a measure of discrepancy between F_n and L is the Kolmogorov–Smirnov statistic:

$$D_n = \sup\{L(x) - F_n(x)\}.$$

The rejection region of H_0 is then of the form $D_n > \epsilon$ where ϵ is a given constant (in $(0, 1)$).

This type of test was originally developed for continuous distributions L. It is precisely the situation discussed here. Then, D_n is distribution-free under H_0, so $L(x)$ can be taken as d.f. of a uniform random variable on the interval $(0, 1)$ (i.e. $L(x) = x$ on $(0, 1)$). We are going to determine in this case the probability of the non-rejection region:

$$\mathbb{P}(D_n < \epsilon) \quad (= 1 - \mathbb{P}(D_n > \epsilon)). \qquad [1.33]$$

1.7.2. *Non-rejection probability*

Let $X_1, \ldots, X_n$ be a sample of n i.i.d. $(0,1)$-uniform random variables. We first derive the desired probability in terms of A–G polynomials defined from a truncated affine family U.

LEMMA 1.2.– *For* $\epsilon \in (0,1)$*:*

$$\mathbb{P}(D_n < \epsilon) = (-1)^n \, n! \, G_n(0|V), \qquad [1.34]$$

where $V = \{v_i,\, 0 \le i \le n-1\}$, *with* $v_i = \min(\epsilon + i/n, 1)$.

Proof. Denote $X_{(1)}, \ldots, X_{(n)}$ the order statistics of the starting sample. By definition, $F_n(X_{(i)}) = i/n$, thus the probability [1.33] is equivalent to:

$$\mathbb{P}(D_n < \epsilon) = \mathbb{P}\big(X_{(1)} < \epsilon, X_{(2)} < \epsilon + 1/n, \ldots, X_{(n)} < \epsilon + (n-1)/n\big),$$

that is, using the notation $v_i = \min(\epsilon + i/n, 1)$ above:

$$\mathbb{P}(D_n < \epsilon) = \mathbb{P}(X_{(i+1)} < v_i, \; for \, i = 0, 1, \ldots, n-1).$$

This can be rewritten as the multiple integral:

$$\mathbb{P}(D_n < \epsilon) = n! \int_0^{v_0} dx_0 \int_{x_0}^{v_1} dx_1 \ldots \int_{x_{n-2}}^{v_{n-1}} dx_{n-1},$$

which is equivalent to:

$$\mathbb{P}(D_n < \epsilon) = (-1)^n \, n! \int_{v_0}^{0} dx_0 \int_{v_1}^{x_0} dx_1 \ldots \int_{v_{n-1}}^{x_{n-2}} dx_{n-1}. \qquad [1.35]$$

Using the representation [1.8] of an A–G polynomial, [1.35] gives the desired result [1.34]. □

Because of the special form of v_i, we can continue the calculation by doing some algebra with the A–G polynomials. Let us introduce the families U and W with $u_i = \epsilon + i/n$ and $w_i = 1$ for $0 \le i \le n-1$. Then:

$$v_i = u_i \, \mathbf{1}(i \le k) + w_i \, \mathbf{1}(i > k), \quad \text{where } k \equiv \max\{i : \epsilon + i/n \le 1\}. \qquad [1.36]$$

PROPOSITION 1.4.– *For* $\epsilon \in (0,1)$*:*

$$\mathbb{P}(D_n < \epsilon) = \epsilon \sum_{i=k+1}^{n} \binom{n}{i} (1-u_i)^{n-i} (u_i)^{i-1}, \qquad [1.37]$$

where $u_i = \epsilon + i/n$, $k+1 \leq i \leq n-1$, *with* k *defined in [1.36].*

Proof. By proposition 1.1, any polynomial R of degree n admits an A–G expansion, hence:

$$R(0) = \sum_{i=0}^{n} R^{(i)}(u_i)\, G_i(0|U). \qquad [1.38]$$

In [1.38], let us choose for U the family V defined in [1.36] and for $R(x)$ the polynomial $G_n(x|V)$. This yields:

$$G_n(0|V) = \sum_{i=0}^{n} G_n^{(i)}(u_i|V)\, G_i(0|U). \qquad [1.39]$$

As seen in [1.13]:

$$G_n^{(i)}(u_i|V) = G_{n-i}(u_i|E^i V),$$

so that [1.39] becomes:

$$G_n(0|V) = \sum_{i=0}^{n} G_{n-i}(u_i|E^i V)\, G_i(0|U). \qquad [1.40]$$

Now, from [1.36], when $i \leq k$, then $v_i = u_i$ and:

$$G_{n-i}(u_i|E^i V) = G_{n-i}(v_i|v_i, \ldots) = \delta_{n,i},$$

by virtue of [1.12] and since $i \leq k \leq n-1$. When $i > k$, then $v_i = w_i$ and $E^i V = E^i W = \{1, 1, \ldots\}$. Therefore, [1.40] is simplified in:

$$G_n(0|V) = \sum_{i=k+1}^{n} G_{n-i}(u_i|\{1, 1, \ldots\})\, G_i(0|U). \qquad [1.41]$$

The u_i having an affine expression, $G_i(0|U)$ reduces to an Abel polynomial. From [1.10], we then have:

$$G_i(0|U) = \frac{(-1)^i \epsilon(\epsilon + i/n)^{i-1}}{i!}.$$

Moreover, it is straightforward that:

$$G_{n-i}(u_i|\{1, 1, \ldots\}) = \frac{(u_i - 1)^{n-i}}{(n-i)!}.$$

By substitution in [1.41] and then in [1.34], we obtain the formula [1.37]. □

Other results are possible in the same direction. Thus, lemma 1.2 can be extended to any model, whether continuous or not. It can also be generalized to the computation of $\mathbb{P}(D_n < \epsilon|F = K)$ where K is a d.f. fixed for F. Moreover, the alternative one-tailed test $H_1 : F(x) > L(x)$ for some x is also tractable using this time the Kolmogorov–Smirnov statistic $\sup\{F_n(x) - L(x)\}$.

1.8. Extension to pseudopolynomials

The previous three examples show that A–G polynomials appear frequently in stochastic models and in natural but unexpected ways. They also indicate where and why the effectiveness of these polynomials may prove to be limited, and the kind of convenient generalizations that would be worth making.

In the first meeting problem, a natural extension would consist of replacing the Poisson process by a compound Poisson process. The general analysis is only slightly modified but the canonical A–G polynomials turn out to be totally insufficient. The situation would become more complex in other cases, notably for processes such as those of the birth or death type or when the boundary itself is the trajectory of a counting process.

In the epidemic context, the introduction of more general functional infection rates would also require a more elaborate mathematical tool. Also, multipopulation models are necessary to take into account the variable degrees of infectivity or resistance to the disease (according to sex, age, family, professions, etc.). This would imply working, at least, with a multivariate version of the polynomials A–G.

In non-parametric statistics, the complementary question of the exact distributions of rank statistics for two independent samples could be addressed through the use of generalized A–G polynomials.

Most of the applications we have in mind, however, rely on a general structure that is roughly similar in shape. Thus, we can expect that new mathematical objects, more general than the A–G polynomials, will allow us to tackle the problems they raise.

The aim of this book is to show that such objects do indeed exist: they will no longer necessarily be polynomials but functions reminiscent of the A–G polynomials. For this reason, we will call them *A–G pseudopolynomials*. The general theory on these pseudopolynomials will first be developed in several chapters, then it will be illustrated by different problems in applied probability. The material presented is directly inspired by our previous work starting with Picard and Lefèvre (1996) (see the list of our papers on A–G pseudopolynomials). The presentation adopted here is however more mathematically sound.

2

Abel-Gontcharoff Pseudopolynomials

2.1. General framework: $\mathfrak{D}, \mathcal{E}, \mathfrak{F}, \Delta$

The definition of pseudopolynomials will require the broader framework presented in the following.

Let $\mathfrak{D}$ be a non-empty set. An arbitrary element of $\mathfrak{D}$ is written as x. This set $\mathfrak{D}$ can be $\mathbb{R}$, $\mathbb{R}_+$, $\mathbb{C}$, $\mathbb{R}^k$, $\mathbb{R}^{\mathbb{N}_0}$ or even $\mathbb{N}_0$, $\mathbb{Z}$, for instance. However, the algebraic form of $\mathfrak{D}$ will not matter much for obtaining many results.

Consider a countable, possibly finite, family $\mathcal{E}$ of functions $e_n : \mathfrak{D} \to \mathbb{R}$ (or $\mathbb{C}$), $n \in \mathbb{N}_0$, which are linearly independent. Such a family $\mathcal{E}$ is said to be basic. It should be noted that these e_n can be, in particular, (linearly independent) polynomials.

The real vector space generated by $\mathcal{E}$ is denoted $\mathfrak{F}$. An element of $\mathfrak{F}$ is a finite sum of expression $f_n \equiv \sum_{i=0}^{n} a_i e_i$, for some $n \in \mathbb{N}_0$. The coefficients a_i can of course depend on n, which will be explicitly indicated if necessary. When $a_n \neq 0$, f_n is called a *pseudopolynomial* of degree n.

Of course, $\mathfrak{F}$ may be extended by introducing the formal series $\sum_{i=0}^{\infty} a_i e_i$. This extension depends on $\mathcal{E}$ (see the remark below), so the vector space of these formal series is denoted $\mathfrak{F}_{\mathcal{E}}$.

REMARK.– Other basic families can be used to generate $\mathfrak{F}$. For example, let $\hat{\mathcal{E}}$ be a family related to $\mathcal{E}$ by the following formulas:

$$\hat{e}_n = \sum_{i=0}^{n} \beta_{n,i} e_i \text{ where } \beta_{n,n} \neq 0, \quad \text{and} \quad e_n = \sum_{i=0}^{n} \gamma_{n,i} \hat{e}_i \text{ where } \gamma_{n,n} \neq 0,$$

Any pseudopolynomial will have the same degree with respect to $\mathcal{E}$ and $\hat{\mathcal{E}}$, but the vector spaces $\mathfrak{F}_{\mathcal{E}}$ and $\mathfrak{F}_{\hat{\mathcal{E}}}$ are not necessarily the same. If we start with $f = \sum_{i=0}^{\infty} a_i e_i \in \mathfrak{F}_{\mathcal{E}}$, all a_i are then assumed to be finite, while

$$f = \sum_{i=0}^{\infty} a_i e_i = \sum_{i=0}^{\infty} a_i \Big(\sum_{j=0}^{i} \gamma_{i,j} \hat{e}_j \Big)$$

is an element of $\mathfrak{F}_{\hat{\mathcal{E}}}$ only if the coefficients $\sum_{i=0}^{\infty} a_i \gamma_{i,j}$ are all finite, which is not guaranteed.

Now, we define a generalized "differentiation" operator Δ on $\mathcal{E}$ as follows.

DEFINITION 2.1.– *On $\mathcal{E}$:*

$$\Delta e_{n+1} = e_n, \quad n \in \mathbb{N}_0, \quad \textit{and} \quad \Delta e_0 = 0. \qquad [2.1]$$

This operator Δ is directly extended linearly to $\mathfrak{F}$. As usual, the iterate Δ^{j+1}, $j \geq 1$, is defined recursively from $\Delta^{j+1} = \Delta(\Delta^j)$ with $\Delta^1 = \Delta$, and Δ^0 is the identity operator.

REMARK.– The family $\{\lambda_n e_n, n \in \mathbb{N}_0\}$ of coefficients $\lambda_n \neq 0$ also generates $\mathfrak{F}$ but it does not provide the same Δ. This is also the case for the e_n, which would be taken in a modified order.

At first glance, this whole presentation seems abstract and leads to inevitable questions like which families $\mathcal{E}$ or sets $\mathfrak{D}$ are worth considering. This is discussed below.

2.1.1. *Previous simple situation*

Our starting point is the historical case in which Abel and Gontcharoff worked, namely, $\mathfrak{D} = \mathbb{R}$ (or $\mathbb{C}$) and $e_n(x) = x^n/n!$, $n \in \mathbb{N}_0$, which provides a range of convenient properties. In any extension of the theory, some of these properties will necessarily be lost. Sometimes this loss will only pose a difficulty to overcome, but often it can turn out to be a complete obstruction.

A wise precaution is to list these nice features first:

1) Each function e_n is a polynomial of degree n.

2) The operator Δ is the simple differentiation D with its usual properties (e.g. linearity and Leibniz formula).

3) The resulting integration operator is positive ($f \geq 0 \to \int_u^x f(\xi)d\xi \geq 0$ when $x \geq u$).

4) The so-called binomial (or convolutional) identity applies and asserts that:

$$e_n(x+y) = \sum_{i=0}^{n} e_i(x)e_{n-i}(y).$$

5) The differentiation commutes with the shift operator $S^a : e_n(x) \to e_n(x+a)$ for any a:

$$D\big(S^a e_n(x)\big) = D\frac{(x+a)^n}{n!} = \frac{(x+a)^{n-1}}{(n-1)!} = S^a\big(De_n(x)\big).$$

6) The multiplicative property is satisfied in the sense that for any λ:

$$e_n(\lambda x) - \lambda^n e_n(x).$$

7) A less referenced relationship:

$$\binom{n}{k} e_n(x) = e_k(x)e_{n-k}(x)$$

has a certain robustness as will be seen later.

Note that, for example, the properties (4)–(6) are meaningless without a suitable structure on $\mathfrak{D}$: for (4) and (5), an operation of addition is required, and for (6), an external product $(\lambda, x) \mapsto \lambda x$ must be defined when $\lambda \in \mathbb{R}$ and $x, \lambda x \in \mathfrak{D}$.

2.1.2. *Choices of $\mathfrak{D}$ and $\mathcal{E}$*

In practice, $\mathfrak{D}$ and $\mathcal{E}$ are closely related and must be defined simultaneously. Three different strategies are possible.

i) By starting with *explicit definitions* of these sets. Here are four examples:

a) $\mathfrak{D} = \mathbb{N}_0$ and $e_n(x) = \binom{x}{n}$;

b) $\mathcal{D} = \mathbb{Z}$ and $e_n(x) = \binom{x+n-1}{n}$;

c) $\mathfrak{D} = \mathbb{R}$ and $e_n(x) = e^{nx}/n!$;

d) $\mathfrak{D} = \mathbb{R}^{\mathbb{N}}$ and $e_n(\mathbf{x}) = x_1 x_2 \ldots x_n/n!$.

Note that property (6) above is lost in cases (a)–(c) but not in (d). In cases (a), (b) and (d), $e_n(x+a)$ is a real number but the function $x \mapsto e_n(x+a)$ is not an element of $\mathfrak{F}$. In case (c), $e_n(x+a) = e^{na}e_n(x)$ and $x \mapsto e_n(x+a)$ is an element of $\mathfrak{F}$. The binomial identity (4) is kept in cases (a) and (b) only. Other comments in the same vein are immediate.

ii) By using a *(formal) generating function* of the form $\sum_{n \geq 0} e_n(x)s^n$, $s \in \mathbb{R}$ (or $\mathbb{C}$). Thus, the generating function for the $x^n/n!$ is given by e^{xs} and for the $\binom{x}{n}$ by $(1+s)^x$. More generally, we can consider a function of the form $exp(xg(s))$, or $exp(\theta(x)s)$, or even $exp(g(x,s))$ where g is an analytic function (or a formal sum) in s. The first case is frequently used, and the third is especially interesting when $g(x,s) = \sum_j \theta_j(x) g_j(s)$.

iii) By taking for $\mathcal{E}$ an *abstract family*, while postulating specific simple properties like each e_n is a polynomial of degree n, or $D = h(\Delta)$ with h a formal series. In this case, the e_n can have complicated expressions.

2.2. Copies $\hat{\mathcal{E}}$ and standard families

2.2.1. *Copies $\hat{\mathcal{E}}$ of $\mathcal{E}$*

To any sequence of real (or complex) numbers $\{d_i,\ i \geq 0\}$ with $d_0 = 1$, we can associate a formal linear operator $\mathcal{T}$ given by:

$$\mathcal{T} = \sum_{i=0}^{\infty} d_i \Delta^i. \qquad [2.2]$$

This allows us to introduce a new family of functions $\hat{e}_n$ as:

DEFINITION 2.2.–

$$\hat{e}_n = \mathcal{T} e_n, \quad n \in \mathbb{N}_0. \qquad [2.3]$$

The whole family $\hat{\mathcal{E}}$ of the $\hat{e}_n$ is called a copy of $\mathcal{E}$ for the following reasons. First, we will show that Δ acts in the same way on $\mathcal{E}$ and $\hat{\mathcal{E}}$.

PROPERTY 2.1.– *As for [2.1]:*

$$\Delta \hat{e}_{n+1} = \hat{e}_n, \; n \in \mathbb{N}_0, \quad \textit{and} \quad \hat{e}_0 = 1. \qquad [2.4]$$

Proof. Obviously, $\mathcal{T}$ permutes with Δ (i.e. $\Delta\mathcal{T} = \mathcal{T}\Delta$). From [2.1]–[2.3], we have $\hat{e}_0 = d_0 e_0 = 1$ and for $n \in \mathbb{N}_0$:

$$\Delta \hat{e}_{n+1} = \Delta \mathcal{T} e_{n+1} = \mathcal{T} \Delta e_{n+1} = \mathcal{T} e_n = \hat{e}_n,$$

as announced. □

Let us then check that the inverse operator $\mathcal{T}^{-1}$ of $\mathcal{T}$ is well defined.

PROPERTY 2.2.–

$$\mathcal{T}^{-1} = \sum_{i=0}^{\infty} c_i \, \Delta^i, \qquad [2.5]$$

where the numerical sequence $\{c_i, \, i \geq 0\}$ *satisfies the relations:*

$$\sum_{i=0}^{n} d_i \, c_{n-i} = \delta_{n,0}, \quad n \in \mathbb{N}_0. \qquad [2.6]$$

Therefore, the set of all $\mathcal{T}$ *forms an abelian group, and* $\mathcal{E} = \mathcal{T}^{-1}\hat{\mathcal{E}}$ *is a copy of* $\hat{\mathcal{E}}$.

Proof. As [2.6] gives recursively all the $c_j, j \geq 0$ (starting from $c_0 = d_0 = 1$), we can consider the operator $\tilde{\mathcal{T}}$ defined as $\sum_{i=0}^{\infty} c_i \Delta^i$. We see that:

$$\begin{aligned} \tilde{\mathcal{T}}\mathcal{T} e_n &= \sum_{i=0}^{\infty} c_i \Delta^i \Big(\sum_{j=0}^{\infty} d_j \Delta^j \Big) e_n \\ &= \sum_{i,j=0}^{\infty} c_i d_j \Delta^{i+j} e_n = \sum_{k=0}^{\infty} \Big(\sum_{i+j=k}^{\infty} c_i d_j \Big) \Delta^k e_n, \end{aligned}$$

and by virtue of [2.6]:

$$\tilde{\mathcal{T}}\mathcal{T}e_n = \sum_{k=0}^{\infty} \delta_{k,0}\, e_{n-k} = e_n.$$

So, $\tilde{\mathcal{T}}$ exactly matches the inverse operator [2.5]. □

PROPOSITION 2.1.– *$\hat{\mathcal{E}}$ is a linearly independent family that generates Δ (as does $\mathcal{E}$).*

Proof. Suppose there are constants λ_i such that $\sum_{i\in J} \lambda_i \hat{e}_i = 0$ where $J \subset \mathbb{N}_0$. Since $\mathcal{T}^{-1}$ exists:

$$\sum_{i\in J} \lambda_i e_i = \sum_{i\in J} \lambda_i (\mathcal{T}^{-1}\hat{e}_i) = \mathcal{T}^{-1}\Big(\sum_{i\in J} \lambda_i \hat{e}_i\Big) = \mathcal{T}^{-1}(0) = 0.$$

By the linear independence of $\mathcal{E}$, all the λ_i are then equal to 0. □

We can thus conclude that $\hat{\mathcal{E}}$ plays an equivalent role to $\mathcal{E}$.

2.2.2. Pseudopolynomials of Appell type

Let us make the link with families of Appell pseudopolynomials, $\{A_n,\, n \in \mathbb{N}_0\}$ say. As for the polynomial case (see [1.11]), we say that

DEFINITION 2.3.– *$\mathcal{A}$ is a family of Appell pseudopolynomials (with respect to Δ) if:*

$$\Delta A_{n+1} = A_n, \quad n \in \mathbb{N}_0, \quad \textit{and} \quad \mathrm{A}_0 = 1. \qquad [2.7]$$

This condition [2.7] implies a simple expansion for an Appell family.

PROPOSITION 2.2.– *Every Appell pseudopolynomial A_n can be expressed as:*

$$A_n = \sum_{i=0}^{n} a_{n-i}\, e_i, \quad n \in \mathbb{N}_0, \qquad [2.8]$$

for some coefficients a_{n-i} with $a_0 = 1$.

Proof. By definition, any pseudopolynomial A_n can be expressed as:

$$A_n = \sum_{i=0}^{n} a_{n,i}\, e_i, \quad n \in \mathbb{N}_0, \qquad [2.9]$$

for appropriate coefficients $a_{n,i}$ with $a_{n,0} = 1$. Thus, the Appell condition [2.7] is equivalent to:

$$\sum_{i=1}^{n+1} a_{n+1,i}\, e_{i-1} = \sum_{j=0}^{n} a_{n,j}\, e_j,$$

that is

$$\sum_{i=0}^{n} a_{n+1,i+1}\, e_i = \sum_{j=0}^{n} a_{n,j}\, e_j, \quad n \in \mathbb{N}_0.$$

Therefore, $a_{n+1,i+1} = a_{n,i}$ for $0 \leq i \leq n$, and by iteration, $= a_{n-1,i-1} = \ldots = a_{n-i,0}$. So, [2.9] can be rewritten as:

$$A_n - \sum_{i=0}^{n} a_{n-i,0}\, e_i,$$

hence the simplified formula [2.8]. □

It easily follows that there is identity between the copies of $\mathcal{E}$ and the Appell families.

PROPOSITION 2.3.– *Any copy $\hat{\mathcal{E}}$ of $\mathcal{E}$ is an Appell family, and conversely.*

Proof. By [2.4], we know that $\hat{\mathcal{E}}$ is an Appell family. Conversely, we have from [2.8]:

$$A_n = \sum_{j=0}^{n} a_j\, e_{n-j} = \Big(\sum_{j=0}^{n} a_j \Delta^j\Big) e_n,$$

which means with [2.2] and [2.3] that $\mathcal{A}$ corresponds to a particular copy of $\mathcal{E}$. □

Note that there exist basic families generating $\mathfrak{F}$ that are not copies of $\mathcal{E}$. This is the case, for instance, for the family $\{e_n - \delta_{n,1}e_0, n \in \mathbb{N}_0\}$ or for the family $\mathcal{G}$ considered later.

EXAMPLE 2.1.– Let $\mathfrak{D} = \mathbb{R}$ and define $\mathcal{E}$ via the generating function:

$$\sum_{n=0}^{\infty} e_n(x)s^n = e^{xg(s)}, \quad \text{where } g(s) = \sum_{i=1}^{\infty} \alpha_i s^i \text{ with } \alpha_1 \neq 0. \qquad [2.10]$$

First, given real numbers x, a, the identity $e^{(x+a)g(s)} = e^{xg(s)}e^{ag(s)}$ gives from [2.8]:

$$\sum_{n=0}^{\infty} e_n(x+a)s^n = \sum_{i=0}^{\infty} e_i(x)s^i \sum_{j=0}^{\infty} e_j(a)s^j,$$

which implies

$$e_n(x+a) = \sum_{i=0}^{n} e_i(a)e_{n-i}(x) = \Big(\sum_{i=0}^{\infty} e_i(a)\Delta^i\Big)e_n(x).$$

Thus, we see that the operator $S^a \equiv \sum_{i=0}^{\infty} e_i(a)\Delta^i$ provides a copy of $\mathcal{E}$, while the operator $S^a - \Delta^0 = \sum_{i=1}^{\infty} e_i(a)\Delta^i$ does not.

Second, differentiating [2.10] with respect to x gives:

$$\sum_{n=0}^{\infty} e_n^{(1)}(x)s^n = g(s)e^{xg(s)} = \sum_{i=1}^{\infty} \alpha_i s^i \sum_{k=0}^{\infty} e_k(x)s^k,$$

so that

$$e_n^{(1)}(x) = \sum_{k=1}^{n} \alpha_k e_{n-k} = \Big(\sum_{k=1}^{\infty} \alpha_k \Delta^k\Big)e_n = g(\Delta)e_n,$$

that is, $D = g(\Delta)$. Then, we see that the operator D does not give a copy of $\mathcal{E}$, while the operator $\Delta^0 + D$ does and leads to $\hat{e}_n = e_n + e_n^{(1)}$.

2.2.3. *Standard families $\mathcal{E}$ and $\hat{\mathcal{E}}$*

Recall that for the A-G polynomials discussed in Chapter 1, $e_n(x) = x^n/n!$, $n \in \mathbb{N}_0$, and thus $e_n(0) = 0$ for $n \geq 1$. This property is generalized as follows.

DEFINITION 2.4.– *The family $\mathcal{E}$ is standard at point ω if there exists $\omega \in \mathfrak{D}$ such that:*

$$e_n(\omega) = \delta_{n,0}, \quad n \in \mathbb{N}_0. \qquad [2.11]$$

So, under this condition, any pseudopolynomial $f_n = \sum_{i=0}^{n} a_i e_i$ satisfies:

$$f_n(\omega) = a_0, \quad n \in \mathbb{N}_0. \qquad [2.12]$$

Let us see the effect of [2.11] on a copy $\hat{\mathcal{E}}$ of $\mathcal{E}$.

PROPOSITION 2.4.– *When $\mathcal{E}$ is standard at $\omega \in \mathfrak{D}$, then:*

$$\hat{e}_n = \sum_{i=0}^{n} \hat{e}_{n-i}(\omega) e_i, \quad \text{and} \quad \mathcal{T} = \sum_{j=0}^{\infty} \hat{e}_j(\omega)\Delta^j. \qquad [2.13]$$

Moreover, a copy $\hat{\mathcal{E}}$ cannot be standard at this ω without being identical to $\mathcal{E}$.

Proof. From [2.2], [2.3] and using [2.11]:

$$\hat{e}_n(\omega) = \sum_{i=0}^{n} d_i\, e_{n-i}(\omega) = \sum_{i=0}^{n} d_i\, \delta_{n-i,0} = d_n,$$

which implies the expression of $\hat{e}_n$ in [2.13] and thus also that of the operator $\mathcal{T}$.

Suppose $\hat{\mathcal{E}}$ standard at ω, that is, $\hat{e}_n(\omega) = \delta_{n,0}$ for all n. From [2.13]:

$$\hat{e}_n = \sum_{i=0}^{n} \delta_{n-i,0}\, e_i = e_n,$$

hence $\hat{\mathcal{E}}$ and $\mathcal{E}$ are identical. □

We cannot guarantee in advance that $\mathcal{E}$ is standard at some point ω. One reason for this is that $\mathfrak{D}$ may not be big enough. For example, if $\mathfrak{D} = \mathbb{N}$, a priori $\mathcal{E} = \{e_n(x) = x_{[n]}/n!, n \in \mathbb{N}_0\}$ could be standard at point $\omega = 0$ except that $0 \notin \mathfrak{D}$. In fact, $\mathcal{E}$ is not standard on $\mathfrak{D}$.

As shown below, however, there is always a copy of $\mathcal{E}$ that is standard.

PROPOSITION 2.5.– *A copy $\hat{\mathcal{E}}$ standard at any point ω' exists and is unique.*

Proof. From [1.9], $\hat{\mathcal{E}}$ standard at ω' requires that:

$$\delta_{n,0} = \hat{e}_n(\omega') = \sum_{i=0}^{n} d_{n-i}\, e_i(\omega') = d_n + \sum_{i=0}^{n-1} d_{n-i}\, e_i(\omega'), \quad n \in \mathbb{N}_0,$$

which recursively gives all the d_n starting at $d_0 = 1$. □

Note that $\mathcal{E}$ (or any $\hat{\mathcal{E}}$) may be standard at ω and $\omega' \neq \omega$. This occurs, for instance, if the e_n are periodic functions (of period independent of n).

According to proposition 2.5, when $\mathcal{E}$ is not standard at ω, there is a (unique) copy which is standard at ω. It will be denoted as $\hat{\mathcal{E}}^{(\omega)}$.

2.3. An integration operator I_u

After the "differentiation" operator Δ defined in [2.1], it is natural to introduce in parallel a generalized "integration" operator.

DEFINITION 2.5.– *Given any $u \in \mathfrak{D}$, the integral operator $I_u : \mathcal{E} \to \mathfrak{F}$ is defined by:*

$$I_u e_n = e_{n+1} - e_{n+1}(u)\, e_0, \quad n \in \mathbb{N}_0, \tag{2.14}$$

and more generally, by:

$$I_u\Big(\sum_{i=0}^{\infty} a_i\, e_i\Big) = \sum_{i=0}^{\infty} a_i\, e_{i+1} - \Big(\sum_{i=0}^{\infty} a_i\, e_{i+1}(u)\Big)\, e_0, \tag{2.15}$$

at least when the numerical series is convergent.

EXAMPLE 2.2.– Consider the A–G polynomial case where $\mathfrak{D} = \mathbb{R}$ and $e_n(x) = x^n/n!, n \in \mathbb{N}_0$. Then,

$$I_u e_n(x) = \frac{x^{n+1}}{(n+1)!} - \frac{u^{n+1}}{(n+1)!} = \int_u^x \frac{z^n}{n!} dz = \int_u^x e_n(z) dz, \qquad [2.16]$$

that is, I_u is the standard integral computed from u.

Let us show that I_u affects any copy $\hat{\mathcal{E}}$ in the same way as $\mathcal{E}$.

PROPOSITION 2.6.–

$$I_u \hat{e}_n = \hat{e}_{n+1} - \hat{e}_{n+1}(u)\, \hat{e}_0, \quad n \in \mathbb{N}_0. \qquad [2.17]$$

So, $\hat{\mathcal{E}}$ generates I_u as does $\mathcal{E}$.

Proof. From [2.2], [2.3] and using [2.14]:

$$I_u \hat{e}_n = I_u \Big(\sum_{i=0}^{n} d_{n-i}\, e_i \Big) = \sum_{i=0}^{n} d_{n-i} \left(e_{i+1} - e_{i+1}(u) e_0 \right)$$

$$= \sum_{j=1}^{n+1} d_{n+1-j} \left(e_j - e_j(u) e_0 \right). \qquad [2.18]$$

As the sum [2.18] can be taken from $j = 0$, we get the desired formula [2.17]. □

From [2.1] and [2.14], we see that Δ and I_u do not permute: for example, $I_u \Delta e_0 = 0$ while $\Delta I_u e_0 = \Delta(e_1 - e_1(u) e_0) = e_0 = 1$. In fact, we obviously have $\Delta I_u = \Delta^0$, and the inverse is given by [2.19].

PROPOSITION 2.7.– *For any $f \in \mathfrak{F}$:*

$$(I_u f)(u) = 0, \quad \textit{and} \quad I_u \Delta f = f - f(u). \qquad [2.19]$$

Proof. Obviously, [2.15] implies $(I_u f)(u) = 0$. In addition:

$$I_u \Delta f = I_u \Delta \sum_{i=0}^{\infty} a_i\, e_i = I_u \sum_{i=1}^{\infty} a_i\, e_{i-1} = \sum_{i=1}^{\infty} a_i \left(e_i - e_i(u) e_0 \right),$$

which is equivalent to $f - f(u)$ since the term for $i = 0$ reduces to 0. □

Note that by [2.19], $I_\omega \Delta e_n = e_n - e_n(\omega)$ for $n \geq 0$, whatever $\omega \in \mathcal{D}$. Therefore, if $\mathcal{E}$ is standard at ω, then $I_\omega \Delta e_n = e_n$ is only valid for $n \geq 1$.

The second formula in [2.19] can be rewritten as:

$$f(u) = f - I_u \Delta f = (\Delta^0 - I_u \Delta) f.$$

This leads to defining a new operator δ_u as follows.

DEFINITION 2.6.– $\delta_u = \Delta^0 - I_u \Delta$ *is the operator on* $\mathfrak{F}$ *such that* $f \mapsto f(u)$.

REMARK.– The integration operator makes it possible to generalize [2.16], which comes from the elementary theory of integration. All the beautiful properties of this theory are however not retained in this mimicry.

For example, when $\mathcal{D} = \mathbb{R}$, I_u does not necessarily keep the usual positivity of the integration. Here is an example where $f \geq 0$ for $x > 0$ and simultaneously $I_u f < 0$ for some $x > u$. Let $\mathcal{D} = \mathbb{R}$ and a family $\mathcal{E}$ such that for $n = 0, 1, 2, 3$:

$$e_0 = 1,\ e_1(x) = x,\ e_2(x) = x^2 + ax,\ e_3(x) = x^3 + bx^2 + cx, \ldots,$$

where a, b, c are real constants. Consider the function $f(x) = x^2\,(\geq 0)$. As $x^2 = e_2(x) - ae_1(x)$, we have $f = e_2 - ae_1 \in \mathfrak{F}$ and from [2.14]:

$$I_0 f(x) = I_0\big(e_2(x) - ae_1(x)\big) = e_3(x) - ae_2(x) = x^3 + (b-a)x^2 + (c-a)x,$$

which may be negative in $x = 1\,(> u = 0)$ if $1 + b - a + c - a < 0$. This disagrees with $\int_0^x z^2 dz = x^3/3 > 0$.

The operator Δ which is a mimicry of D leads to similar comments. Note that, a priori, $(e_1)^2$ could be outside $\mathfrak{F}$, but this is not the case here since $(e_1)^2 = e_2 - ae_1 \in \mathfrak{F}$. We then see that $\Delta(e_1 e_1)$ gives $\Delta(e_1)^2 = e_1 - ae_0$, which is not equal to $2e_1 = e_1 e_0 + e_0 e_1 = e_1 \Delta(e_1) + \Delta(e_1) e_1$. Thus, the Leibniz formula valid for D does not apply to Δ.

2.4. A-G pseudopolynomials $\bar{G}_n(\ |U)$

By similarity with the family of A–G polynomials $\{G_n(x|U),\ n \in \mathbb{N}_0\}$, we start from an integral representation of the family of *A–G pseudopolynomials* $\{\bar{G}_n(\ |U),\ n \in \mathbb{N}_0\}$ of degree n.

DEFINITION 2.7.– *Any countable (or finite) family* $U = \{u_i \in \mathfrak{D},\ i \in \mathbb{N}_0\}$ *is associated with the family* $\mathcal{G}(U)$ *of A–G pseudopolynomials defined by:*

$$\begin{cases} \bar{G}_0(\ |U) = e_0, \\ \bar{G}_n(\ |U) = I_{u_0} I_{u_1} \dots I_{u_{n-1}} e_0, \quad n \geq 1, \end{cases} \qquad [2.20]$$

that is, given recursively by:

$$\bar{G}_n(\ |U) = I_{u_0} \bar{G}_{n-1}(\ |EU), \quad n \geq 1, \qquad [2.21]$$

where EU *is the shifted family* $\{u_{1+i} \in \mathfrak{D},\ i \in \mathbb{N}_0\}$.

Of course, $\bar{G}_n(x|U)$ is the value of $\bar{G}_n(\ |U)$ at point $x \in \mathfrak{D}$. For convenience, we will sometimes put $\bar{G}_n(\ |U) = 0$ when $n < 0$.

EXAMPLE 2.3.– Consider again the A–G polynomial case. Using [2.16], we have:

$$\bar{G}_n(x|U) = \int_{u_0}^{x} dx_1 \int_{u_1}^{x_1} dx_2 \dots \int_{u_{n-1}}^{x_{n-1}} dx_n, \qquad [2.22]$$

which provides the integral representation [1.8] of the A–G polynomials $G_n(x|U)$.

PROPOSITION 2.8.– *Similarly to [1.12] and [1.13]:*

$$\Delta^m \bar{G}_n(\ |U) = \bar{G}_{n-m}(\ |E^m U), \quad 0 \leq m \leq n, \qquad [2.23]$$

where $E^m U = \{u_{m+i},\ i \in \mathbb{N}_0\}$, *and*

$$\Delta^m \bar{G}_n(u_m|U) = \delta_{m,n}, \quad 0 \leq m \leq n. \qquad [2.24]$$

Proof. Since $\Delta I_u = \Delta^0$, we get:

$$\Delta^m \bar{G}_n(\ |U) = \Delta^m I_{u_0} I_{u_1} \dots I_{u_m} \dots I_{u_{n-1}}\, e_0 = I_{u_m} \dots I_{u_{m+(n-m)-1}}\, e_0,$$

hence [2.23]. The formula [2.24] follows from the property $(I_u f)(u) = 0$ stated in [2.19]. □

COROLLARY 2.1.–

$$\Delta I_{u_0}\bar{G}_n(\ |U) = \bar{G}_n(\ |U) = I_{u_0}\Delta\bar{G}_n(\ |U), \quad n \geq 1. \tag{2.25}$$

Proof. The first identity of [2.25] is immediate from [2.23]. The second comes from:

$$I_{u_0}\Delta\big(I_{u_0}I_{u_1}\dots e_0\big) = I_{u_0}(\Delta I_{u_0})I_{u_1}\dots e_0 = I_{u_0}\Delta^0 I_{u_1}\dots = I_{u_0}I_{u_1}\dots e_0,$$

as announced. For another proof, [2.19] says that $I_{u_0}\Delta f = f - f(u_0)$, and here $f(u_0) = \bar{G}_n(u_0)(u_0|U) = 0$ by [2.24]. □

PROPOSITION 2.9.– *$\mathcal{G}(U)$ is a basic family for $\mathfrak{F}$.*

Proof. From [2.20] with [2.14], $\bar{G}_n(\ |U)$ is a pseudopolynomial of the form:

$$\bar{G}_n(\ |U) = e_n + \sum_{i=0}^{n-1} c_i\, e_i, \tag{2.26}$$

where the c_i are constant coefficients. Therefore, e_n can be expressed as:

$$e_n = \bar{G}_n(\ |U) + \sum_{i=0}^{n-1} a_i\bar{G}_i(\ |U) \tag{2.27}$$

for some appropriate constants a_i. From [2.26] and [2.27], $\mathcal{G}(U)$ is linearly independent and generates $\mathfrak{F}$ as does $\mathcal{E}$. □

Note that $\mathcal{G}(U)$ is not a copy of $\mathcal{E}$ except for the case of $EU = U$, that is, $U = \{u, u, u, \dots\}$.

First A–G pseudopolynomials. For any family $\mathcal{E}$:

$$\begin{aligned}
\bar{G}_1(\ |U) &= I_{u_0}e_0 = e_1 - e_1(u_0)e_0,\\
\bar{G}_2(\ |U) &= I_{u_0}\bar{G}_1(\ |EU) = I_{u_0}\{e_1 - e_1(u_1)e_0\}\\
&= e_2 - e_2(u_0) - e_1(u_1)[e_1 - e_1(u_0)e_0]\\
&= e_2 - e_1(u_1)e_1 + [e_1(u_1)e_1(u_0) - e_2(u_0)]e_0,
\end{aligned}$$

$$
\begin{aligned}
\bar{G}_3(\ |U) &= I_{u_0}\bar{G}_2(\ |EU) \\
&= I_{u_0}\{e_2 - e_1(u_2)e_1 + [e_1(u_2)e_1(u_1) - e_2(u_1)]e_0\} \\
&= e_3 - e_3(u_0) - e_1(u_2)[e_2 - e_2(u_0)] + [e_1(u_2)e_1(u_1) \\
&- e_2(u_1)][e_1 - e_1(u_0)] \\
&= e_3 - e_1(u_2)e_2 + [e_1(u_2)e_1(u_1) - e_2(u_1)]e_1 \\
&- \{e_3(u_0) - e_1(u_2)e_2(u_0) + [e_1(u_2)e_1(u_1) \\
&- e_2(u_1)]e_1(u_0)\}e_0.
\end{aligned}
$$

In particular, for the Abel case where $e_n(x) = x^n/n!$ and $u_i = \alpha + \beta i$ for $n, i \in \mathbb{N}_0$, this yields, as expected:

$$
\begin{aligned}
G_1(x|U) &= x - \alpha, \\
G_2(x|U) &= (x^2 - \alpha^2)/2 - (\alpha + \beta)(x - \alpha) \\
&= (x - \alpha)[x - (\alpha + 2\beta)]/2, \\
G_3(x|U) &= (x^3 - \alpha^3)/3! - (\alpha + 2\beta)(x^2 - \alpha^2)/2 \\
&+ [(\alpha + 2\beta)(\alpha + \beta) - (\alpha + \beta)^2/2](x - \alpha) \\
&= (x - \alpha)\{x^2 + \alpha x + \alpha^2 - 3(\alpha + 2\beta)(x + \alpha) \\
&+ 3(\alpha + \beta)[2(\alpha + 2\beta) - (\alpha + \beta)]\}/3! \\
&= (x - \alpha)[x - (\alpha + 3\beta)]^2/3!.
\end{aligned}
$$

Let us point out that in the notation $\bar{G}_n(\ |U)$, we found it unnecessary to specify the basic family $\mathcal{E}$ used. This can be explained based on the following result.

PROPOSITION 2.10.– *Let $\mathcal{E}$ and $\hat{\mathcal{E}}$ be two basic families that generate $\mathfrak{F}$ through the operators Δ and $\hat{\Delta}$, respectively. Then:*

$$\bar{G}_n(\ |\mathcal{E}, U) = \bar{G}_n(\ |\hat{\mathcal{E}}, U), \quad \textit{for all } U \textit{ and } n \in \mathbb{N}_0, \qquad [2.28]$$

if and only if $\hat{\mathcal{E}}$ is a copy of $\mathcal{E}$.

Proof. By proposition 2.6, we already know that the family A–G does not change when working with a copy $\hat{\mathcal{E}}$ of $\mathcal{E}$. We now show the converse, namely that [2.28] implies that $\hat{\mathcal{E}}$ is a copy of $\mathcal{E}$. To do this, we apply the operator Δ to $\bar{G}_n(\ |\mathcal{E}, U)$ and successively obtain:

$$\begin{aligned}\Delta\bar{G}_n(\ |\mathcal{E}, U) &= \bar{G}_{n-1}(\ |\mathcal{E}, EU) \quad \text{(from [2.23])}\\ &= \bar{G}_{n-1}(\ |\hat{\mathcal{E}}, EU) \quad \text{(using [2.28])}\\ &= \hat{\Delta}\bar{G}_n(\ |\hat{\mathcal{E}}, U) \quad \text{(by definition of } \hat{\Delta})\\ &= \hat{\Delta}\bar{G}_n(\ |\mathcal{E}, U) \quad \text{(using again [2.28]).}\end{aligned}$$

Thus, $\Delta = \hat{\Delta}$ on $\mathfrak{F}$, which means that $\Delta\hat{e}_{n+1} = \hat{e}_n$, $n \in \mathbb{N}_0$. By [2.7], $\hat{\mathcal{E}}$ is an Appell type family, and therefore a copy of $\mathcal{E}$. □

Note that if $\hat{\mathcal{E}}$ is not a copy, the situation would be different. This is shown below.

REMARK.– As a first simple example, let $\mathfrak{D} = \mathbb{R}$ and choose $e_n(x) = x^n/n!$ and $\tilde{e}_n(x) = \binom{x}{n}$, $n \in \mathbb{N}_0$. Then, $\tilde{\mathcal{E}}$ is a basic family that generates $\mathfrak{F}$ but is not a copy of $\mathcal{E}$. Let us now explicitly add the basic family in the writing of $\bar{G}_n$. Since:

$$\tilde{e}_1(x) = e_1(x), \quad \text{and} \quad \tilde{e}_2(x) = e_2(x) - e_1(x)/2,$$

we have $\bar{G}_1(\ |\tilde{\mathcal{E}}, U) = \bar{G}_1(\ |\mathcal{E}, U)$, but:

$$\begin{aligned}\bar{G}_2(\ |\tilde{\mathcal{E}}, U) &= [\tilde{e}_2 - \tilde{e}_2(u_0)] - \tilde{e}_1(u_1)[\tilde{e}_1 - \tilde{e}_1(u_0)]\\ &= [e_2 - e_2(u_0)] - e_1(u_1)[e_1 - e_1(u_0)] - [e_1 - e_1(u_0)]/2\\ &= \bar{G}_2(\ |\mathcal{E}, U) - \bar{G}_1(\ |\mathcal{E}, U)/2,\end{aligned}$$

the difference between the two $\bar{G}_2$ being exactly known.

As a second example, take $\mathfrak{D} = \mathbb{R}$ and the same $\mathcal{E}$, but choose $\breve{e}_n(x) = x^{2n}/n!$, $n \in \mathbb{N}_0$. Then:

$$\breve{e}_n(x) = e_n(x^2) = [(2n)!/n!]e_{2n}(x),$$

thus $\breve{\mathcal{E}}$ does not generate $\mathfrak{F}$ but only a subspace $\breve{\mathfrak{F}}$ of $\mathfrak{F}$. In $\breve{\mathfrak{F}}$, we can write:

$$\begin{aligned}\bar{G}_1(x|\breve{\mathcal{E}}, U) &= \breve{e}_1(x) - \breve{e}_1(u_0) = e_1(x^2) - e_1({u_0}^2)\\ &= \bar{G}_1(x^2|\mathcal{E}, \{{u_0}^2, {u_1}^2, \ldots\} \neq \bar{G}_1(x|\mathcal{E}, U),\end{aligned}$$

so the two $\bar{G}_n$ differ as soon as $n \geq 1$.

2.5. Expansions of A–G type

In [1.15], we considered the expansion of a function with respect to a family of A–G polynomials. A similar type of expansion is valid here too using a family of A–G pseudopolynomials.

PROPOSITION 2.11.– *Let $\mathcal{G}(U)$ be a family of A–G pseudopolynomials. Then, any pseudopolynomial $f_n \in \mathfrak{F}$, $n \in \mathbb{N}_0$, admits an expansion, said of Abel-Gontcharoff type, given by:*

$$f_n(\ ) = \sum_{i=0}^{n} \Delta^i f_n(u_i)\, \bar{G}_i(\ |U). \qquad [2.29]$$

This formula extends as a formal series inside the generalized vector space $\mathfrak{F}_{\mathcal{E}}$.

Proof. By proposition 2.9, $\mathcal{G}(U)$ is a basic family for $\mathfrak{F}$, so that f_n can be expressed as:

$$f_n(\ ) = \sum_{i=0}^{n} c_i\, \bar{G}_i(\ |U). \qquad [2.30]$$

Apply now Δ^r to [2.30] and then choose $u_r \in \mathfrak{D}$, for all $0 \leq r \leq n$. Using [2.14], we then obtain:

$$\Delta^r f_n(u_r) = \sum_{i=0}^{n} c_i\, \Delta^r \bar{G}_i(u_r|U) = \sum_{i=0}^{n} c_i\, \delta_{r,i} = c_r,$$

which, after insertion into [2.30], provides the announced expansion [2.29]. □

The A-G type expansion formula [2.29] is a central result of this monography. It is important to emphasize that this result has a kind of universal validity because special assumptions on $\mathfrak{D}$ and $\mathcal{E}$ have not been introduced so far. The situation is different with a Taylor type formula which requires at least defining an addition operator on $\mathfrak{D}$. This exceptional robustness of A-G type expansions probably explains why they appear in so many fields of application, sometimes in unexpected ways.

2.5.1. *Some simple consequences*

i) From [2.24] and [2.29], we easily deduce the following characterization of the A–G pseudopolynomials.

PROPOSITION 2.12.– *A pseudopolynomial f_n corresponds to $\bar{G}_n(\ |U)$ iff f_n satisfies the conditions:*

$$\Delta^m f_n(u_m) = \delta_{m,n}, \quad 0 \leq m \leq n. \tag{2.31}$$

ii) Let us derive a recursion for the A–G pseudopolynomials. From [2.29] with [2.1]:

$$e_n(\ ) = \sum_{i=0}^{n} e_{n-i}(u_i)\, \bar{G}_i(\ |U). \tag{2.32}$$

Thanks to [2.32], $\bar{G}_n(\ |U)$ can therefore be expressed as:

$$\bar{G}_n(\ |U) = e_n(\ ) - \sum_{i=0}^{n-1} e_{n-i}(u_i)\, \bar{G}_i(\ |U), \tag{2.33}$$

which gives a recursive way to evaluate $\bar{G}_n(\ |U)$, $n \geq 1$, starting with $\bar{G}_0(\ |U) = 1$.

iii) Let U and V be two distinct families in $\mathfrak{D}$. From [2.29] with [2.21], we get:

$$\bar{G}_n(\ |V) = \sum_{i=0}^{n} \bar{G}_{n-i}(u_i|E^i V)\, \bar{G}_i(\ |U),$$

$$\bar{G}_n(\ |U) = \sum_{i=0}^{n} \bar{G}_{n-i}(v_i|E^i U)\, \bar{G}_i(\ |V), \tag{2.34}$$

which show how two families of A–G pseudopolynomials are actually related.

iv) Consider the special case where $U = \{\omega, \omega, \ldots\}$ with $\omega \in \mathbb{R}$. Then, $EU = U$ so that [2.23] implies $\Delta\bar{G}_n(\ |U) = \bar{G}_{n-1}(\ |U)$, $n \geq 1$, which is recognized as the Appell property with respect to Δ. Therefore, the family of

$\bar{G}_n(\ |\{\omega, \omega, \ldots\})$ is a copy of $\mathcal{E}$. It is in fact the copy which is standard at ω, that is, $\hat{\mathcal{E}}^{(\omega)} = \mathcal{G}(\{\omega, \omega, \ldots\})$. Moreover, [2.29] then becomes:

$$f_n(\ ) = \sum_{i=0}^{n} \Delta^i f_n(\omega)\, \bar{G}_i(\ |\{\omega, \omega, \ldots\}), \qquad [2.35]$$

whose structure is of the Maclaurin formula type.

Note that regardless of already having [2.32], it is tempting and would be convenient to obtain e_n directly in the form of a $\bar{G}_n$. When this occurs, it most often involves using:

$$e_n(\ ) = \bar{G}_n(\ |\{0, 0, \ldots\}),$$

which is however not a universal identity but requires $\mathcal{E}$ to be standard at 0 (see corollary 2.2 for more details).

v) Continuing the remark of section 2.4, let us return to $\mathfrak{D} = \mathbb{R}$, $e_n(x) = x^n/n!$ and $\breve{e}_n(x) = x^{2n}/n!$, $n \in \mathbb{N}_0$. From the binomial identity:

$$\breve{e}_n(x) = (x^2 - u^2 + u^2)^n/n! = \sum_{i=0}^{n} \left[(u^2)^{n-i}/(n-i)!\right] \left[(x^2 - u^2)^i/i!\right],$$

we see that each $\breve{e}_n(x)$ can be expressed as:

$$\breve{e}_n(x) = \sum_{i=0}^{n} \breve{e}_{n-i}(u)\, (x^2 - u^2)^i/i!, \qquad [2.36]$$

which corresponds to an A–G type expansion.

In fact, the expansion of $\breve{e}_n(x)$ with respect to $\bar{G}_i(x|\breve{\mathcal{E}}, \{u, u, \ldots\})$ is given by [2.29]:

$$\breve{e}_n(x) = \sum_{i=0}^{n} \breve{e}_{n-i}(u)\, \bar{G}_i(x|\breve{\mathcal{E}}, \{u, u, \ldots\}). \qquad [2.37]$$

By subtracting [2.37] from [2.36], we then get:

$$0 = \sum_{i=0}^{n} \breve{e}_{n-i}(u)\left[(x^2 - u^2)^i/i! - \bar{G}_i(x|\breve{\mathcal{E}}, \{u, u, \ldots\})\right], \quad n \in \mathbb{N}_0. \quad [2.38]$$

Consider [2.38] successively for $n = 0, 1, \ldots$. Since for $n = 0$, the term $[\ldots]$ vanishes, we see by recursion that all the terms $[\ldots]$ in [2.38] are equal to 0. In other words, we find that:

$$\bar{G}_n(x|\breve{\mathcal{E}}, \{u, u, \ldots\}) = (x^2 - u^2)^n/n!, \quad n \in \mathbb{N}_0. \qquad [2.39]$$

vi) More generally, for any family $U = \{u_i \in \mathfrak{D},\ i \in \mathbb{N}_0\}$, [2.32] implies:

$$\breve{e}_n(x) = \sum_{i=0}^{n} \breve{e}_{n-i}(u_i)\, \bar{G}_i(x|\breve{\mathcal{E}}, U),$$

that is, since $\breve{e}_n(x) = e_n(x^2)$:

$$e_n(x^2) = \sum_{i=0}^{n} e_{n-i}(u_i^2)\, \bar{G}_i(x|\breve{\mathcal{E}}, U). \qquad [2.40]$$

On the other hand, from [2.32] applied to $e_n(x^2)$ and using the family $\{u_i^2,\ i \in \mathbb{N}_0\}$, we know that:

$$e_n(x^2) = \sum_{i=0}^{n} e_{n-i}(u_i^2)\, \bar{G}_i(x^2|\mathcal{E}, \{u_0^2, u_1^2, \ldots\}). \qquad [2.41]$$

By comparing [2.40] and [2.41], we then obtain the identity:

$$\bar{G}_n(x|\breve{\mathcal{E}}, \{u_0, u_1, \ldots\}) = \bar{G}_n(x|\mathcal{E}, \{u_0^2, u_1^2, \ldots\}), \quad n \in \mathbb{N}_0.. \qquad [2.42]$$

Note that when U is affine, the l.h.s. of (2.42) is not a standard Abel polynomial because $\breve{\mathcal{E}} \neq \mathcal{E}$. Also, the difference operator defined by $\breve{\mathcal{E}}$ is not shift-invariant as defined in section 2.8, so the explicit formula derived in section 3.1 will not be valid for these pseudopolynomials.

2.5.2. *Back to the A-G polynomials*

Let us return to the historical case where $\mathfrak{D} = \mathbb{R}$ and $e_n(x) = x^n/n!$, $n \in \mathbb{N}_0$. Suppose now that the family U is a sequence of non-increasing reals u_i (i.e. $u_i \geq u_{i+1}$). For some applications, it is then natural to introduce a family U_+ defined by:

$$U_+ = \{(u_i)_+,\ i \in \mathbb{N}_0\}, \quad \text{where} \quad (\mathrm{u_i})_+ = \max\{0, \mathrm{u_i}\}.$$

We are going to show that the corresponding A–G polynomials $G_n(x|U_+)$ can be expressed in terms of the simpler A–G polynomials $G_n(x|U)$.

PROPERTY 2.3.–

$$G_n(x|U_+) = \frac{x^n}{n!} - \sum_{i=0}^{n-1} \frac{[(u_i)_+]^{n-i}}{(n-i)!} G_i(x|U), \quad n \in \mathbb{N}_0. \qquad [2.43]$$

Proof. Indeed, an A–G expansion of $x^n/n!$, $n \geq 0$, with respect to the family $\mathcal{G}(\mathcal{U}_+)$ gives:

$$\frac{x^n}{n!} = \sum_{i=0}^{n} \frac{[(u_i)_+]^{n-i}}{(n-i)!} G_i(x|U_+). \qquad [2.44]$$

In the sum of [2.44], U can be substituted for U_+ for all the terms of $i = 0, \ldots, n-1$ since either $(u_i)_+ = 0$ and the substitution has no effect, or $(u_i)_+ \neq 0$ and then $u_i > 0$ with thus $u_0, \ldots, u_{i-1}$ also because U is assumed to be non-increasing. So, we can write $G_i(x|U)$ instead of $G_i(x|U_+)$ for all $i < n$. When $i = n$, such a substitution is not allowed. Therefore, [2.43] follows directly from [2.44]. □

An equivalent formula is possible which this time uses the parameters $(u_i)_- = \max\{0, -u_i\}$.

PROPERTY 2.4.–

$$G_n(x|U_+) = \sum_{i=0}^{n} \frac{[-(u_i)_-]^{n-i}}{(n-i)!} G_i(x|U), \quad n \in \mathbb{N}_0. \qquad [2.45]$$

Proof. Indeed, let us denote L the sum in the r.h.s. of [2.45]. Subtracting L from [2.43], we obtain:

$$G_n(x|U_+) - L = \frac{x^n}{n!} - \sum_{i=0}^{n-1} \frac{[(u_i)_+]^{n-i} + [-(u_i)_-]^{n-i}}{(n-i)!} G_i(x|U) - G_n(x|U). \qquad [2.46]$$

Now, for any $a \in \mathbb{R}$, we have the identity $(a_+)^{n-i} + (-a_-)^{n-i} = a^{n-i}$. So, [2.46] becomes:

$$G_n(x|U_+) - L = \frac{x^n}{n!} - \sum_{i=0}^{n} \frac{[(u_i)]^{n-i}}{(n-i)!} G_i(x|U),$$

whose r.h.s. is equal to 0 due to the A–G expansion of $x^n/n!$ with respect to $\mathcal{G}(U)$. In other words, $L = G_n(x|U_+)$ as announced in [2.45]. □

REMARK.– A–G type expansions can be considered as generalizations of other expansions encountered in calculus. Since $\mathfrak{F}$ is a vector space, a more geometric presentation including the definition of a metric is possible. In geometry, finding the components of a given vector on the principal basis of a vector space is a standard operation, the task being generally easier if the basis has been ortho-normalized. However, in the context of pseudopolynomials, things are different. Here, no canonical basis really exists, and one can look up the components of any $B \in \mathfrak{F}$ on the family $\mathcal{E}$, one of its copies, the family $\mathcal{G}(U)$, or other families $\mathcal{G}(V), \mathcal{G}(W)$, because all these families will generate $\mathfrak{F}$. Such operations will only perform A–G type expansions of B, and a metric seems pointless, if not useless.

2.6. A shift operator S^a

Let us suppose that an abelian group structure is defined on $\mathfrak{D}$, with 0 as the neutral element. Given any $a \in \mathfrak{D}$, the functions $e_n(. + a)$ then have a meaning, and we may consider the family:

$$\mathcal{E}^a \equiv \{e_n(. + a),\, n \in \mathbb{N}_0\}.$$

Note that the familly $\mathcal{E}^a$ is linearly independent. Indeed, if there were constants λ_i for i in a subset K such that:

$$\sum_{i \in K} \lambda_i\, e_i(. + a) = 0,$$

then at $x - a$, that would give $\sum_{i \in K} \lambda_i\, e_i(x) = 0$, implying that all λ_i are zero by the linear independence of $\mathcal{E}$. Thus, the family $\mathcal{E}^a$ generates a vector space, denoted $\mathfrak{F}^a$, which is similar but generally not identical to $\mathfrak{F}$. This naturally leads us to introduce a shift operator as follows.

DEFINITION 2.8.– *Any $a \in \mathfrak{D}$ is associated with the shift operator S^a defined on $\mathcal{E}$ by:*

$$S^a : e_n(.) \to e_n(. + a), \quad n \in \mathbb{N}_0. \qquad [2.47]$$

Of course, $S^a S^b = S^b S^a$ for $a, b \in \mathfrak{D}$. As indicated before, S^a changes the family $\mathcal{E}$ to $\mathcal{E}^a$ and the space $\mathfrak{F}$ to $\mathfrak{F}^a$. By construction, $\mathcal{E}^a$ is a basic family of $\mathfrak{F}^a$, so if f_n is a pseudopolynomial of degree n in $\mathfrak{F}$, $S^a f_n$ will be a pseudopolynomial of degree n in $\mathfrak{F}^a$. However, if it happens that $S^a f_n \in \mathfrak{F}$, it is not guaranteed that as an element of $\mathfrak{F}$, its degree is always equal to n.

Moreover, S^a transforms the operator Δ on $\mathcal{E}$ into an operator Δ_a on $\mathcal{E}^a$ whose characteristic is obtained below.

PROPOSITION 2.13.– *On $\mathcal{E}^a$:*

$$\Delta_a = S^a \Delta S^{-a}. \qquad [2.48]$$

Proof. By [2.47], Δ_a acts on $\mathcal{E}^a$ in such a way that:

$$\Delta_a(S^a e_n) = S^a e_{n-1}, \;\; n \geq 1, \;\; \text{and} \;\; \Delta_a(S^a e_0) = 0. \qquad [2.49]$$

Using [2.1], we see from [2.49] that $\Delta_a(S^a e_n) = S^a(\Delta e_n)$ for $n \geq 1$ and $\Delta_a(S^a)e_0 = S^a(\Delta e_0) = 0$. This shows that $\Delta_a S^a = S^a \Delta$ on $\mathcal{E}$, and thus $\Delta_a = S^a \Delta S^{-a}$ on $\mathcal{E}^a$, that is, [2.48]. □

Observe that $S^a \Delta$ still makes sense but ΔS^a does not since Δ is defined on $\mathfrak{F}$ and not on $\mathfrak{F}^a$. When $\mathfrak{F}$ contains $\mathfrak{F}^a$, the two operators S^a and Δ are however well defined, but they do not necessarily commute, so $\Delta_a \neq \Delta$.

EXAMPLE 2.4.–

1) Take $\mathfrak{D} = \mathbb{R}^{\mathbb{N}}$ with generic elements $\mathbf{x} = \{x_j, \, j \geq 1\}$, and define $e_n(\mathbf{x}) = \prod_{j=1}^{n}(1/n!) \exp(\alpha_j x_j)$, $n \geq 1$, for some reals α_j, with $e_0 = 1$. Then,

$$e_n(\mathbf{x} + \mathbf{a}) = n! e_n(\mathbf{a}) e_n(\mathbf{x}), \quad n \in \mathbb{N}_0,$$

and since $n! e_n(\mathbf{a}) \in \mathbb{R}$, we see that $S^{\mathbf{a}} e_n \in \mathfrak{F}$. In fact, $\mathcal{E}^{\mathbf{a}} \subset \mathfrak{F}$, so $\mathfrak{F}^{\mathbf{a}} \subset \mathfrak{F}$, but also $\mathfrak{F} \subset \mathfrak{F}^{\mathbf{a}}$, hence $\mathfrak{F}^{\mathbf{a}} = \mathfrak{F}$. The situation here is exceptionally simple.

2) Again let $\mathfrak{D} = \mathrm{IR}^{\mathrm{IN}}$, but define $e_n(\mathbf{x}) = (1/n!)x_1x_2\ldots x_n$, $n \geq 1$, with $e_0 = 1$. Then:

$$S^{\mathbf{a}}e_1(\mathbf{x}) = e_1(\mathbf{x}+\mathbf{a}) = e_1(x_1) + a_1 = e_1(\mathbf{x}) + a_1e_0(\mathbf{x}),$$

thus $S^{\mathbf{a}}e_1(\mathbf{x}) \in \mathfrak{F}$, but:

$$\begin{aligned} S^{\mathbf{a}}e_2(\mathbf{x}) &= (x_1+a_1)(x_2+a_2)/2! = x_1x_2/2 + (a_2/2)x_1 \\ &\quad + (a_1/2)x_2 + (a_1a_2)/2 \\ &= e_2(\mathbf{x}) + (a_2/2)e_1(\mathbf{x}) + (a_1/2)x_2 + e_2(\mathbf{a}), \end{aligned}$$

where the first, second and fourth terms are in $\mathfrak{F}$ but not the third, thus $S^{\mathbf{a}}e_2(\mathbf{x}) \notin \mathfrak{F}$.

Let us highlight a simple effect of a shift $a \in \mathfrak{D}$ on the A–G pseudopolynomials. For this, the notation $\bar{G}_n^a$ will indicate that the A–G pseudopolynomial is then defined on the set $\mathfrak{D}^a$; an explicit reference to $\mathcal{E}^a$ will also be added. Note that $\bar{G}_n \in \mathfrak{F}$ while $\bar{G}_n^a \in \mathfrak{F}^a$, they are thus distinct although being linked between them by their images in IR (or $\mathbb{C}$).

PROPOSITION 2.14.–

$$\bar{G}_n(x+a|\mathcal{E}, U+a) = \bar{G}_n^a(x|\mathcal{E}^a, U), \quad n \in \mathrm{IN}_0, \tag{2.50}$$

where $U + a = \{u_i + a,\ i \in \mathrm{IN}_0\} \subset \mathfrak{D}$.

Proof. In $\mathfrak{F}$, the expansion [2.32] with respect to $U + a$ gives:

$$e_n(\;) = \sum_{i=0}^{n} e_{n-i}(u_i+a)\,\bar{G}_i(\;|\mathcal{E}, U+a),$$

and thus at point $x + a$:

$$e_n(x+a) = \sum_{i=0}^{n} e_{n-i}(u_i+a)\,\bar{G}_i(x+a|\mathcal{E}, U+a). \tag{2.51}$$

In $\mathfrak{F}^a$, [2.32] with respect to U implies that:

$$S^a e_n(\;) = \sum_{i=0}^{n} \left[\Delta_a^i(S^a e_n)\right](u_i)\,\bar{G}_i^a(\;|\mathcal{E}^a, U),$$

in which, as with [2.49]:

$$\Delta_a^i(S^a e_n) = S^a e_{n-i} = e_{n-i}(\ +a),$$

so that at point x:

$$e_n(x+a) = \sum_{i=0}^{n} e_{n-i}(u_i+a)\, \bar{G}_i^a(x|\mathcal{E}^a, U). \qquad [2.52]$$

By subtracting [2.52] from [2.51], we have:

$$0 = \sum_{i=0}^{n} e_{n-i}(u_i+a)\,[\bar{G}_i(x+a|\mathcal{E}, U+a) - \bar{G}_i^a(x|\mathcal{E}^a, U)] \equiv \sum_{i=0}^{n} e_{n-i}(u_i+a)\,\gamma_i,$$

after an obvious notation, and therefore:

$$\gamma_n = -\sum_{i=0}^{n-1} e_{n-i}(u_i+a)\,\gamma_i. \qquad [2.53]$$

Starting from $\gamma_0 = 0$, we deduce by recursion that all the γ_i are null, hence [2.50]. □

REMARK.– A special case of [2.50] is when all the u_i are null, which gives:

$$\bar{G}_n(x+a|\mathcal{E}, \{a, a, \ldots\}) = \bar{G}_n^a(x|\mathcal{E}^a, \{0, 0, \ldots\}), \quad n \in \mathbb{N}_0. \qquad [2.54]$$

This is the value at x of the copy of the standard $S^a e_n$ at point 0, which is denoted $\widehat{(S^a e_n)}^{(0)}(x)$ according to section 2.2.

Note that from [2.51] and [2.54], we can write:

$$e_n(x+a) = \sum_{i=0}^{n} e_{n-i}(a)\, \widehat{(S^a e_i)}^{(0)}(x),$$

and more generally, for any $f_n \in \mathfrak{F}$:

$$f_n(x+a) = \sum_{i=0}^{n} (\Delta^i f_n)(a)\, \widehat{(S^a e_i)}^{(0)}(x).$$

These two formulas have a certain similarity with the binomial identity and the Taylor formula, respectively. However, to achieve this identification, we will show in section 2.8 that a property of shift invariance (S.I.) is necessary.

2.7. A multiplication operator M^λ

Consider for $\mathfrak{D}$ a real vector space with the associated external product:

$$\mathbb{R} \times \mathfrak{D} \to \mathfrak{D} : (\lambda, \mathrm{x}) \to \lambda\mathrm{x}.$$

DEFINITION 2.9.– *Any $\lambda \in \mathbb{R}$ is associated with the multiplication operator M^λ defined on $\mathcal{E}$ by:*

$$M^\lambda : e_n(x) \to e_n(\lambda x), \quad n \in \mathbb{N}_0.$$

The family $\mathcal{E}$ has the so-called multiplication property when:

$$M^\lambda e_n(x) = e_n(\lambda x) \;\equiv\; \lambda^n e_n(x), \quad n \in \mathbb{N}_0. \qquad [2.55]$$

Of course, [2.55] gives $M^\lambda \mathfrak{F} = \mathfrak{F}$. In the following, we will always assume that $\mathcal{E}$ has the multiplication property. First, here are some direct implications

PROPOSITION 2.15.– *Suppose [2.55] is verified.*

a) $\mathcal{E}$ is standard at 0.

b) If $\mathfrak{D} = \mathbb{R}$, $e_n(x) = e_n(1)\, x^n$.

c) A copy $\hat{e}_n$ of e_n satisfies $\hat{e}_n(\lambda x) = \lambda^n\, \hat{e}_n^\lambda(x)$ where $\hat{e}_n^\lambda$ is another copy of e_n depending on λ.

Proof. For (a), $e_n(\lambda 0) = e_n(0) = \lambda^n e_n(0)$ from the assumption [2.55]. Thus, $(1 - \lambda^n) e_n(0) = 0$, so that $e_n(0) = \delta_{n,0}$. The property (b) is straightforward. For (c), the definitions [2.2] and [2.3] and the multiplication assumption give:

$$\hat{e}_n(\lambda x) = \sum_{i=0}^{n} d_i\, e_{n-i}(\lambda x) = \lambda^n \sum_{i=0}^{n} (d_i/\lambda^i)\, e_{n-i}(x),$$

hence the announced result. □

In parallel with [2.50], we obtain the relation below concerning the A–G pseudopolynomials.

PROPOSITION 2.16.–

$$\bar{G}_n(\lambda x|\lambda U) = \lambda^n \bar{G}_n(x|U), \quad n \in \mathbb{N}_0, \tag{2.56}$$

where $\lambda U = \{\lambda u_i,\ i \in \mathbb{N}_0\}$.

Proof. From the A–G expansion [2.30], we have, for any $n \in \mathbb{N}_0$:

$$\bar{G}_n(\ |U) = e_n(\ ) - \sum_{i=0}^{n-1} e_{n-i}(u_i)\,\bar{G}_i(\ |U), \tag{2.57}$$

$$\bar{G}_n(\ |\lambda U) = e_n(\ ) - \sum_{i=0}^{n-1} e_{n-i}(\lambda u_i)\,\bar{G}_i(\ |\lambda U). \tag{2.58}$$

Take [2.57] at x with multiplication by λ^n, and substract [2.58] at λx. Using [2.55], we get:

$$\lambda^n \bar{G}_n(x|U) - \bar{G}_n(\lambda x|\lambda U) = \alpha_n - \sum_{i=0}^{n-1} e_{n-i}(\lambda u_i)[\lambda^i \bar{G}_i(x|U) - \bar{G}_i(\lambda x|\lambda U)] \tag{2.59}$$

where $\alpha_n \equiv \lambda^n e_n(x) - e_n(\lambda x)$ which reduces to 0. Therefore, [2.59] can be rewritten as:

$$\beta_n = -\sum_{i=0}^{n-1} e_{n-i}(\lambda u_i)\,\beta_i, \tag{2.60}$$

where $\beta_i \equiv \lambda^i \bar{G}_i(x|U) - \bar{G}_i(\lambda x|\lambda U)$, $0 \le i \le n$. Since $\beta_0 = 0$, we deduce from [2.60], by recursion on n, that all the β_i vanish, hence the identity [2.56]. □

We finish by examining a possible permutation with the previous operators Δ and S^a.

PROPOSITION 2.17.–

$$\Delta M^\lambda = \lambda M^\lambda \Delta, \quad \textit{and} \quad DM^\lambda = \lambda M^\lambda D \textit{ when } e_n^{(1)} \textit{ exists}, \tag{2.61}$$

$$\textit{but} \quad S^a M^\lambda \neq M^\lambda S^a. \tag{2.62}$$

Proof. For [2.61], we have from [2.1] and [2.55]:

$$M^\lambda \Delta e_n(x) = M^\lambda e_{n-1}(x) = e_{n-1}(\lambda x) = \lambda^{n-1} e_{n-1}(x)$$
$$= (1/\lambda)\lambda^n \Delta e_n(x) = (1/\lambda)\Delta e_n(\lambda x) = (1/\lambda)\Delta M^\lambda e_n(x),$$

and if e_n' exists:

$$DM^\lambda e_n(x) = De_n(\lambda x) = \lambda e_n^{(1)}(\lambda x) = \lambda M^\lambda De_n(x).$$

For [2.62], we get from [2.47] and [2.55]:

$$S^a M^\lambda e_n(x) = S^a e_n(\lambda x) = e_n(\lambda x + a),$$

while

$$M^\lambda S^a e_n(x) = M^\lambda e_n(x + a) = e_n(\lambda(x + a)). \qquad \square$$

2.8. Shift invariance property

Let us return to section 2.6 which concerns the shift operator S^a. It can happen that $\mathfrak{F}^a \subset \mathfrak{F}$ for all $a \in \mathfrak{D}$. If this is true, since $\mathfrak{F}^{-a} \subset \mathfrak{F}$ implies $\mathfrak{F} \subset \mathfrak{F}^a$, we are sure that $\mathfrak{F}^a = \mathfrak{F}$. This means that any $S^a e_n$ can be written in the form:

$$S^a e_n = \sum_{i \in J} c_i^{(n)}(a)\, e_i, \qquad [2.63]$$

for appropriate coefficients $c_i^{(n)}(a)$. We know nothing, however, about the subset J of $\mathbb{N}_0$.

DEFINITION 2.10.– *There is an S.I. property when the operators Δ and S^a commute with each other, that is, if:*

$$\Delta S^a = S^a \Delta, \quad \textit{for all } a \in \mathfrak{D} \textit{ with } \mathfrak{F}^a \subset \mathfrak{F}. \qquad [2.64]$$

Note that in this case, $\Delta_a = \Delta$ by virtue of [2.48]. In the remainder of the section, this S.I. property is assumed to hold.

PROPOSITION 2.18.–

$$\Delta e_{n+1}(x+a) = e_n(x+a), \quad n \in \mathbb{N}_0, \tag{2.65}$$

and $\mathcal{E}^a$ is a copy of $\mathcal{E}$ which is standard at $-a$ if $\mathcal{E}$ is standard at 0.

Proof. First, [2.65] follows easily from [2.64] since:

$$\Delta e_{n+1}(x+a) = \Delta S^a e_{n+1}(x) = S^a \Delta e_{n+1}(x) = S^a e_n(x) = e_n(x+a).$$

Second, we will show that $\mathcal{E}^a$ is a copy of $\mathcal{E}$. By the S.I. and from [2.1]:

$$\Delta^{n+1} S^a e_n = S^a \Delta^{n+1} e_n = 0,$$

while using [2.63]:

$$S^a \Delta^{n+1} e_n = \sum_{i\in J} c_i^{(n)}(a)\, e_{i-n-1} = \sum_{i\in J\cap\{n+1,n+2,\ldots\}} c_i^{(n)}(a)\, e_{i-n-1},$$

since $c_{i-n-1} = 0$ when $i \le n$. Consequently, all coefficients $c_i^{(n)}(a)$, $i \in J \cap \{n+1,\ldots\}$, must cancel. By writing $c_i^{(n)}(a) = b_{n-i}^{(n)}(a)$ for $0 \le i \le n$, [2.63] can then be simplified to:

$$S^a e_n = \sum_{i=0}^{n} b_{n-i}^{(n)}(a)\, e_i. \tag{2.66}$$

Note that

$$b_0^{(0)}(a) = b_0^{(0)}(a)\, e_0 = S^a e_0 = 1. \tag{2.67}$$

Applying Δ in [2.66], we get:

$$\Delta S^a e_n = \sum_{i=1}^{n} b_{n-i}^{(n)}(a)\, e_{i-1} = \sum_{j=0}^{n-1} b_{n-1-j}^{(n)}(a)\, e_j, \tag{2.68}$$

and by the S.I.:

$$\Delta S^a e_n = S^a e_{n-1},$$

where [2.66] with $n-1$ substituted for n gives:

$$S^a e_{n-1} = \sum_{j=0}^{n-1} b^{(n-1)}_{n-1-j}(a)\, e_j. \qquad [2.69]$$

The identities [2.68] and [2.69] together imply that:

$$b^{(n)}_{n-1-j}(a) = b^{(n-1)}_{n-1-j}(a), \quad 0 \leq j \leq n-1,$$

that is, $b_r^{(n)}(a) = b_r^{(n-1)}(a)$ for $0 \leq r \leq n-1$. By recursion, we see that:

$$b_r^{(n)}(a) = b_r^{(n-1)}(a) = \ldots = b_r^{(r)}(a) \equiv b_r(a). \qquad [2.70]$$

Inserting [2.70] in [2.66] then leads to:

$$S^a e_n = \sum_{i=0}^{n} b_{n-i}(a)\, e_i = \sum_{j=0}^{n} b_j(a)\, e_{n-j} = \Big(\sum_{j=0}^{\infty} b_j(a)\, \Delta^j\Big)\, e_n. \qquad [2.71]$$

From [2.71] with [2.67], we conclude that $\mathcal{E}^a$ is a copy of $\mathcal{E}$ through the operator $\mathcal{T}^a = \sum_{j=0}^{\infty} b_j(a)\Delta^j$ (see [2.2]).

Finally, if $\mathcal{E}$ is standard at 0, then $e_n(0) = \delta_{n,0}$ by [2.11], so that:

$$S^a e_n(-a) = e_n(a-a) = e_n(0) = \delta_{n,0}.$$

In other words, $S^a e_n$ is standard in $-a$ as announced. □

Proposition 2.14 can now take the simpler form given below.

PROPOSITION 2.19.–

$$\bar{G}_n(x+a|U+a) = \bar{G}_n(x|U), \quad n \in \mathbb{N}_0. \qquad [2.72]$$

Proof. We know that $\Delta_a = \Delta$ and $\mathcal{E}^a$ is a copy of $\mathcal{E}$. As indicated in section 2.4, we can then write that:

$$\bar{G}^a_n(x|\mathcal{E}^a, U) = \bar{G}_n(x|\mathcal{E}, U),$$

and [2.72] becomes immediate from [2.50]. □

2.8.1. *Binomial-type identities*

The property of S.I., possibly supplemented, is closely linked to different identities which recall the traditional binomial (or convolution) form.

PROPOSITION 2.20.– *S.I. is equivalent to the generalized binomial identity:*

$$\sum_{i=0}^{n} e_{n-i}(0)\, e_i(x+a) = \sum_{i=0}^{n} e_{n-i}(a)\, e_i(x), \quad n \in \mathbb{N}_0. \qquad [2.73]$$

Proof. For convenience, we rewrite [2.73] as $L_n = H_n$ say.

i) First suppose [2.73] is true. Since the member H_n is a copy of e_n, so is the member L_n. Thus, $\Delta L_{n+1} = L_n$ by virtue of [2.4], that is, explicitly:

$$\Delta\Big(\sum_{i=0}^{n+1} e_{n+1-i}(0)\, S^a e_i\Big) = \sum_{i=0}^{n} e_{n-i}(0)\, S^a e_i. \qquad [2.74]$$

By permutating $\sum$ and Δ in the l.h.s., and putting $i = j-1, e_{j-1} = \Delta e_j$ in the r.h.s., [2.74] becomes:

$$\sum_{i=0}^{n+1} e_{n+1-i}(0)\, \Delta S^a e_i = \sum_{j=1}^{n+1} e_{n+1-j}(0)\, S^a \Delta e_j,$$

and since $S^a \Delta e_0 = 0$, in compact form:

$$\sum_{i=0}^{n+1} e_{n+1-i}(0)\, \big(\Delta S^a - S^a \Delta\big)\, e_j = 0. \qquad [2.75]$$

Starting from $(\Delta S^a - S^a\Delta)e_0 = 0$, we see by recursion of [2.75] that all the terms $(\Delta S^a - S^a\Delta)e_n$ cancel, so the S.I. property is proven.

ii) Now suppose that S.I. holds and consider the difference $L_n - H_n$. Due to the S.I. hypothesis, we have:

$$\Delta(L_{n+1} - H_{n+1}) = L_n - H_n, \qquad [2.76]$$

and changing i in $n-i$ in L_n, we see that:

$$(L_n - H_n)(0) = 0. \qquad [2.77]$$

Thus, if $L_n - H_n = 0$, we get from [2.76] and [2.77] that:

$$L_{n+1} - H_{n+1} = \text{constant} = 0.$$

Starting from $L_0 - H_0 = 0$, we then deduce by recursion that all the differences $L_n - H_n$ cancel, so [2.73] is true. □

We will reinforce S.I. by also assuming that $\mathcal{E}$ is standard at 0.

PROPOSITION 2.21.– *S.I. with $\mathcal{E}$ standard at 0 is equivalent to the classical binomial identity:*

$$e_n(x+a) = \sum_{i=0}^{n} e_{n-i}(a)\, e_i(x), \quad n \in \mathbb{N}_0. \qquad [2.78]$$

Proof. On one side, when $\mathcal{E}$ is standard at 0, [2.73] (true under S.I.) reduces to [2.78]. On the other side, the binomial identity [2.78] with $a = 0$ gives:

$$e_n(x) = \sum_{i=0}^{n} e_{n-i}(0)\, e_i(x),$$

which implies $e_j(0) = \delta_{j,0}$, $j \in \mathbb{N}_0$, so $\mathcal{E}$ is standard at 0. The identities [2.73] and [2.78] are then equivalent, so S.I. is true. □

2.8.2. *Taylor-type formula*

Under S.I., a pseudopolynomial can also be expanded through a Taylor-type formula.

PROPOSITION 2.22.– *S.I. with $\mathcal{E}$ standard at 0 imply that a Taylor-type formula applies to any f_n in $\mathcal{F}$, namely:*

$$f_n(x+a) = \sum_{i=0}^{n} (\Delta^i f_n)(a)\, e_i(x), \quad n \in \mathbb{N}_0. \qquad [2.79]$$

Proof. Obviously, the binomial identity [2.78] can be rewritten as:

$$e_j(x+a) = \sum_{i=0}^{j} (\Delta^i e_j)(a)\, e_i(x), \quad j \in \mathbb{N}_0. \qquad [2.80]$$

By using linear combinations, [2.80] then leads to the formula [2.79]. □

The formula [2.79] generalizes to the case where $\mathcal{E}$ is standard in a general sense.

COROLLARY 2.2.– *S.I. with $\mathcal{E}$ standard at $\omega \in \mathfrak{D}$ lead to:*

$$\begin{aligned} f_n(x+a) &= \sum_{i=0}^{n} (\Delta^i f_n)(a)\, e_i(x+\omega) \\ &= \sum_{i=0}^{n} (\Delta^i f_n)(a-\omega)\, e_i(x), \quad n \in \mathbb{N}_0. \end{aligned} \qquad [2.81]$$

Proof. Note that $\mathcal{E}$ standard at ω means that $\mathcal{E}^\omega = S^\omega \mathcal{E}$ is standard at 0. Since $\Delta_\omega = \Delta$ by S.I., the first equality in [2.81] comes from [2.79]; the second equality directly follows. □

2.8.3. *Regarding integration*

To close, let us look at a possible permutation of the shift and integration operators.

PROPOSITION 2.23.– *S.I. with $\mathcal{E}$ standard at 0 implies:*

$$I_u e_n(x-u) = e_{n+1}(x-u). \qquad [2.82]$$

In particular,

$$\bar{G}_n(x|\{u,u,\ldots\}) = e_n(x-u), \quad n \in \mathbb{N}_0. \qquad [2.83]$$

Proof. If $\mathcal{E}$ is standard at 0, [2.14] gives $I_0 e_n = e_{n+1}$, so that from [2.20]:

$$\bar{G}_n(x|\{0,0,\ldots\}) = e_n(x). \qquad [2.84]$$

By the S.I., [2.72] is of application, thus we get from [2.84]:

$$I_u e_n(x-u) = I_u \bar{G}_n(x-u|\{0,0,\dots\}) = I_u \bar{G}_n(x|\{u,u,\dots\}). \quad [2.85]$$

Moreover, from [2.21], [2.72] and using [2.84], we see that:

$$\begin{aligned} I_u \bar{G}_n(x|\{u,u,\dots\}) &= \bar{G}_{n+1}(x|\{u,u,\dots\}) = \bar{G}_{n+1}(x-u|\{0,0,\dots\}) \\ &= e_{n+1}(x-u). \end{aligned} \quad [2.86]$$

Therefore, [2.85] with [2.86] give the first formula [2.82].

Furthermore, the S.I. yields:

$$\bar{G}_n(x|\{u,u,\dots\}) = \bar{G}_n(x-u|\{0,0,\dots\}),$$

while from [2.20] with [2.82], we have:

$$\bar{G}_n(x-u|\{0,0,\dots\}) = e_n(x-u),$$

hence the second formula [2.83]. □

3

General Theory and Explicit Results

3.1. Return to the shift invariance

Let us return to the notion of shift invariance defined in [2.64] and characterized by the generalized binomial identity [2.73]. This time, we will use the formal generating functions of $\mathcal{E}$, denoted as:

$$\psi(x, s) = \sum_{n=0}^{\infty} e_n(x)\, s^n, \quad x \in \mathfrak{D},$$

where s plays the role of a real (or complex) parameter. Note that $\psi(x, 0) = e_0(x) = 1$ and $\psi(0, s) = 1$ when $\mathcal{E}$ is standard at 0.

In fact, the S.I. condition stated in [2.73] means that:

$$\psi(x + a, s)\, \psi(0, s) = \psi(x, s)\, \psi(a, s). \qquad [3.1]$$

Defining

$$\phi(x, s) = \psi(x, s)/\psi(0, s), \qquad [3.2]$$

this relation becomes

$$\phi(x + a, s) = \phi(x, s)\, \phi(a, s), \qquad [3.3]$$

with $\phi(0, s) = 1$. This is a well-known functional equation, which is discussed below when $\mathfrak{D}$ represents the set of integers or reals.

3.1.1. *The case where* $x \in \mathfrak{D} = \mathbb{Z}$ *(or* $\mathbb{N}_0$*)*

Then, any solution to [3.3] is such that:

$$\phi(x,s) = [\phi(1,s)]^x. \tag{[3.4]}$$

Indeed, when $x \in \mathbb{N}$, this is immediate by taking $a = 1$ in [3.3]. When $-x \in \mathbb{N}$, [3.3] with $x = -a$ gives, using the previous result:

$$\phi(x,s) = \phi(0,s)/\phi(-x,s) = 1/[\phi(1,s)]^{-x} = [\phi(1,s)]^x.$$

From [3.2] and [3.4], we then get:

$$\psi(x,s) = \psi(0,s)\,\phi(x,s) = \psi(0,s)\,[\phi(1,s)]^x, \tag{[3.5]}$$

where $\psi(0,s)$ and $\phi(1,s)$ are arbitrary functions such that $\phi(1,0) = 1$ and the linear independence in $\mathcal{E}$ is preserved. Explicitly, [3.5] is written as:

$$\sum_{n=0}^{\infty} e_n(x)\, s^n = \Big(\sum_{n=0}^{\infty} e_n(0)\, s^n\Big)\Big(1 + \sum_{i=1}^{\infty} \beta_i s^i\Big)^x, \tag{[3.6]}$$

for appropriate coefficients β_i. Note that $\beta_1 \neq 0$ necessarily, otherwise e_1 would be a constant that contradicts the linear independence of $\mathcal{E}$.

Actually, we prefer to introduce the notation that will be adopted later, namely:

$$\phi(1,s) = e^{g(s)} = 1 + \sum_{k=1}^{\infty} \frac{[g(s)]^k}{k!}, \tag{[3.7]}$$

where $g(s)$ is the series $\sum_{i=1}^{\infty} \alpha_i s^i$ in which the coefficients α_i are obtained from the identity:

$$\sum_{i=1}^{\infty} \beta_i s^i = \sum_{k=1}^{\infty} \frac{1}{k!}\Big(\sum_{i=1}^{\infty} \alpha_i s^i\Big)^k.$$

By identification of both members, we get $\beta_1 = \alpha_1$ and for $n \geq 2$:

$$\beta_n = \alpha_n + \text{terms in } \alpha_i \alpha_{n-i} + \text{terms in } \alpha_i \alpha_j \alpha_{n-i-j} + \dots,$$

so that by recursion, all the α_r are expressed in terms of the β_i for $i \leq r$.

By making the notation change [3.7], the relation [3.6] simply becomes:

$$\sum_{n=0}^{\infty} e_n(x)\, s^n = \Big(\sum_{n=0}^{\infty} e_n(0)\, s^n\Big)\, e^{xg(s)}. \qquad [3.8]$$

Note that $g(0) = 0$ and $g^{(1)}(0) = \alpha_1 = \beta_1 \neq 0$. Of course, [3.8] provides us with an expression for each e_n. For that, we introduce the convolution notation:

$$\Big(\sum_{i=1}^{\infty} \alpha_i s^i\Big)^k = \sum_{i=k}^{\infty} \alpha_i^{*k}\, s^i, \qquad [3.9]$$

showing, in particular, that $\alpha_i^{*0} = \delta_{i,0}$, $\alpha_i^{*1} = \alpha_i$ and $\alpha_k^{*k} = \alpha_1^k$. Then,

$$e^{xg(s)} = \sum_{k=0}^{\infty} \frac{x^k}{k!} \Big(\sum_{i=1}^{\infty} \alpha_i s^i\Big)^k = \sum_{k=0}^{\infty} \frac{x^k}{k!} \Big(\sum_{i=k}^{\infty} \alpha_i^{*k}\, s^i\Big),$$

so that using [3.8]:

$$e_n(x) = \sum_{0 \leq k \leq i \leq n} e_{n-i}(0)\, \frac{x^k}{k!}\, \alpha_i^{*k}. \qquad [3.10]$$

We will continue the analysis below assuming that $\mathcal{E}$ is standard at 0. In this case, $\psi(0, s) = 1$, therefore:

$$\psi(x, s) = \phi(x, s) = e^{xg(s)}. \qquad [3.11]$$

PROPOSITION 3.1.– *Suppose $\mathcal{E}$ is standard at* 0.

a) Each polynomial e_n is explicitly given by:

$$e_n(x) = \sum_{k=0}^{n} \frac{x^k}{k!}\, \alpha_n^{*k}, \quad n \in \mathbb{N}_0, \qquad [3.12]$$

with, for example, $e_1(x) = \alpha_1 x$.

b) The operator S^a has the simple expression:

$$S^a = \sum_{i=0}^{\infty} e_i(a)\, \Delta^i. \qquad [3.13]$$

c) Defining an operator O by $O \equiv \sum_{i=1}^{\infty} i\alpha_i \Delta^i$:

$$Oe_n(x) = \frac{ne_n(x)}{x}, \tag{3.14}$$

and

$$\Delta\Big(\frac{ne_n(x)}{x}\Big) = \frac{(n-1)e_{n-1}(x)}{x}, \quad n \geq 1. \tag{3.15}$$

d) For any $b \neq 0$:

$$\frac{ne_n(x+b)}{x+b} = \sum_{i=1}^{n} \frac{ie_i(b)}{b} e_{n-i}(x), \quad n \geq 1. \tag{3.16}$$

Proof. The formula [3.12] of $e_n(x)$ directly follows from [3.10].

For S^a, we have from [2.78]:

$$S^a e_n(x) = e_n(x+a) = \sum_{i=0}^{n} e_i(a)\, e_{n-i}(x) = \Big(\sum_{i=0}^{n} e_i(a)\, \Delta^i\Big)\, e_n(x),$$

hence the representation [3.13].

Differentiating [3.11] with respect to s gives:

$$\begin{aligned}\sum_{n=1}^{\infty} ne_n(x)\, s^{n-1} &= e^{xg(s)} x\big(g(s)\big)^{(1)} \\ &= xe^{xg(s)} \sum_{i=1}^{\infty} i\alpha_i\, s^{i-1},\end{aligned}$$

which implies that:

$$\frac{ne_n(x)}{x} = \sum_{i=1}^{n} i\alpha_i\, e_{n-i}(x), \tag{3.17}$$

that is, [3.14] is true and $ne_n(x)/x \in \mathfrak{F}$. The operator Δ may thus be applied to [3.17], yielding:

$$\Delta\Big(\frac{ne_n(x)}{x}\Big) = \sum_{i=1}^{n-1} i\alpha_i\, e_{n-1-i},$$

which is equal, by virtue of [3.17], to $(n-1)e_{n-1}(x)/x$ as announced by [3.15].

Finally, let us differentiate the function $\exp\big((x+b)g(s)\big)$ with respect to s. Since:

$$\big(e^{(x+b)g(s)}\big)^{(1)} = \frac{x+b}{b}\big(e^{bg(s)}\big)^{(1)} e^{xg(s)},$$

we get this time:

$$\sum_{n=0}^{\infty} \frac{ne_n(x+b)}{x+b} s^{n-1} = \sum_{i=1}^{\infty} \frac{ie_i(b)}{b} s^{i-1} e^{xg(s)},$$

which leads to the desired formula [3.16]. □

Here are some other, more specific, properties.

PROPOSITION 3.2.– *Suppose the generating function $\psi(x,s)$ converges.*

a) For $s \in \mathbb{R}$:

$$|\sum_{m=0}^{\infty} e_{2m+1}(x)\, s^{2m+1}| < \sum_{m=0}^{\infty} e_{2m}\, s^{2m}. \qquad [3.18]$$

b) When there exist real numbers b and s_b verifying $b = g(s_b)$:

$$\Delta\big(e^{bx}\big) = s_b\,\big(e^{bx}\big). \qquad [3.19]$$

c) When $\tilde{\mathcal{E}}$ is defined by [3.8] with now the factor $e^{x(g\circ\bar{g})(s)}$ where $\bar{g}(s) \equiv \sum_{i=1}^{\infty} \bar{\gamma}_i s^i$:

$$\tilde{e}_n(x) = \sum_{k=0}^{n} e_k(x)\, \bar{\gamma}_n^{*k}. \qquad [3.20]$$

The family $\tilde{\mathcal{E}}$ is still standard at 0 but it is not a copy of $\mathcal{E}$.

Proof. By [3.11], $\psi(x,s) = \exp(xg(s))$ which is positive for any real s. Thus, considering s and $-s$, we deduce the inequality [3.18].

Moreover, we have:

$$\Delta\left(e^{xg(s)}\right) = \Delta\Big(\sum_{n=0}^{\infty} e_n(x)\, s^n\Big) = \sum_{n=1}^{\infty} e_{n-1}(x)\, s^n = se^{xg(s)}.$$

By setting $s = s_b$ and since $b = g(s_b)$ by hypothesis, we obtain [3.19].

Finally, we write:

$$\begin{aligned}\sum_{n=0}^{\infty} \tilde{e}_n(x)\, s^n &= e^{x(g\circ\bar{g})(s)} \\ &= \sum_{k=0}^{\infty} e_k(x)\, [\bar{g}(s)]^k = \sum_{k=0}^{\infty} e_k(x) \Big(\sum_{n=k}^{\infty} \gamma_n^{*k}\, s^n\Big),\end{aligned}$$

hence the formula [3.20] for $\tilde{e}_n(x)$. □

3.1.2. *The case where* $x \in \mathfrak{D} = \mathbb{R}$

We start from the practical assumption that $\psi(x,s)$ is the sum of a convergent series, that is, it is a function of x defined on a real interval. The solutions of [3.3] are then known to be either extremely regular or extremely irregular. In fact, any regularity condition imposed on the function $\psi(x,s)$, such as continuity, differentiability, monotonicity and measurability, is sufficient to eliminate most solutions. The set of acceptable solutions is thus considerably reduced, and only constants (of no interest) and exponential functions $\exp(xg(s))$ will remain. All this has been well established since the work of Cauchy[1] on the functional equation $f(x+a) = f(x) + f(a)$. Surprisingly, all the regularity conditions mentioned above have the same impact and there is no gain in choosing one rather than another.

In the present study, we focus on manageable families $\mathcal{E}$, and we are led to accept only generating functions of the form:

1 Cauchy, A.-L. (1821). *Analyse algébrique*. De Bure Frères, Paris.

$$\begin{aligned}\psi(x,s) &= \theta(s)\, e^{xg(s)} \\ &= \Big(\sum_{n=0}^{\infty} e_n(0)\, s^n\Big)\, e^{x\sum_{i=1}^{\infty}\alpha_i\, s^i},\end{aligned} \qquad [3.21]$$

where $\theta(s)$ and $g(s)$ are sums of convergent series in s, independently of x. As in the previous case (1), $g(0) = 0$, $g^{(1)}(0) \neq 0$ and $\theta(s) = 1$ when $\mathcal{E}$ is standard at 0.

Note that convergence hypotheses are not in the spirit of the text since only polynomials and formal series are generally considered. What would happen if $\psi(x,s)$ were only a formal series? In this case, [3.1] would have more solutions, most of them very irregular, in addition to the few interesting previous ones. To select these manageable solutions, additional regularity conditions should certainly be added. An effective assumption for this is the differentiability of each $e_n(x)$: it will be introduced below.

PROPOSITION 3.3.– *Suppose the functions $e_n(x)$ are differentiable. Then, the formal series $\psi(x,s) = \sum_{n=0}^{\infty} e_n(x)s^n$ is of the form [3.21] where $g(s)$ is a formal series if and only if the derivative operator D can be expressed as:*

$$D = \sum_{i=1}^{\infty} \alpha_i\, \Delta^i. \qquad [3.22]$$

If this is true, each $e_n(x)$ is necessarily a polynomial of degree n in x.

Proof. Consider $\psi(x,s)$ as a formal series [3.21]. By differentiation with respect to x, we have:

$$D\psi(x,s) = \sum_{n=0}^{\infty} De_n(x)\, s^n = g(s)\,\theta(s)\, e^{xg(s)} = g(s)\,\psi(x,s),$$

so that

$$De_n(x) = \sum_{i=1}^{n} \alpha_i\, e_{n-i} = \Big(\sum_{i=1}^{n} \alpha_i\, \Delta^i\Big)\, e_n, \qquad [3.23]$$

which is equivalent to [3.22].

Conversely, by reasoning recursively from [3.23], we observe that, as announced, each e_n is indeed a polynomial of degree n. Moreover, its Maclaurin expansion gives:

$$e_n(x) = \sum_{r=0}^{n} \frac{x^r}{r!} \, (D^r e_n)(0), \tag{3.24}$$

in which, by virtue of [3.22]:

$$D^r = \Big(\sum_{i=1}^{\infty} \alpha_i \, \Delta^i\Big)^r = \sum_{i=r}^{\infty} \alpha_i^{*r} \, \Delta^i, \tag{3.25}$$

with the same notation as for [3.9]. Inserting [3.25] into [3.24], we obtain:

$$e_n(x) = \sum_{r=0}^{n} \frac{x^r}{r!} \Big(\sum_{i=r}^{\infty} \alpha_i^{*r} \, e_{n-i}(0)\Big), \tag{3.26}$$

thus taking the generating function of [3.26]:

$$\begin{aligned} \psi(x,s) &= \sum_{r=0}^{\infty} \frac{x^r}{r!} \Big(\sum_{i=r}^{\infty} \alpha_i^{*r} \, s^i\Big) \Big(\sum_{n-i=0}^{\infty} e_{n-i}(0) \, s^{n-i}\Big) \\ &= \sum_{r=0}^{\infty} \frac{x^r}{r!} \Big(\sum_{j=1}^{\infty} \alpha_j \, s^j\Big)^r \Big(\sum_{k=0}^{\infty} e_k(0) \, s^k\Big), \end{aligned}$$

which corresponds to [3.21]. □

PROPOSITION 3.4.– *Furthermore, the S.I. property is equivalent to the representation [3.22] of the derivative operator D.*

Proof. Suppose [3.22] applies. By proposition 3.3, we can then write:

$$\begin{aligned} \psi(x,s) \, \psi(y,s) &= \theta(s) \, e^{xg(s)} \, \theta(s) \, e^{yg(s)} \\ &= \theta(s) \, \theta(s) \, e^{(x+y)g(s)} = \theta(s) \sum_{r=0}^{\infty} e_r(x+y) \, s^r. \end{aligned}$$

This means that:

$$\sum_{i=0}^{n} e_{n-i}(x)\, e_i(y) = \sum_{r=0}^{n} e_{n-r}(0)\, e_r(x+y), \qquad [3.27]$$

that is, the generalized binomial identity [2.70] is satisfied, hence the S.I. property as well.

Under S.I., the identity [3.27] is true. By differentiation with respect to y and setting $y = 0$, we then obtain:

$$\sum_{i=0}^{n} e_{n-i}(x)\, De_i(0) = \sum_{r=0}^{n} e_{n-r}(0)\, De_r(x). \qquad [3.28]$$

Let us discuss separately the cases where $\mathcal{E}$ is standard at 0 or not.

If yes, [3.28] reduces to:

$$\sum_{i=1}^{n} e_{n-i}(x)\, De_i(0) = De_n(x),$$

implying that:

$$D = \sum_{i=1}^{n} De_i(0)\, \Delta^i,$$

which is of the form [3.22].

If no, the result is similar but less direct. First, [3.28] implies that:

$$De_n(x) = \sum_{i=1}^{n} De_i(0)\, e_{n-i}(x) - \sum_{r=1}^{n-1} e_{n-r}(0)\, De_r(x),$$

that is, recursively, for $n = 1$ and 2:

$$De_1(x) = De_1(0)\, e_0(x),$$

$$De_2(x) = \sum_{i=1}^{2} De_i(0)\, e_{2-i}(x) - e_1(0)\, De_1(x) = \sum_{i=1}^{2} De_i(0)\, e_{2-i}(x)$$
$$- e_1(0)\, \big(De_1(0)\, e_0(x)\big),$$

and more generally,

$$De_n(x) = \sum_{j=1}^{n} \beta_j^{(n)} \, e_{n-j}(x), \qquad [3.29]$$

for appropriate coefficients $\beta_j^{(n)}$ which are well-defined and could a priori depend on n. Actually, they are independent of n. Indeed, let us rewrite [3.28] as:

$$\sum_{i=0}^{n} e_i(x) \, De_{n-i}(0) = \sum_{r=0}^{n} e_{n-r}(0) \, De_r(x), \qquad [3.30]$$

and inside both sides, let us use [3.29] with simple coefficients β_j, that is,

$$De_n(x) = \sum_{j=1}^{n} \beta_j \, e_{n-j}(x) = \sum_{i=0}^{n-1} \beta_{n-i} \, e_i(x).$$

We then obtain:

$$\sum_{i=0}^{n} e_i(x) \Big(\sum_{j=1}^{n} \beta_j \, e_{n-i-j}(0) \Big) = \sum_{r=0}^{n} e_{n-r}(0) \Big(\sum_{i=0}^{r-1} \beta_{r-i} \, e_i(x) \Big),$$

or equivalently, for any i:

$$\sum_{j=1}^{n} \beta_j \, e_{n-i-j}(0) = \sum_{r=i+1}^{n} e_{n-r}(0) \, \beta_{r-i},$$

which is an obvious identity (although the β_j are not explicitly given). Since the $\beta_j^{(n)}$ are well defined, they necessarily identify with the β_j, so that [3.29] leads to [3.22]. □

PROPOSITION 3.5.– *In addition, suppose $\mathcal{E}$ is standard at* 0*. Then, the results of propositions 3.1 and 3.2 are valid. Moreover, defining an operator S by $S \equiv \sum_{i=2}^{\infty} (i/2) \alpha_i^{*2} \Delta^i$:*

$$Se_n(x) = \frac{n}{x} \Big(De_n(x) - \frac{e_n(x)}{x} \Big), \qquad [3.31]$$

and

$$\Delta\Big[\frac{n}{x}\Big(De_n(x)-\frac{e_n(x)}{x}\Big)\Big]=\frac{n-1}{x}\Big(De_{n-1}(x)-\frac{e_{n-1}(x)}{x}\Big),\quad n\geq 1. \tag{[3.32]}$$

Proof. We first establish [3.31]. By differentiating the generating function $\psi(x,s)$ of $\mathcal{E}$ twice, one with respect to x and the other with respect to s, we get from [3.8] when $\mathcal{E}$ standard at 0, and using the definition of $g(s)$:

$$\begin{aligned}\sum_{n=0}^{\infty} nDe_n(x)\,s^{n-1} &= D\big[e^{xg(s)}\,xg'(s)\big]=e^{xg(s)}\,\big[g'(s)+xg(s)g'(s)\big]\\ &= e^{xg(s)}\sum_{i=1}^{\infty} i\alpha_i\,s^{i-1}+e^{xg(s)}\,(x/2)\sum_{i=2}^{\infty} i\alpha_i^{*2}\,s^{i-1},\end{aligned} \tag{[3.33]}$$

since $(gg')(s)=(1/2)(g^2)'(s)=(1/2)(\sum_{i=2}^{\infty}\alpha_i^{*2}s^i)'$.

Let us work on the r.h.s. of [3.33]. For the first sum, we recourse to [3.16], and for the second sum, to the expansion of the exponential. This transforms [3.33] into:

$$\sum_{n=1}^{\infty} nDe_n(x)\,s^{n-1}-\sum_{n=1}^{\infty}\frac{ne_n(x)}{x}\,s^{n-1}+(x/2)\sum_{r=0}^{\infty}e_r(x)\,s^r\sum_{i=2}^{\infty} i\alpha_i^{*2}\,s^{i-1},$$

that is, equivalently:

$$\sum_{n=1}^{\infty}\Big(\frac{nDe_n(x)}{x}-\frac{ne_n(x)}{x^2}\Big)\,s^{n-1}=(1/2)\sum_{n=2}^{\infty}s^{n-1}\sum_{i=2}^{\infty} i\alpha_i^{*2}\,e_{n-i}(x). \tag{[3.34]}$$

From [3.34], we deduce that:

$$\frac{n}{x}\Big(De_n(x)-\frac{e_n(x)}{x}\Big)=(1/2)\sum_{i=2}^{n} i\alpha_i^{*2}\,e_{n-i}(x), \tag{[3.35]}$$

which gives [3.31] as desired and also shows that the l.h.s. is an element of $\mathcal{F}$. Thus, Δ may be applied to [3.35] and this directly leads to [3.32]. □

3.2. The higher dimensional case

In this section, we continue the discussion of the S.I. property by examining the case where $\mathcal{D}$ is the set of higher dimensional reals (finite or not). We will end by pointing out cases of exponential generating functions outside the S.I. context.

3.2.1. *The case where* $\mathbf{x} \in \mathfrak{D} = \mathrm{I\!R}^k$

The results are quite similar to the previous ones. Let $\mathbf{x} = (x_1, \dots, x_k)$. We will now suppose that the partial derivatives $\partial_r e_n(\mathbf{x}) \equiv (\partial/\partial x_r) e_n(\mathbf{x})$ exist for $r = 1, \dots, k$. Given two vectors $\mathbf{a}, \mathbf{b} \in \mathrm{I\!R}^k$, denote $< \mathbf{a}, \mathbf{b} >$ the standard scalar product $\sum_{r=1}^{k} a_r b_r$.

PROPOSITION 3.6.– *Each partial derivative operator has the representation:*

$$\partial_r = \sum_{i=1}^{\infty} \alpha_{r,i} \, \Delta^i \qquad [3.36]$$

if and only if

$$\sum_{n=0}^{\infty} e_n(\mathbf{x}) \, s^n = \theta(s) \, e^{<\mathbf{x},\mathbf{g}(s)>}, \qquad [3.37]$$

where $\mathbf{g}(s) = \big(g_1, \dots, g_k\big)(s)$, *and each* $g_r(s)$ *is a power series of the form:*

$$g_r(s) = \sum_{i=1}^{\infty} \alpha_{r,i} \, s^i, \qquad [3.38]$$

with, of course, $\theta(s) = \sum_{n=0}^{\infty} e_n(\mathbf{0}) s^n$.

Proof. First, suppose [3.37] is true. Then:

$$\partial_r \Big[\sum_{n=0}^{\infty} e_n(\mathbf{x}) \, s^n \Big] = g_r(s) \theta(s) \, e^{<\mathbf{x},\mathbf{g}(s)>},$$

or equivalently,

$$\sum_{n=0}^{\infty} \partial_r e_n(\mathbf{x}) \, s^n = g_r(s) \sum_{n=0}^{\infty} e_n(\mathbf{x}) \, s^n.$$

Using [3.38], this implies that for all r:

$$\partial_r e_n(\mathbf{x}) = \sum_{i=1}^{n} \alpha_{r,i}\, e_{n-i}(\mathbf{x}),$$

hence the desired representation [3.36].

Now, suppose [3.36] is true. The partial derivative of order i_r is then of expression:

$$\partial_r^{i_r} = \Big(\sum_{i=1}^{\infty} \alpha_{r,i}\, \Delta^i\Big)^{i_r} = \sum_{j_r=i_r}^{\infty} \alpha_{r,j_r}^{*i_r}\, \Delta^{j_r},$$

by virtue of the convolution definition. The Maclaurin formula applied to any $e_n(\mathbf{x})$ thus gives:

$$\begin{aligned} e_n(\mathbf{x}) &= \sum_{i_1\geq 0,\ldots,i_k\geq 0} \frac{x_1^{i_1}}{i_1!}\cdots\frac{x_k^{i_k}}{i_k!}\, \partial_1^{i_1}\ldots\partial_k^{i_k} e_n(\mathbf{0}) \\ &= \sum_{i_1\geq 0,\ldots,i_k\geq 0} \frac{x_1^{i_1}}{i_1!}\cdots\frac{x_k^{i_k}}{i_k!} \sum_{j_1\geq i_1,\ldots,j_k\geq i_k} \alpha_{1,j_1}^{*i_1}\cdots\alpha_{k,j_k}^{*i_k}\, e_{n-(j_1+\ldots+j_k)}(\mathbf{0}), \end{aligned}$$

with $j_1+\ldots+j_k \leq n$ of course. For their generating function, we so obtain:

$$\sum_{n=0}^{\infty} e_n(\mathbf{x})\, s^n = \sum_{i_1\geq 0,\ldots,i_k\geq 0} \frac{x_1^{i_1}}{i_1!}\cdots\frac{x_k^{i_k}}{i_k!} \sum_{j_1\geq i_1,\ldots,j_k\geq i_k} \alpha_{1,j_1}^{*i_1}\, s^{j_1}\ldots\alpha_{k,j_k}^{*i_k}\, s^{j_k} \Big(\sum_{n=j_1+\ldots+j_k}^{\infty} e_{n-(j_1+\ldots+j_k)}(\mathbf{0})\, s^{n-(j_1+\ldots+j_k)}\Big). \qquad [3.39]$$

Note that the sum $(\ldots)$ in [3.39] corresponds precisely to the term $\theta(s)$ in [3.36]. By eliminating the convolution, [3.39] becomes:

$$\begin{aligned} \sum_{n=0}^{\infty} e_n(\mathbf{x})\, s^n &= \theta(s) \sum_{i_1\geq 0,\ldots,i_k\geq 0} \frac{x_1^{i_1}}{i_1!}\cdots\frac{x_k^{i_k}}{i_k!} \Big(\sum_{r_1\geq 0} \alpha_{1,r_1} s^{r_1}\Big)^{i_1} \\ &\quad \ldots\Big(\sum_{r_k\geq 0} \alpha_{k,r_k} s^{r_k}\Big)^{i_k} \\ &= \theta(s)\, \exp\Big(x_1 \sum_{r_1\geq 0} \alpha_{1,r_1} s^{r_1}\Big)\ldots\exp\Big(x_k \sum_{r_k\geq 0} \alpha_{k,r_k} s^{r_k}\Big) \\ &= \theta(s)\, \exp\big(x_1 g_1(s)\big)\ldots\exp\big(x_k g_k(s)\big), \end{aligned}$$

hence the formula [3.37]. □

PROPOSITION 3.7.– *Furthermore, the S.I. property is equivalent to the representation [3.36] of the partial derivative operators* ∂_r, $1 \leq r \leq k$.

Proof. The reasoning is similar to that followed for proposition 3.4. If [3.36] applies, we get here too:

$$\sum_{i=0}^{n} e_{n-i}(\mathbf{x})\, e_i(\mathbf{y}) = \sum_{r=0}^{n} e_{n-r}(\mathbf{0})\, e_r(\mathbf{x}+\mathbf{y}), \qquad [3.40]$$

which is the S.I. condition.

Conversely, suppose S.I. is true and consider the case where $\mathcal{E}$ is standard at $\mathbf{0}$. If this is not the case, the result remains valid but the proof is left to the reader for the sake of brevity. Now, from [3.40], we get for all r:

$$\sum_{i=0}^{n} e_{n-i}(\mathbf{x})\, \partial_r e_i(\mathbf{0}) = \partial_r e_n(\mathbf{x}),$$

or equivalently,

$$\partial_r e_n(\mathbf{x}) = \Big(\sum_{i=0}^{n} \partial_r e_i(\mathbf{0})\, \Delta^i\Big) e_n(\mathbf{x}).$$

This leads to the representation [3.36] of ∂_r in which $\alpha_{r,i} = \partial_r e_i(\mathbf{0})$. □

REMARK.– When $\mathcal{E}$ is standard at $\mathbf{0}$, suppose that each $g_i(s)$ is of the special form $\gamma_i h(s)$ with the same function $h(s)$. Then:

$$\sum_{n=0}^{\infty} e_n(\mathbf{x})\, s^n = e^{<\mathbf{x},\gamma>h(s)},$$

so that

$$\sum_{n=0}^{\infty} n e_n(\mathbf{x})\, s^{n-1} = <\mathbf{x},\gamma>\, h^{(1)}(s)\, e^{<\mathbf{x},\gamma>h(s)}$$
$$= <\mathbf{x},\gamma>\, h^{(1)}(s) \sum_{n=0}^{\infty} e_n(\mathbf{x})\, s^n.$$

Let us show that as a consequence:

$$\Delta\Big(\frac{ne_n(\mathbf{x})}{<\mathbf{x},\gamma>}\Big) = \frac{(n-1)e_{n-1}(\mathbf{x})}{<\mathbf{x},\gamma>}, \qquad [3.41]$$

a result reminiscent of [3.14]. Indeed, we then obtain:

$$\sum_{n=0}^{\infty}\Delta\Big(\frac{ne_n(\mathbf{x})}{<\mathbf{x},\gamma>}\,s^{n-1}\Big) = h^{(1)}(s)\sum_{n=1}^{\infty}e_{n-1}(\mathbf{x})\,s^n = s\,h^{(1)}(s)\sum_{n=0}^{\infty}e_n(\mathbf{x})\,s^n$$

$$= s\,\frac{1}{<\mathbf{x},\gamma>}\sum_{n=0}^{\infty}ne_n(\mathbf{x})\,s^{n-1} = \sum_{n=0}^{\infty}\frac{ne_n(\mathbf{x})}{<\mathbf{x},\gamma>}\,s^n,$$

hence the announced identity [3.41].

3.2.2. *The case where* $\mathrm{x} \in \mathfrak{D} = \mathrm{I\!R}^{\mathrm{I\!N}}$

In most cases, the functions $e_n(\mathbf{x})$ do not depend on the entire sequence of components of $\mathbf{x} = \{x_j, j \geq 1\}$ but only on a finite number of them. Therefore, instead of working on $\mathrm{I\!R}^{\mathrm{I\!N}}$, it is often enough to work on $\mathrm{I\!R}^{\mathrm{k}}$ where k is sufficiently large.

For instance, suppose that:

$$\sum_{n=0}^{\infty}e_n(\mathbf{x})\,s^n = \exp\Big(\sum_{j=1}^{\infty}x_j\,s^j\Big)$$

$$= \prod_{j=1}^{\infty}e^{x_j s^j} = \prod_{j=1}^{\infty}\Big(\sum_{i_j=0}^{\infty}\frac{(x_j s^j)^{i_j}}{i_j!}\Big).$$

This yields:

$$e_n(\mathbf{x}) = \sum_{\substack{i_1,i_2\dots,i_n\geq 0:\\ i_1+2i_2+\dots+ni_n=n}}\frac{x_1^{i_1}}{i_1!}\,\frac{x_2^{i_2}}{i_2!}\cdots\frac{x_n^{i_n}}{i_n!}, \qquad [3.42]$$

and thus depends only on $(x_1,\dots,x_n)$. Note that the S.I. condition is satisfied since:

$$\prod_{j=1}^{\infty}e^{(x_j+a_j)s^j} = \prod_{j=1}^{\infty}e^{x_j s^j}\prod_{j=1}^{\infty}e^{a_j s^j}.$$

Another example mentioned in section 2.1 is where:

$$e_n(\mathbf{x}) = x_1 \dots x_n/n!, \quad n \geq 1.$$

This time, however, the S.I. property is no longer true.

COMMENT (out of context of the S.I.)– Exponential generating functions for $\mathcal{E}$ can exist even if the S.I. property does not hold. Below we give two simple illustrations when $\mathfrak{D} = \mathrm{I\!R}$.

Let us start with a family $\tilde{\mathcal{E}}$ standard at 0 which is defined by:

$$\sum_{n=0}^{\infty} \tilde{e}_n(x)\, s^n = e^{\rho(x) g(s)},$$

for some differentiable function $\rho(x)$ with $\rho(0) = 0$. This implies that:

$$\tilde{e}_n(x) = \sum_{k=0}^{n} \frac{[\rho(x)]^k}{k!} \alpha_n^{*k} = e_n[\rho(x)],$$

the last equality following from [3.12]. Instead of [3.22], D can here be expressed as:

$$D = \rho' \sum_{i=1}^{\infty} \alpha_i \tilde{\Delta}^i, \quad \text{where} \quad \tilde{\Delta}\big(e^{\rho(x) g(s)}\big) = s\, e^{\rho(x) g(s)}.$$

Moreover, the identity [3.15] is modified as:

$$\tilde{\Delta}\Big(\frac{n\tilde{e}_n(x)}{\rho(x)}\Big) = \frac{(n-1)\tilde{e}_{n-1}(x)}{\rho(x)}, \quad n \geq 1.$$

Suppose now that:

$$\psi(x, s) = \exp\Big(\sum_{j=1}^{\infty} e^{a_j x}\, s^j\Big)$$

$$= \prod_{j=1}^{\infty} \exp\big(e^{a_j x}\, s^j\big) = \prod_{j=1}^{\infty} \Big(\sum_{i_j=0}^{\infty} \frac{\exp(a_j x i_j)\, s^{j i_j}}{i_j!}\Big).$$

so that

$$e_n(x) = \sum_{\substack{i_1,i_2\ldots,i_n\geq 0:\\ i_1+2i_2+\ldots+ni_n=n}} \frac{\exp\left(x\sum_{j=1}^{\infty} a_j i_j\right)}{i_1!i_2!\ldots i_n!}.$$

Note that $\mathcal{E}$ is not standard at 0 and also not in agreement with S.I.

3.3. Calculation formulas for $\bar{G}_n(\ |U)$

In this section, we discuss three possible methods for determining A–G pseudopolynomials in general. They rely directly on key properties satisfied by these $\bar{G}_n$. Alternative approaches, however, could be proposed.

3.3.1. *From an A-G type expansion of* e_n

As seen in [2.32], each e_n can be written in terms of the $\bar{G}_i$, $0 \leq i \leq n$, as:

$$e_n(\ ) = \sum_{i=0}^{n} e_{n-i}(u_i)\,\bar{G}_i(\ |U), \quad n > 1. \qquad [3.43]$$

Conversely, we would here like to express each $\bar{G}_n(\ |U)$ in terms of the e_i, $0 \leq i \leq n$.

a) *A first method*: by definition, $\bar{G}_n$ is of the form:

$$\bar{G}_n(\ |U) = \sum_{i=0}^{n} g_{n,i}(U)\,e_i(\ ), \quad n \geq 1. \qquad [3.44]$$

Moreover, by virtue of [2.24], taking Δ^j of [3.44] at u_j yields:

$$\Delta^j \bar{G}_n(u_j|U) = \sum_{i=j}^{n} g_{n,i}(U)\,e_{i-j}(u_j) = \delta_{j,n}, \quad 0 \leq j \leq n$$

Therefore, the coefficients $g_{i,j}$ in [3.44] satisfy the system:

$$\begin{cases} \sum_{i=j}^{n} g_{n,i}\,a_{i,j} = \delta_{j,n}, \\ \sum_{i=j}^{n} a_{n,i}\,g_{i,j} = \delta_{j,n}, \end{cases} \qquad [3.45]$$

after denoting:

$$a_{i,j} \equiv e_{i-j}(u_j). \qquad [3.46]$$

At this stage, it may be useful to use a matricial approach. For $n \in \mathbb{N}_0$, the system [3.43] can be rewritten as:

$$\mathbf{e} = \mathcal{A}\,\bar{\mathbf{G}},$$

where $\mathbf{e}$ (respectively, $\bar{\mathbf{G}}$) is the column vector of elements $e_i, i \geq 0$ (respectively, $\bar{G}_j, j \geq 0$), and $\mathcal{A}$ is the square matrix of elements $a_{i,j} = e_{i-j}(u_j)$. These vectors and matrix are of infinite or large (say N) dimension. $\mathcal{A}$ is a lower triangular matrix with $a_{i,i} = 1$ for all i and $\det(\mathcal{A}) = 1$. So, the inverse matrix $\mathcal{A}^{-1}$ of elements $g_{i,j}$ has the same structure: triangular with $g_{i,i} = 1$ for all i and $\det(\mathcal{A}^{-1}) = 1$. For $g_{i,j} \neq 0$ (i.e. when $i > j$), its value is:

$$g_{i,j} = (-1)^{i+j} \det(\mathcal{A}_{i,j}), \qquad [3.47]$$

where $\mathcal{A}_{i,j}$ is the square matrix of elements $a_{i',j'}$ with $j < i' \leq i$ and $j \leq j' < i$.

When $\mathcal{A}$ is of finite dimension N, we may write $\mathcal{A} = \mathbf{1} + \mathcal{B}$ where $\mathbf{1}$ is the identity matrix and with $\mathcal{B}$ such that $\mathcal{B}^k = 0$ for $k \geq N$. Thus:

$$\mathcal{A}^{-1} = (\mathbf{1} + \mathcal{B})^{-1} = \sum_{k=0}^{N-1} (-\mathcal{B})^k,$$

which provides the corresponding pseudopolynomials $\bar{G}_0, \ldots, \bar{G}_{N-1}$.

Application to U periodic: suppose the u_i are of period ν. Then, in the notation [3.46]:

$$a_{i+\nu,j+\nu} = e_{i+\nu-(j+\nu)}(u_{j+\nu}) = e_{i,j}(u_j) = a_{i,j}. \qquad [3.48]$$

By [3.48], $\mathcal{A}$ becomes a lower triangular matrix of the form:

$$\mathcal{A} = \begin{pmatrix} \mathcal{A}_0 & & & \\ \mathcal{A}_1 & \mathcal{A}_0 & & \\ \mathcal{A}_2 & \mathcal{A}_1 & \mathcal{A}_0 & \\ \cdots & \cdots & \cdots & \mathcal{A}_0 \end{pmatrix},$$

where $\mathcal{A}_i$, $i \geq 0$, are the submatrices:

$$\mathcal{A}_0 = \begin{pmatrix} a_{0,0} & \dots & a_{0,\nu-1} \\ \dots & \dots & \dots \\ a_{\nu-1,0} & \dots & a_{\nu-1,\nu-1} \end{pmatrix} = \begin{pmatrix} 1 & \dots & 0 \\ \dots & \dots & \dots \\ a_{\nu-1,0} & \dots & 1 \end{pmatrix},$$

$$\mathcal{A}_i = \begin{pmatrix} a_{i\nu,0} & \dots & a_{i\nu,\nu-1} \\ \dots & \dots & \dots \\ a_{i\nu+\nu-1,0} & \dots & a_{i\nu+\nu-1,\nu-1} \end{pmatrix}, \quad i \geq 1.$$

The matrix $\mathcal{A}^{-1}$ has the same triangular structure with submatrices $\mathcal{B}_i$, say, in place of $\mathcal{A}_i$. Let us determine these $\mathcal{B}_i$.

For that, consider the product $\mathcal{A}\mathcal{A}^{-1}$ which reduces to a diagonal matrix with unit submatrices $\mathbf{1}_i \equiv \mathbf{1}$ on the diagonal. In this product, the term on any line $n+r$, column r, yields:

$$\sum_{\alpha \geq r}^{n+r} \mathcal{A}_{n+r-\alpha}\, \mathcal{B}_{\alpha-r} = \mathbf{1}\, \delta_{n+r,r}, \quad n, r \geq 1.$$

Equivalently, $\mathcal{A}_0\mathcal{B}_0 = \mathbf{1}$, and:

$$\mathcal{A}_n\mathcal{B}_0 + \mathcal{A}_{n-1}\mathcal{B}_1 + \ldots + \mathcal{A}_1\mathcal{B}_{n-1} + \mathcal{A}_0\mathcal{B}_n = \mathbf{0}, \quad n \geq 1.$$

Setting $\mathcal{A}_0\mathcal{B}_i = \tilde{\mathcal{B}}_i$ and $\mathcal{A}_i\mathcal{A}_0{}^{-1} = \tilde{\mathcal{A}}_i$ for $i \geq 0$, this becomes $\tilde{\mathcal{B}}_0 = \tilde{\mathcal{A}}_0 = \mathbf{1}$, and:

$$\tilde{\mathcal{A}}_n + \tilde{\mathcal{A}}_{n-1}\tilde{\mathcal{B}}_1 + \ldots + \tilde{\mathcal{A}}_1\tilde{\mathcal{B}}_{n-1} + \tilde{\mathcal{B}}_n = \mathbf{0}, \quad n \geq 1.$$

The $\tilde{\mathcal{B}}_i$ are then obtained by recursion. For example:

$$\tilde{\mathcal{B}}_1 = -\tilde{\mathcal{A}}_1,\ \tilde{\mathcal{B}}_2 = -\tilde{\mathcal{A}}_2 + \tilde{\mathcal{A}}_1^2,\ \tilde{\mathcal{B}}_3 = -\tilde{\mathcal{A}}_3 + (\tilde{\mathcal{A}}_2\tilde{\mathcal{A}}_1 + \tilde{\mathcal{A}}_1\tilde{\mathcal{A}}_2) - \tilde{\mathcal{A}}_1^3,$$

$$\tilde{\mathcal{B}}_4 = -\tilde{\mathcal{A}}_4 + (\tilde{\mathcal{A}}_3\tilde{\mathcal{A}}_1 + \tilde{\mathcal{A}}_2\tilde{\mathcal{A}}_2 + \tilde{\mathcal{A}}_1\tilde{\mathcal{A}}_3)$$

$$- (\tilde{\mathcal{A}}_2\tilde{\mathcal{A}}_1^2 + \tilde{\mathcal{A}}_1\tilde{\mathcal{A}}_2\tilde{\mathcal{A}}_1 + \tilde{\mathcal{A}}_1^2\tilde{\mathcal{A}}_2) + \tilde{\mathcal{A}}_1^4.$$

More generally, we get:

$$\tilde{\mathcal{B}}_n - \sum_{k=1}^{n} (-1)^k\, \sigma_{n,k}, \qquad [3.49]$$

where $\sigma_{n,k}$ denotes the sum of all terms $\tilde{A}_{i_1}\tilde{A}_{i_2}\dots\tilde{A}_{i_k}$ with k factors such as $i_1 + i_2 + \dots + i_k = n$ (the i_j being non necessarily distinct). From [3.49], we then have $\tilde{\mathcal{B}}_n = H_n(\tilde{A}_1, \dots, \tilde{A}_n)$, and therefore:

$$\mathcal{B}_n = \mathcal{A}_0^{-1} H_n(\mathcal{A}_1\mathcal{A}_0^{-1}, \dots, \mathcal{A}_n\mathcal{A}_0^{-1}).$$

Finally, to deduce the $g_{i,j} \neq 0$ given in [3.47], we express j as $j \equiv k\nu + j_0$, $0 \leq j_0 < \nu$ (Euclidean algorithm). Thus:

$$g_{i,j} = g_{i-k\nu,j-k\nu} = g_{i-k\nu,j_0},$$

which is precisely a term of B_r where r denotes the integer part of $(i - k\nu)/\nu$.

b) *A second method*: it suffices to rewrite [3.43]:

$$\bar{G}_n(\ |U) = e_n(\ ) - \sum_{i=0}^{n-1} e_{n-i}(u_i)\,\bar{G}_i(\ |U), \quad n \geq 1, \qquad [3.50]$$

which provides recursively $\bar{G}_n$, $n \geq 1$, starting from $\bar{G}_0 = 1$.

If $\mathcal{E}$ is standard at ω: In this case, the calculation via [3.50] can be made partly explicit. Indeed, as $e_n(\omega) = 0$, $n \geq 1$, [3.50] evaluated at ω yields:

$$\bar{G}_n(\omega|U) = -\sum_{i=0}^{n-1} e_{n-i}(u_i)\,\bar{G}_i(\omega|U), \quad n \geq 1. \qquad [3.51]$$

After successive elimination of all the $\bar{G}_i(\omega|U)$ in the r.h.s. of [3.51], we then obtain for $\bar{G}_n(\omega|U)$ an expression independent of ω which is the sum of products having the form:

$$\pi_{n;i_0,\dots,i_k} \equiv \big(-e_{n-i_0}(u_{i_0})\big)\big(-e_{i_0-i_1}(u_{i_1})\big)\dots\big(-e_{i_{k-1}-i_k}(u_{i_k})\big), \qquad [3.52]$$

where $n > i_0 > i_1 > \dots > i_{k-1} > i_k = 0$.

A suggestive description of this method is to represent it as performed along a path from n to 0 by $k + 1$ steps (with k intermediate stops in $\{i_0, i_1, \dots, i_{k-1}\} \subseteq \{n-1, n-2, \dots, 1\}$). The structure of $\bar{G}_n(\omega|U)$ is thus given by:

$$\bar{G}_n(\omega|U) = \sum_{paths} \prod_{steps} \pi_{n;i_0,\dots,i_k}. \qquad [3.53]$$

Note that the number of paths in [3.53] is equal to 2^{n-1}: paths of $j+1$ steps (i.e. with j intermediate stops) are in number $\binom{n-1}{j}$, and summing these numbers for all $j = 0, \dots n-1$ gives 2^{n-1}. Observe also that each path ends on some factor $e_s(u_0)$, $0 \leq s \leq n$, and does not contain any other factor in u_0.

For any element of $\mathcal{D}$: Having obtained the $\bar{G}_{n-i}(\omega|U)$, we can deduce $\bar{G}_n(\ |U)$ from [2.79], that is, by the Taylor-type formula:

$$\bar{G}_n(\ |U) = \sum_{i=0}^{n} \bar{G}_{n-i}(\omega|E^i U)\, e_i(\ ), \quad n \geq 1.$$

A more efficient method is however possible, which has the advantage of not depending on the existence of ω. We note that [3.50] can be rewritten as:

$$\bar{G}_n(\ |U) = [e_n(\ ) - e_n(u_0)] - \sum_{i=1}^{n-1} e_{n-i}(u_i)\, \bar{G}_i(\ |U), \quad n \geq 1. \quad [3.54]$$

It is therefore sufficient to proceed iteratively as above for $\bar{G}_n(\omega|U)$, after having substituted $e_r(u_0) - e_r(\ )$ for each $e_r(u_0)$.

3.3.2. *From the integral definition of $\bar{G}_n$*

As seen in [2.20], $\bar{G}_n$, $n \geq 1$, is defined using the integration operator I_u defined by [2.14] as:

$$\bar{G}_n(\ |U) = I_{u_0} I_{u_1} \dots I_{u_{n-1}}\, e_0, \quad n \geq 1. \quad [3.55]$$

Let us now introduce a forward operator J by $Je_n = e_{n+1}$, and remember the operator $\delta_u : f \in \mathcal{F} \to f(u)$ (see definition 2.6). Then,

$$I_u e_n = e_{n+1} - e_{n+1}(u) e_0 = (1 - \delta_u) e_{n+1} = (1 - \delta_u) J e_n,$$

so that [3.55] becomes:

$$\bar{G}_n(\ |U) = [(1-\delta_{u_0})J][(1-\delta_{u_1})J] \dots [(1-\delta_{u_{n-1}})J]\, e_0, \quad n \geq 1. \quad [3.56]$$

By virtue of [3.56], $\bar{G}_n$ is a sum of 2^n terms, each of them being a product of n terms in J preceded by a factor that is either 1 or $-\delta_{u_i}$, in short by $(-\delta_{u_i})^{a_i}$ with $a_i = 0$ or 1. More explicitly:

$$\bar{G}_n(\ |U) = \sum_{a_0,a_1,\ldots a_{n-1}} C_{a_0,a_1,\ldots a_{n-1}}, \qquad [3.57]$$

where

$$C_{a_0,a_1,\ldots a_{n-1}} = (-\delta_{u_0})^{a_0} J\, (-\delta_{u_1})^{a_1} J \ldots (-\delta_{u_{n-1}})^{a_{n-1}} J\, e_0.$$

For instance:

$$C_{0,0,\ldots,0} = J^n e_0 = e_n, \qquad C_{1,1,\ldots,1} = [-e_1(u_0)] \ldots [-e_1(u_{n-1})],$$

when all $a_i = 0$ except for $i = r$ with $0 \le r \le n-1$:

$$C_{0,\ldots,0,1,0,\ldots,0} = J^r\, (-\delta_{u_r})\, J^{n-r}\, e_0 = -e_r\, e_{n-r}(u_r),$$

and if $a_i = 0$ except for $i = r_1, r_2$ with $0 \le r_1 < r_2 \le n-1$:

$$\begin{aligned} C_{0,\ldots,0,1,0,\ldots,0,1,0,\ldots,0} &= J^{r_1}\, (-\delta_{u_{r_1}})\, J^{r_2-r_1}\, (-\delta_{u_{r_2}})\, J^{n-r_2}\, e_0 \\ &= e_{r_1}\, e_{r_2-r_1}(u_{r_1})\, e_{n-r_2}(u_{r_2}). \end{aligned}$$

In general, if $a_i = 0$ except for $i = r_1, r_2, \ldots, r_s$ with $0 \le r_1 < r_2 < \ldots < r_s \le n-1$, we have:

$$C_{a_0,a_1,\ldots a_{n-1}} = (-1)^s\, e_{r_1}\, e_{r_2-r_1}(u_{r_1}) \ldots e_{n-r_s}(u_{r_s}). \qquad [3.58]$$

To conclude, $\bar{G}_n(\ |U)$ is the sum [3.57] of 2^n terms having the form [3.58].

Application to $\partial_{u_i}(\bar{G}_n(\ |U))$*:* Suppose U is a family of numbers and the functions e_n are differentiable. To measure the influence of any u_i, we will determine the partial derivative $\partial_{u_i} \equiv (\partial/\partial u_i)$ of the pseudopolynomial $\bar{G}_n(\ |U)$.

PROPOSITION 3.8.– *For* $1 \le i \le n-1$*:*

$$\partial_{u_i}\big[\bar{G}_n(\ |U)\big] = \bar{G}_i(\ |U)\, \partial_{u_i}\big[\bar{G}_{n-i}(\ |E^i U)\big]. \qquad [3.59]$$

Proof. Inside [3.57], we distinguish the term dependent on u_i from the others, hence:

$$\begin{aligned}\bar{G}_n(\ |U) &= I_{u_0} \dots I_{u_{i-1}} (-\delta_{u_i}) J\, I_{u_{i+1}} \cdots I_{u_{n-1}}\, e_0 \\ &\quad + \text{terms independent of } u_i \\ &= \bar{G}_i(\ |U)\, (-\delta_{u_i}) J\, \bar{G}_{n-i-1}(\ |E^{i+1}U) \\ &\quad + \text{terms independent of } u_i.\end{aligned}$$

By differentiation, we then obtain:

$$\begin{aligned}\partial_{u_i}\big[\bar{G}_n(\ |U)\big] &= \bar{G}_i(\ |U)\, \partial_{u_i}\big[(-\delta_{u_i}) J\, \bar{G}_{n-i-1}(\ |E^{i+1}U)\big] \\ &= \bar{G}_i(\ |U)\, \partial_{u_i}\big[(1-\delta_{u_i}) J\, \bar{G}_{n-i-1}(\ |E^{i+1}U) - J\, \bar{G}_{n-i-1}(\ |E^{i+1}U)\big].\end{aligned} \quad [3.60]$$

Now, the first term in [...] of [3.60] corresponds to $\bar{G}_{n-i}(\ |E^iU)$ (see [2.21]), while the second is independent of u_i. Thus, [3.60] yields [3.59] as desired. □

REMARK.– By definition, $\bar{G}_n(x|U)$, $n \geq 1$, can be expressed as $f(x) - f(u_0)$ for some function $f \in \mathcal{F}$. This implies that $\partial_{u_0}\big[\bar{G}_n(x|U)\big]$ is independent of x. Consequently, [3.59] can be rewritten as:

$$\partial_{u_i}\big[\bar{G}_n(x|U)\big] = \bar{G}_i(x|U)\ \partial_{u_i}\big[\bar{G}_{n-i}(a|E^iU)\big], \quad [3.61]$$

where a denotes any constant.

For the A–G polynomials $G_n(x|U)$, the same formula [3.59] follows directly from the integral representation [1.8].

3.3.3. *From the classical binomial identity*

We here assume S.I. and $\mathcal{E}$ standard at 0, so that the binomial identity [2.78] applies. We also recall that the identity [2.82] is then true, namely:

$$I_u e_0(x) = I_u e_0(x-u) = e_1(x-u), \text{ and } I_u e_n(x-u) = e_{n+1}(x-u), \quad n \geq 1. \quad [3.62]$$

Now, consider the family $U = \{a_0, \ldots, a_0, a_1, \ldots, a_1\}$, where there are n_0 parameters a_0 and n_1 parameters a_1. From [2.20] and using [3.62], $\bar{G}_{n_0+n_1}(\ |U)$ is given by:

$$\begin{aligned}\bar{G}_{n_0+n_1}(x|U) &= I_{a_0} \ldots I_{a_0}\, I_{a_1} \ldots I_{a_1}\, e_0(x) \\ &= I_{a_0} \ldots I_{a_0}\, e_{n_1}(x - a_1).\end{aligned}$$

By applying [2.78] to $e_{n_1}(x - a_1)$, this gives:

$$\begin{aligned}\bar{G}_{n_0+n_1}(x|U) &= I_{a_0} \ldots I_{a_0} \sum_{i_1=0}^{n_1} e_{i_1}(x - a_0)\, e_{n_1-i_1}(a_0 - a_1) \\ &= \sum_{i_1=0}^{n_1} e_{n_0+i_1}(x - a_0)\, e_{n_1-i_1}(a_0 - a_1).\end{aligned} \quad [3.63]$$

After iterations, we are going to derive the following general result.

PROPOSITION 3.9.– *Suppose S.I. satisfied and $\mathcal{E}$ standard at* 0*. Let:*

$$U = \{a_0, \ldots, a_0, a_1, \ldots, a_1, \ldots, a_{r-1}, \ldots, a_{r-1}\},$$

where a_j, $0 \leq j \leq r-1$, are real (complex) numbers repeated n_j (> 0) times. Then:

$$\bar{G}_{n_0+\ldots+n_{r-1}}(x|U) = \sum_{i_0,\ldots,i_{r-1}\in\mathbb{N}_0} e_{i_0}(x) \left(\prod_{j=0}^{r-1} e_{w_j}(b_j)\right), \quad r \geq 1, \quad [3.64]$$

after denoting

$$b_j = a_{j-1} - a_j, \quad \textit{and} \quad w_j = n_j + i_{j+1} - i_j,$$

where $a_{-1} \equiv 0$, $i_r \equiv 0$ and $e_{w_j} \equiv 0$ if $w_j < 0$. When $r \geq 2$, an equivalent formula is:

$$\bar{G}_{n_0+\ldots+n_{r-1}}(x|U) = \sum_{i_1,\ldots,i_{r-1}\in\mathbb{N}_0} e_{n_0+i_1}(x - a_0) \left(\prod_{j=1}^{r-1} e_{w_j}(b_j)\right). \quad [3.65]$$

Proof. First, notice that when $r \geq 2$, [3.64] is identical to [3.65] since:

$$\sum_{i_0 \in \mathbb{N}_0} e_{i_0}(x)\, e_{w_0}(b_0) = \sum_{i_0=0}^{n_0+i_1} e_{i_0}(x)\, e_{n_0+i_1-i_0}(-a_0) = e_{n_0+i_1}(x-a_0),$$

by virtue of [2.78].

When $r = 1$ (i.e. $U = \{a_0, \dots, a_0\}$), then $i_1 = 0$, $b_0 = -a_0$ and $w_0 = n_0 - i_0$, so that [3.64] becomes:

$$\bar{G}_{n_0}(x|U) = \sum_{i_0=0}^{n_0} e_{i_0}(x)\, e_{n_0-i_0}(-a_0) = e_{n_0}(x-a_0),$$

which is true since $u_i = a_0$ for all i (see [2.83]). When $r = 2$, then $i_2 = 0$, $b_1 = a_0 - a_1$, $w_1 = n_1 - i_1$, and [3.64] is identical to [3.65] which has been established by [3.63].

For the other cases, we will proceed by recurrence. Suppose that [3.64] (respectively, [3.65]) has been proven for $r = 1, \dots, k$ (respectively, $r = 2, \dots, k$), and then consider the identity (see [2.23]):

$$\Delta^{n_0} \bar{G}_{n_0+\ldots+n_k}(x|U) = \bar{G}_{n_1+\ldots+n_k}(x|E^{n_0}U). \qquad [3.66]$$

We express the r.h.s. of [3.66] using [3.65] for $r = k$ but in which $n_1, \dots, n_k$ are substituted for $n_0, \dots, n_{k-1}$, also $a_1, \dots, a_k$ for $a_0, \dots, a_{k-1}$ (thus, $b_1, \dots, b_k$ for $b_0, \dots, b_{k-1}$) and, for convenience, the indices $i_2, \dots, i_k$ for $i_1, \dots, i_{k-1}$ (thus, $w_1, \dots, w_k$ for $w_0, \dots, w_{k-1}$). This gives:

$$\begin{aligned}
\bar{G}_{n_1+\ldots+n_k}(x|E^{n_0}U) &= \sum_{i_2,\dots,i_k \in \mathbb{N}_0} e_{n_1+i_2}(x-a_1) \left(\prod_{j=2}^{k} e_{w_j}(b_j)\right) \\
&= \sum_{i_2,\dots,i_k \in \mathbb{N}_0} \left[\sum_{i_1 \in \mathbb{N}_0} e_{i_1}(x-a_0)\, e_{n_1+i_2-i_1}(a_0-a_1)\right] \left(\prod_{j=2}^{k} e_{w_j}(b_j)\right) \\
&= \sum_{i_1,\dots,i_k \in \mathbb{N}_0} e_{i_1}(x-a_0) \left(\prod_{j=1}^{k} e_{w_j}(b_j)\right). \qquad [3.67]
\end{aligned}$$

Now, let us invert [3.66] by applying the operator I_{a_0} successively n_0 times. Thanks to [3.67], we deduce that:

$$\bar{G}_{n_0+\ldots+n_k}(x|U) = \sum_{i_1,\ldots,i_k \in \mathbb{N}_0} e_{n_0+i_1}(x-a_0)\left(\prod_{j=1}^{k} e_{w_j}(b_j)\right),$$

that is, [3.65] when $r = k+1$. □

The previous results are easily generalized when $\mathcal{E}$ is standard at ω instead of 0.

COROLLARY 3.1.– *Suppose S.I. satisfied and $\mathcal{E}$ standard at ω. Then, [3.65] simply becomes:*

$$\bar{G}_{n_0+\ldots+n_{r-1}}(x|U) = \sum_{i_1,\ldots,i_{r-1} \in \mathbb{N}_0} e_{n_0+i_1}(x-a_0+\omega)\left(\prod_{j=1}^{r-1} e_{w_j}(b_j+\omega)\right), \quad r \geq 2.$$

Proof. Define $\hat{e}_n(.) = e_n(.+\omega)$, so that $\hat{\mathcal{E}}$ is standard at 0. According to the binomial formula [2.78], it is a copy of $\mathcal{E}$. By [2.28], we thus have:

$$\bar{G}_{n_0+\ldots+n_{r-1}}(x|\mathcal{E}, U) = \bar{G}_{n_0+\ldots+n_{r-1}}(x|\hat{\mathcal{E}}, U),$$

and using [3.65]:

$$\bar{G}_{n_0+\ldots+n_{r-1}}(x|\hat{\mathcal{E}}, U) = \sum_{i_1,\ldots,i_{r-1} \in \mathbb{N}_0} \hat{e}_{n_0+i_1}(x-a_0)\left(\prod_{j=1}^{r-1} \hat{e}_{w_j}(b_j)\right),$$

hence the announced result. The adaptation of [3.64] is quite similar. □

3.4. Geometric or affine form for U

The A–G pseudopolynomials can sometimes be made more explicit. Below we examine three particular situations where this is feasible, namely, when the $u_i \in \mathcal{D}$ are constant and, more generally, in geometric progression or of affine form.

3.4.1. *When* $U = \{u_i \equiv a,\ i \in \mathbb{N}_0\}$

As pointed out in section 2.2 (see definition 2.3) and section 2.5 (by the consequence (iv)), the family of $\bar{G}_n(x|U)$ in that case has an Appell structure and is a copy of $\mathcal{E}$ that is standard at a. In fact, by virtue of [2.83], we know that $\bar{G}_n(x|\{a, a, \ldots\}) = e_n(x - a)$ under the S.I. and $\mathcal{E}$ standard at 0.

Hereafter, we consider the general situation, that is, without any assumption on $\mathcal{E}$. Since all $u_i = a$, the A-G expansion [2.32] of e_n becomes:

$$e_n(x) = \sum_{i=0}^{n} e_{n-i}(a)\, \bar{G}_i(x|U).$$

In terms of generating functions, we then obtain:

$$\sum_{n=0}^{\infty} e_n(x)\, s^n = \sum_{j=0}^{\infty} e_j(a)\, s^j \sum_{n=0}^{\infty} \bar{G}_n(x|U)\, s^n,$$

which gives:

$$\sum_{n=0}^{\infty} \bar{G}_n(x|U)\, s^n = \frac{\sum_{n=0}^{\infty} e_n(x)\, s^n}{\sum_{n=0}^{\infty} e_n(a)\, s^n}. \qquad [3.68]$$

Let us list a few simple ways to deduce $\bar{G}_n(x|U)$.

a) Writing $\sum_{n=0}^{\infty} e_n(a)\, s^n = 1 + \sum_{n=1}^{\infty} e_n(a)\, s^n$, we have:

$$\frac{1}{\sum_{n=0}^{\infty} e_n(a)\, s^n} = \sum_{k=0}^{\infty} (-1)^k \left(\sum_{n=1}^{\infty} e_n(a)\, s^n\right)^k = \sum_{k=0}^{\infty} (-1)^k \sum_{i=k}^{\infty} e_i^{*k}(a)\, s^i, \qquad [3.69]$$

using the convolution notation (see [3.9]). Remember that the $e_i^{*k}(a)$ can be determined recursively from:

$$e_i^{*1}(a) = e_i(a), \quad \text{and} \quad e_i^{*(k+1)}(a) = \sum_{j=k}^{i-1} e_j^{*k}(a)\, e_{i-j}(a), \quad i \geq k+1.$$

Therefore, by inserting [3.69] into [3.68], we get:

$$\bar{G}_n(x|U) = \sum_{i=0}^{n} e_{n-i}(x) \sum_{k=0}^{i} (-1)^k e_i^{*k}(a).$$

b) Denote $\psi(x,s) = \sum_{n=0}^{\infty} e_n(x)\, s^n$. Taking the nth derivative [3.68] evaluated at $s=0$, we directly see that:

$$\bar{G}_n(x|U) = \frac{1}{n!}\left[\frac{\psi(x,s)}{\psi(a,s)}\right]_{s=0}^{(n)} = \frac{1}{n!}\sum_{i=0}^{n}\binom{n}{i}\left[\psi(x,s)\right]_{s=0}^{(n-i)}\left[\frac{1}{\psi(a,s)}\right]_{s=0}^{(i)}. \quad [3.70]$$

From the definition of $\psi(x,s)$, [3.70] then simplifies to:

$$\bar{G}_n(x|U) = \sum_{i=0}^{n} e_{n-i}(x)\frac{1}{i!}\left[\frac{1}{\psi(a,s)}\right]_{s=0}^{(i)}. \quad [3.71]$$

It remains to determine in [3.71] the factors $(1/i!)[1/\psi(a,s)]_{s=0}^{(i)}$. For that, we recall that $\bar{G}_n(a|U) = \delta_{n,0}$ (by [2.24] for $m=0$). Thus, setting $x=a$ in [3.71] yields:

$$\delta_{n,0} = \sum_{i=0}^{n} e_{n-i}(a)\frac{1}{i!}\left[\frac{1}{\psi(a,s)}\right]_{s=0}^{(i)},$$

which gives recursively all the desired terms.

c) Since $EU = U$, the u_i are of period $\nu = 1$. Therefore, we may follow the reasoning developed in part 1 of section 3.3. Note that as $\nu = 1$, all the involved submatrices will simply be numbers.

d) If the formal series in [3.68] are convergent series for $|s|$ small enough, $\bar{G}_n(x|U)$ can be obtained as a Cauchy integral on a circle of radius $R>0$:

$$\bar{G}_n(x|U) = \frac{1}{2\pi i}\int_{C(0,R)} \frac{\psi(x,z)}{\psi(a,z)}\frac{1}{z^{n+1}}\,dz.$$

e) Under the S.I. and $\mathcal{E}$ standard at 0, [2.35] for $f_n = \bar{G}_n$ and $\omega = 0$ gives, using [2.83]:

$$\bar{G}_n(x|U) = \sum_{i=0}^{n} \bar{G}_{n-i}(0|E^i U)\, e_i(x).$$

Therefore, we can only examine the case where $x = 0$ and consider, instead of [3.68], the relation:

$$\sum_{n=0}^{\infty} \bar{G}_n(0|U)\, s^n = \frac{1}{\sum_{n=0}^{\infty} e_n(a)\, s^n}.$$

3.4.2. *When* $U = \{u_i \equiv aq^i,\ i \in \mathbb{N}_0\}$

For this case, we assume that $a \in \mathcal{D}$, a real vector space, q is any fixed real and $\mathcal{E}$ satisfies the multiplicative property [2.55].

The A–G expansion of $e_n(x)$ gives here:

$$e_n(x) = \sum_{i=0}^{n} e_{n-i}(u_i)\, \bar{G}_i(x|U) = \sum_{i=0}^{n} q^{i(n-i)}\, e_{n-i}(a)\, \bar{G}_i(x|U).$$

Multiplying by $q^{-n^2/2}$, this becomes:

$$q^{-n^2/2}\, e_n(x) = \sum_{i=0}^{n} q^{-(n-i)^2/2}\, e_{n-i}(a)\, q^{-i^2/2}\, \bar{G}_i(x|U). \qquad [3.72]$$

We now introduce a new family $\hat{\mathcal{E}}$ of functions $\hat{e}_n(x)$ defined by:

$$\hat{e}_n(x) = q^{-n^2/2}\, e_n(x), \quad n \geq 0. \qquad [3.73]$$

Note that like $\mathcal{E}$, the family $\hat{\mathcal{E}}$ is linearly independent, and it satisfies the multiplicative property, implying that both families are standard at 0. To $\hat{\mathcal{E}}$ is attached a differentiation operator $\hat{\Delta}$, which is not explicitly indicated to simplify the notation. This will be done later as well: below, for example, an operator $\tilde{\Delta}$ is implicitly attached to the family $\tilde{\mathcal{E}}$ defined by [3.78].

For the rest, it will of course be necessary to indicate in the notation which family, $\mathcal{E}$ or $\hat{\mathcal{E}}$, is used in the formulas. Thus, via the substitution [3.73], the relation [3.72] is rewritten as:

$$\hat{e}_n(x) = \sum_{i=0}^{n} \hat{e}_{n-i}(a)\, q^{-i^2/2}\, \bar{G}_i(x|\mathcal{E}, U). \qquad [3.74]$$

Simultaneously, consider the A–G pseudopolynomials $\bar{G}_i(x|\hat{\mathcal{E}}, \{a, a, \ldots\})$, so that the corresponding A–G expansion of $\hat{e}_n(x)$ is:

$$\hat{e}_n(x) = \sum_{i=0}^{n} \hat{e}_{n-i}(a)\, \bar{G}_i(x|\hat{\mathcal{E}}, \{a, a, \ldots\}). \qquad [3.75]$$

Arguing by recursion on $\hat{e}_n$ for $n \geq 0$, we then deduce from [3.74] and [3.75] the following identification:

$$q^{-i^2/2}\, \bar{G}_i(x|\mathcal{E}, U) = \bar{G}_i(x|\hat{\mathcal{E}}, \{a, a, \ldots\}), \quad i \geq 0.$$

In other words, we have:

$$\bar{G}_i(x|\mathcal{E}, U) = q^{i^2/2}\, \bar{G}_i(x|\hat{\mathcal{E}}, \{a, a, \ldots\}), \quad i \geq 0, \qquad [3.76]$$

where $\bar{G}_i(x|\hat{\mathcal{E}}, \{a, a, \ldots\})$ can be evaluated as explained in part 1 above.

3.4.3. *When* $U = \{u_i \equiv a + bi,\ i \in \mathbb{N}_0\}$*: first model*

Here, $a, b \in \mathcal{D}$, an abelian group, and the S.I. property is assumed to be satisfied.

The A-G expansion of $e_n(x - u_n)$ then gives:

$$e_n(x-u_n) = \sum_{i=0}^{n} e_{n-i}(u_i-u_n)\, \bar{G}_i(x|U) = \sum_{i=0}^{n} e_{n-i}(a-u_{n-i})\bar{G}_i(x|U), \qquad [3.77]$$

since $u_i - u_n = -b(n-i) = a - u_{n-i}$.

As in the previous part, we introduce a new family $\tilde{\mathcal{E}}$ of functions $\tilde{e}_n(x)$ defined by:

$$\tilde{e}_n(x) = e_n(x - u_n), \quad n \geq 0, \tag{3.78}$$

which inherits from $\mathcal{E}$ the property of linear independence. Again, this leads us to add $\mathcal{E}$ or $\tilde{\mathcal{E}}$ in the notation.

On the one hand, [3.77] with [3.78] yields:

$$\tilde{e}_n(x) = \sum_{i=0}^{n} \tilde{e}_{n-i}(a)\, \bar{G}_i(x|\mathcal{E}, U). \tag{3.79}$$

On the other hand, we consider a similar A–G expansion of $\tilde{e}_n$ given by:

$$\tilde{e}_n(x) = \sum_{i=0}^{n} \tilde{e}_{n-i}(a)\, \bar{G}_i(x|\tilde{\mathcal{E}}, \{a, a, \ldots\}). \tag{3.80}$$

From [3.79] and [3.80], we deduce by recursion the identification:

$$\bar{G}_i(x|\mathcal{E}, U) = \bar{G}_i(x|\tilde{\mathcal{E}}, \{a, a, \ldots\}), \quad i \geq 0, \tag{3.81}$$

where the r.h.s. can again be evaluated as in part 1.

3.4.4. *When* $U = \{u_i \equiv a + bi,\ i \in \mathbb{N}_0\}$*: second model*

This time, $a, b \in \mathcal{D} = \mathbb{R}$ or $\mathbb{Z}$, and $\mathcal{E}$ has the S.I. property and is provided by the generating function $\theta(s)\exp(xg(s))$ given in [3.8] (with $\theta(s) = \sum_{n=0}^{\infty} e_n(0)s^n$). We will discuss separately the cases where $\mathcal{E}$ is standard at 0 (i.e. $\theta(s) = 1$) or not.

a) Let us start with the case $\theta(s) = 1$.

PROPOSITION 3.10.– *When $\mathcal{E}$ is standard at* 0*:*

$$\begin{aligned} \bar{G}_n(x|U) &= \left[1 + \frac{e_1(u_n - u_0)}{e_1(x - u_n)}\right] e_n(x - u_n) && [3.82] \\ &= \frac{x - u_0}{x - u_n}\, e_n(x - u_n), \quad n \geq 1. && [3.83] \end{aligned}$$

Proof. Note that as seen in [3.12], $e_1(x) = \alpha_1 x$, thus [3.82] directly reduces to [3.83]. Moreover, the hypotheses made imply [3.15], that is:

$$\Delta\left(\frac{ne_n(x)}{e_1(x)}\right) = \frac{(n-1)e_{n-1}(x)}{e_1(x)}, \quad n \geq 1. \tag{3.84}$$

Now, we denote by $K_n(x|U)$ the r.h.s. of [3.82]. This is an element of $\mathcal{F}$ since $e_n(x)/e_1(x) \in \mathcal{F}$ and $S^{-u_n}\mathcal{F} = \mathcal{F}$. Consequently, applying the operator Δ, we have:

$$\Delta K_n(x|U) = S^{-u_n}\,\Delta\Big\{\Big[1 + \frac{e_1(u_n - u_0)}{e_1(x)}\Big]e_n(x)\Big\}, \tag{3.85}$$

and because of [3.84]:

$$\begin{aligned}\Delta K_n(x|U) &= S^{-u_n}\left\{e_{n-1}(x) + \frac{e_1(u_n-u_0)}{n}\,\Delta\left[\frac{ne_n(x)}{e_1(x)}\right]\right\}\\ &= S^{-u_n}\left\{e_{n-1}(x) + \frac{e_1(u_n-u_0)}{n}\,\frac{(n-1)e_{n-1}(x)}{e_1(x)}\right\}.\end{aligned}$$

Since the u_i are of the form $a + bi$, $(n-1)e_1(u_n - u_0) = ne_1(u_n - u_1)$, so that $\Delta K_n(x|U)$ can be rewritten as:

$$\begin{aligned}\Delta K_n(x|U) &= S^{-u_n}\left\{\left[1 + \frac{e_1(u_n-u_1)}{e_1(x)}\right]e_{n-1}(x)\right\}\\ &= \left[1 + \frac{e_1(u_n-u_1)}{e_1(x-u_n)}\right]e_{n-1}(x-u_n)\\ &= \frac{e_1(x-u_1)}{e_1(x-u_n)}\,e_{n-1}(x-u_n).\end{aligned}$$

Using the usual operator E acting on U (i.e. such that $Eu_i = u_{i+1}$), we obtain:

$$\begin{aligned}\Delta K_n(x|U) &= E\left[\frac{e_1(x-u_0)}{e_1(x-u_{n-1})}\,e_{n-1}(x-u_{n-1})\right]\\ &= EK_{n-1}(x|U),\end{aligned} \tag{3.86}$$

from the r.h.s. of [3.82]. Obviously, $EK_{n-1}(x|U) = K_{n-1}(x|EU)$, so that [3.86] yields $\Delta K_n(x|U) = K_{n-1}(x|EU)$. Since $K_n(u_0|U) = \delta_{n,0}$, we conclude that $K_n(x|U) = \bar{G}_n(x|U)$, that is, the result [3.82]. □

b) Now, let us move on to the case $\theta(s) \neq 1$. We are going to see that the A–G pseudopolynomials then have a more complicated expression. For clarity, they will be denoted $\bar{G}_n(x|\bar{\mathcal{E}}, U)$ where $\bar{\mathcal{E}}$ is the family of functions $\bar{e}_n$ obtained from the generating function $\bar{\psi}(x,s) = \theta(s)\exp(xg(s))$, while $\mathcal{E}$ is provided as in part (a) by $\psi(x,s) = \exp(xg(s))$.

Note that $\bar{\mathcal{E}}$ and $\mathcal{E}$ are copies of each other. We saw in section 2.4 that this implies:

$$\bar{G}_n(x|\bar{\mathcal{E}}, U) = \bar{G}_n(x|\mathcal{E}, U),$$

so that by virtue of [3.83]:

$$\bar{G}_n(x|\bar{\mathcal{E}}, U) = \frac{x - u_0}{x - u_n}\, e_n(x - u_n), \quad n \geq 1. \qquad [3.87]$$

It now remains to go from e_n to the $\bar{e}_i$ in [3.87]. Of course:

$$\begin{aligned}
\psi(x,s) = \sum_{n=0}^{\infty} e_n(x)\, s^n &= \frac{\bar{\psi}(x,s)}{\theta(s)} = \frac{\sum_{n=0}^{\infty} \bar{e}_n(x)\, s^n}{1 + \sum_{j=1}^{\infty} \bar{e}_j(0)\, s^j} \\
&= \sum_{n=0}^{\infty} \bar{e}_n(x)\, s^n \sum_{k=0}^{\infty} (-1)^k \Big(\sum_{j=1}^{\infty} \bar{e}_j(0)\, s^j \Big)^k \\
&= \sum_{n=0}^{\infty} \bar{e}_n(x)\, s^n \sum_{k=0}^{\infty} (-1)^k \sum_{j=k}^{\infty} \left(\bar{e}_j(0)\right)^{*k} s^j,
\end{aligned}$$

using convolution notation. Therefore:

$$e_n(x) = \sum_{k=0}^{n} (-1)^k \sum_{j=k}^{n} \left(\bar{e}_j(0)\right)^{*k} e_{n-j}(x), \qquad [3.88]$$

and inserting [3.88] taken at $x - u_n$ into [3.87] gives $\bar{G}_n(x|\bar{\mathcal{E}}, U)$.

3.5. Extension to special sequences $\mathrm{u}_i = \{u_{i,j}\}$

Let us take $\mathcal{D} = \mathbb{R}^{\mathbb{N}}$ and $e_n(\mathrm{x}) = (1/n!)x_1 \ldots x_n$, $n \geq 1$. Recall that this case has already been encountered in sections 2.1 and 2.6. Inside this frame, we assume in addition that the family of parameters is a sequence of real numbers with double indices, namely:

$$\mathbf{U} = \{\mathbf{u}_i,\ i \in \mathbb{N}_0\},\ \text{ where } \mathbf{u}_i = \{u_{i,j},\ j \in \mathbb{N}\}.$$

Our goal here is to show that the A–G pseudopolynomials can be explicitly derived for three special forms of $\mathbf{U}$.

3.5.1. *When the $u_{i,j}$ do not depend on i*

Then, $E\mathbf{U} = \mathbf{U}$, so that the family $\bar{G}(\mathbf{U})$ is a copy of $\mathcal{E}$. This implies that:

$$\bar{G}_n(\mathbf{x}|\mathbf{U}) = \sum_{i=0}^{n} a_i\, e_{n-i}(\mathbf{x}). \qquad [3.89]$$

By [2.24], $\bar{G}_n(\mathbf{u}_0|\mathbf{U}) = \delta_{0,n}$, therefore:

$$\sum_{i=0}^{n} a_i\, e_{n-i}(\mathbf{u}_0) = \delta_{0,n},$$

that is, using generating functions:

$$\sum_{i=0}^{\infty} a_i\, s^i = 1/\sum_{i=0}^{\infty} e_i(\mathbf{u}_0)\, s^i. \qquad [3.90]$$

In other words, [3.90] provides the unknown a_i, and [3.89] then gives the $\bar{G}_n$.

EXAMPLE 3.1.– Suppose that $u_{i,j} = \alpha(j - \beta)$, $j \geq 1$, where $\alpha, \beta \in \mathbb{R}$. Then:

$$e_n(\mathbf{u}_0) = (1/n!)(-\alpha)^n(\beta-1)(\beta-2)\ldots(\beta-n) = (-\alpha)^n \binom{\beta-1}{n},$$

thus from [3.90]:

$$\sum_{i=0}^{\infty} a_i\, s^i = 1/(1-\alpha s)^{\beta-1} = \sum_{i=0}^{\infty} \binom{1-\beta}{i} (-\alpha)^i\, s^i,$$

directly yielding the a_i. After insertion in [3.89], we obtain:

$$\bar{G}_n(\mathbf{x}|\mathbf{U}) = \sum_{i=0}^{n} \binom{1-\beta}{i} (-\alpha)^i \frac{x_1 \ldots x_{n-i}}{(n-i)!}.$$

3.5.2. *When the $u_{i,j}$ do not depend on j*

Put $u_{i,j} = v_i$ and $V = \{v_i, i \in \mathbb{N}_0\}$. Since $e_n(x,\ldots,x) = x^n/n!$, we see from its A–G expansion [2.33] and proposition 1.1 that:

$$\bar{G}_n(x,\ldots,x|\mathbf{U}) = G_n(x|V), \qquad [3.91]$$

where $G_n(x|V)$ represents a classical A–G polynomial associated with the family V. Now, by definition, $\bar{G}_n(\mathbf{x}|\mathbf{U})$ can be expressed as:

$$\bar{G}_n(\mathbf{x}|\mathbf{U}) = \sum_{i=0}^{n} \chi_i\, e_i(\mathbf{x}), \qquad [3.92]$$

for appropriate coefficients χ_i. Taking $\mathbf{x} = (x, x, \ldots)$ in [3.92], we get from [3.91]:

$$G_n(x|V) = \sum_{i=0}^{n} \chi_i \frac{x^i}{i!},$$

which implies that:

$$\chi_i = G_{n-i}(0|E^i V), \quad 0 \leq i \leq n.$$

From [3.92], we then deduce that:

$$\bar{G}_n(\mathbf{x}|\mathbf{U}) = \sum_{i=0}^{n} G_{n-i}(0|E^i V)\, \frac{x_1 \ldots x_i}{i!}.$$

EXAMPLE 3.2.– Suppose that $u_{i,j} \equiv v_i = \alpha + \beta i$, $i \geq 0$, where $\alpha, \beta \in \mathbb{R}$. By [1.10], $G_n(x|V) = (x - v_0)(x - v_n)^{n-1}/n!$, thus:

$$\begin{aligned} G_{n-i}(0|E^i V) &= G_{n-i}(0|\{\alpha + \beta(i + j), j \geq 0\}) \\ &= G_{n-i}(-v_i|\{\beta j, j \geq 0\}) \\ &= (-v_i)(-v_n)^{n-i-1}/(n-i)!, \quad i \geq 0. \end{aligned}$$

Inserting this in [3.92], we get:

$$G_n(\mathbf{x}|\mathbf{U}) = \sum_{i=0}^{n} (-1)^{n-i} \frac{(\alpha + \beta i)(\alpha + \beta n)^{n-i-1}}{(n-i)!}\, \frac{x_1 \ldots x_i}{i!}.$$

3.5.3. *When the* $u_{i,j}$ *are of the form* $j\gamma_{i+j}/(i+j)(\beta_{i+j}-\beta_i)$

The parameters β_i involved in $\mathbf{U}$ are assumed to be distinct. To begin with, we will establish the following result concerning the functions $F_{n,r}(x)$ of $x \in \mathbb{R}$, for $0 \leq r \leq n$ and fixed n, defined by:

$$\begin{aligned} F_{n,r}(x) &= \prod_{i=r+1}^{n} \frac{\gamma_i}{\beta_i - x} + \frac{\gamma_n}{x-\beta_n} \prod_{i=r+1}^{n-1} \frac{\gamma_i}{\beta_i - \beta_n} \\ &\quad + \sum_{k=r+1}^{n-1} \frac{\gamma_n}{\beta_k - \beta_n} \left(\prod_{i=r+1}^{k} \frac{\gamma_i}{\beta_i - x} \prod_{j=k+1}^{n-1} \frac{\gamma_j}{\beta_j - \beta_n} \right), \end{aligned} \qquad [3.93]$$

and in which, by convention, $\sum_{k=r+1}^{r} \equiv 0$ and $\prod_{k=r+1}^{r} \equiv 1$.

LEMMA 3.1.– *For each* $r = 0, \ldots, n$, $F_{n,r}(\beta_r) = \delta_{n,r}$, *that is:*

$$\prod_{i=r+1}^{n} \frac{\gamma_i}{\beta_i - \beta_r} + \sum_{k=r}^{n-1} \frac{\gamma_n}{\beta_k - \beta_n} \left(\prod_{i=r+1}^{k} \frac{\gamma_i}{\beta_i - \beta_r} \prod_{j=k+1}^{n-1} \frac{\gamma_j}{\beta_j - \beta_n} \right) = \delta_{n,r}. \qquad [3.94]$$

Proof. The reasoning is by backward recursion on r. For $r = n$, we have:

$$F_{n,n}(\beta_n) = 1 + 0 = \delta_{n,n},$$

as desired. Suppose that the formula [3.93] is true when $r = n, n-1, \ldots, r_0 + 1\,(\leq n)$, and let us show that [3.93] is then also valid for $r = r_0$. Clearly, $F_{n,r_0}(x)$ is of the form:

$$F_{n,r_0}(x) = \sum_{i=r_0+1}^{n} \frac{\zeta_i}{x - \beta_i}, \qquad [3.95]$$

so that it is enough to check that all ζ_i, $r_0 + 1 \leq i \leq n$, are zero.

First, we obtain ζ_n by noting from [3.95] with [3.93] that:

$$\zeta_n = \lim_{x \to \beta_n} (x - \beta_n) F_{n,r_0}(x)$$

$$= -\gamma_n \prod_{i=r_0+1}^{n} \frac{\gamma_i}{\beta_i - \beta_n} + \gamma_n \prod_{j=r_0+1}^{n-1} \frac{\gamma_j}{\beta_j - \beta_n} = 0.$$

Second, for ζ_s with $r_0 + 1 \leq s \leq n-1$:

$$\zeta_s = \lim_{x \to \beta_s} (x - \beta_s) F_{n,r_0}(x)$$

$$= -\gamma_s \prod_{\substack{i=r_0+1 \\ i \neq s}}^{n-1} \frac{\gamma_i}{\beta_i - \beta_s}$$

$$- \gamma_s \sum_{k=s}^{n-1} \frac{\gamma_n}{\beta_k - \beta_n} \Big(\prod_{\substack{i=r_0+1 \\ i \neq s}}^{k} \frac{\gamma_i}{\beta_i - \beta_s} \prod_{j=k+1}^{n-1} \frac{\gamma_j}{\beta_j - \beta_n} \Big),$$

Setting l for either k or $n-1$ ($\geq s+1$), we have:

$$\prod_{\substack{i=r_0+1 \\ i \neq s}}^{l} = \prod_{i=r_0+1}^{s-1} \prod_{i=s+1}^{l},$$

so that ζ_s can then be rewritten as:

$$\zeta_s = -\gamma_s \Big(\prod_{i=r_0+1}^{s-1} \frac{\gamma_i}{\beta_i - \beta_s} \Big) F_{n,s}(\beta_s). \qquad [3.96]$$

Thanks to the recurrence hypothesis, $F_{n,s}(\beta_s) = 0$ in [3.96], thus ζ_s too. Using [3.95], we deduce that $F_{n,r_0}(x)$ is identically null, hence the announced result. □

We are ready to derive the compact formula [3.98] below for the A–G pseudopolynomials associated with the family $\mathbf{U} = \{\mathbf{u}_0, \mathbf{u}_1, \ldots\}$ where:

$$\mathbf{u}_i = \Big\{ u_{i,j} \equiv \frac{j}{i+j} \frac{\gamma_{i+j}}{\beta_{i+j} - \beta_i}, \, j \in \mathbb{N} \Big\}, \quad i \in \mathbb{N}_0. \qquad [3.97]$$

PROPOSITION 3.11.– *Under [3.97]:*

$$\bar{G}_n(\mathbf{x}|\mathbf{U}) = \frac{x_1 \ldots x_n}{n!} + \frac{1}{n!}\sum_{k=0}^{n-1} \frac{\gamma_n}{\beta_k - \beta_n}\Big(\prod_{i=k+1}^{n-1} \frac{\gamma_i}{\beta_i - \beta_n}\Big) x_1 \ldots x_k, \qquad [3.98]$$

Proof. Let $L(\mathbf{x})$ denote the r.h.s. of [3.98]. By the consequence (i) given in section 2.5, it suffices to check that:

$$\Delta^m L(\mathbf{u}_m) = \delta_{m,n}, \quad 0 \leq m \leq n. \qquad [3.99]$$

Applying Δ^m to [3.98], we directly have:

$$\begin{aligned}\Delta^m L(\mathbf{u}_m) &= \frac{u_{m,1} \ldots u_{m,n-m}}{(n-m)!} \\ &+ \frac{1}{n!}\sum_{k=m}^{n-1} \frac{\gamma_n}{\beta_k - \beta_n}\Big(\prod_{i=k+1}^{n-1} \frac{\gamma_i}{\beta_i - \beta_n}\Big) k! \frac{u_{m,1} \ldots u_{m,k-m}}{(k-m)!}.\end{aligned}$$

Let us now examine the two terms of the above sum when $u_{i,j}$ are given by [3.97]. After multiplication by $n!$, the first part becomes:

$$\frac{n!}{(n-m)!}\prod_{j=1}^{n-m} \frac{j}{m+j}\frac{\gamma_{m+j}}{\beta_{m+j} - \beta_m} = m! \prod_{i=m+1}^{n} \frac{\gamma_i}{\beta_i - \beta_m}.$$

By comparison with lemma 3.1, it corresponds to the first term of [3.94] up to the multiplicative factor $m!$ (for $r = m$). The second part multiplied by $n!$ is given by:

$$\sum_{k=m}^{n-1} \frac{\gamma_n}{\beta_k - \beta_n}\frac{k!}{(k-m)!}\Big(\prod_{i=k+1}^{n-1} \frac{\gamma_i}{\beta_i - \beta_n}\prod_{j=1}^{k-m} \frac{j}{m+j}\frac{\gamma_{m+j}}{\beta_{m+j} - \beta_m}\Big),$$

where, of course:

$$\frac{k!}{(k-m)!}\prod_{j=1}^{k-m} \frac{j}{m+j} = \frac{k!}{(k-m)!}\frac{(k-m)!m!}{k!} = m!.$$

Thus, it corresponds to the second term in [3.94] again up to the factor $m!$ (for $r = m$). Thanks to lemma 3.1, we deduce that:

$$n!\Delta^m L(\mathbf{u}_m) = m!\delta_{m,n},$$

which is identical to [3.99] because $m!\delta_{m,n} = n!\delta_{m,n}$. □

Note that when $\gamma_i \neq 0$, a more compact expression is:

$$\bar{G}_n(\mathbf{x}|\mathbf{U}) = \frac{1}{n!}\sum_{k=0}^{n} \frac{\gamma_n}{\gamma_k}\left(\prod_{i=k}^{n-1} \frac{\gamma_i}{\beta_i - \beta_n}\right) x_1 \dots x_k. \qquad [3.100]$$

3.6. When $\mathcal{D}$ is the set of integers

As an illustration of $\mathcal{D} = \mathbb{Z}$, we will consider the special case where $\mathcal{E}$ is the family:

$$\mathcal{E} = \{e_n(x) \equiv \binom{x+n-1}{n}, \; n \in \mathbb{N}_0\}.$$

Clearly, $\mathcal{E}$ is standard at 0, its generating function is $(1-s)^{-x}$ and the S.I. property is satisfied.

We first derive the associated operator Δ and simple recurrences for $e_n(x)$.

PROPOSITION 3.12.–

$$e_n(x+1) = \sum_{i=0}^{n} e_i(x), \qquad [3.101]$$

$$\Delta = I - S^{-1}, \qquad [3.102]$$

$$e_{n+1}(x) = \begin{cases} \sum_{j=1}^{x} e_n(j) \;\; when \;\; x \geq 0, \\ -\sum_{j=1+x}^{0} e_n(j) \;\; when \;\; x \leq 0. \end{cases} \qquad [3.103]$$

Proof. Formula [3.101] follows from the binomial identity [2.78] and the fact that $e_n(1) = 1$. For [3.102], we have from [3.101] that $e_n(x+1) = e_n(x) + e_{n-1}(x+1)$, and changing n to $n+1$ and x to $x-1$:

$$e_{n+1}(x) = e_{n+1}(x-1) + e_n(x),$$

so that $\Delta e_{n+1}(x) = e_n(x)$ yields:

$$\Delta e_{n+1}(x) = e_{n+1}(x) - e_{n+1}(x-1) = (1 - S^{-1})e_{n+1}(x),$$

as desired. For [3.103], the first formula expressed in terms of generating function becomes:

$$(1-s)^{-x} - 1 = \sum_{j=1}^{x} s(1-s)^{-j},$$

which is an identity since the r.h.s. is:

$$\frac{s}{1-s}\left[1 + \frac{1}{1-s} + \ldots + \left(\frac{1}{1-s}\right)^{x-1}\right] = \frac{s}{1-s}\,\frac{1-\left(1/(1-s)\right)^{x}}{1-\left(1/(1-s)\right)}$$
$$= -\left[1-(1-s)^{-x}\right].$$

Similarly, the second formula is equivalent to:

$$(1-s)^{-x} - 1 = -s\sum_{j=1+x}^{0} (1-s)^{-j},$$

and since the r.h.s. is:

$$-s\left[1 + (1-s)^{1} + \ldots + (1-s)^{-1+|x|}\right]$$
$$= \frac{-s}{1-(1-s)}\left[1-(1-s)^{|x|}\right] = (1-s)^{|x|} - 1,$$

this is again an identity. □

We now work with the integration operator I_u, which provides the A–G pseudopolynomials.

PROPOSITION 3.13.–

$$I_u e_n(x) = \begin{cases} \sum_{j=1+u}^{x} e_n(j) \text{ when } x \geq u, \\ -\sum_{j=1+x}^{u} e_n(j) \text{ when } x \leq u, \end{cases} \quad [3.104]$$

$$I_u \bar{G}_n(x|U) = \begin{cases} \sum_{j=1+u}^{x} \bar{G}_n(j|U) \text{ when } x \geq u, \\ -\sum_{j=1+x}^{u} \bar{G}_n(j|U) \text{ when } x \leq u, \end{cases} \quad [3.105]$$

$$\bar{G}_n(x|U) = \sum_{j_0=1+u_0}^{x} \sum_{j_1=1+u_1}^{j_0} + \ldots + \sum_{j_{n-1}=1+u_{n-1}}^{j_{n-2}} e_0 \text{ when } x \geq u_0$$

$$\text{and } 1+u_0 \geq u_1, 1+u_1 \geq u_2 \ldots, 1+u_{n-1} \geq u_n. \quad [3.106]$$

Proof. By [2.14], $I_u e_n(x) = e_{n+1}(x) - e_{n+1}(u)$. It thus suffices to insert [3.103] by considering the respective positions of x and u. When $x \geq u$, we obtain that if $u \geq 0$:

$$I_u e_n(x) = \sum_{j=1}^{x} e_n(j) - \sum_{j=1}^{u} e_n(j) = \sum_{j=1+u}^{x} e_n(j),$$

that is, the first formula of [3.104]. The same result is valid if $u \leq 0$ and $x \geq 0$ since:

$$I_u e_n(x) = \sum_{j=1}^{x} e_n(j) + \sum_{j=1+u}^{0} e_n(j),$$

and if $x \leq 0$ since:

$$I_u e_n(x) = -\sum_{j=1+x}^{0} e_n(j) + \sum_{j=1+u}^{0} e_n(j).$$

Moreover, when $x \leq u$, a similar checking leads to the second formula of [3.104].

For [3.105], we write $\bar{G}_n(x|U) = \sum_{i=0}^{n} c_i\, e_i(x)$ (as in [2.26]), so that:

$$I_u \bar{G}_n(x|U) = \sum_{i=0}^{n} c_i\, I_u e_i(x).$$

From [3.104], we then get when $x \geq u$:

$$I_u \bar{G}_n(x|U) = \sum_{i=0}^{n} c_i \sum_{j=1+u}^{x} e_i(j) = \sum_{j=1+u}^{x} \bar{G}_n(j|U),$$

with an analogous result when $x \leq u$.

For [3.106], recall that $\bar{G}_n(x|U) = I_{u_0} \bar{G}_{n-1}(x|EU)$ (see [2.21]). Using [3.105], we obtain for $x \geq u_0$ that if $1 + u_0 \geq u_1$:

$$\bar{G}_n(x|U) = \sum_{j_0=1+u_0}^{x} \bar{G}_{n-1}(j_0|EU),$$

by recursion if $1 + u_1 \geq u_2$:

$$\bar{G}_n(x|U) = \sum_{j_0=1+u_0}^{x} \sum_{j_1=1+u_1}^{j_0} \bar{G}_{n-2}(j_1|E^2U),$$

hence the result after $n - 2$ other iterations. □

4

Further Results and Properties

4.1. A related basic family $\mathcal{E}(b)$

Suppose the S.I. property is satisfied, and consider the special affine case where:

$$u_i = -bi, \quad i \in \mathbb{N}_0.$$

DEFINITION 4.1.– *Let b be any element of $\mathcal{D}$. A new basic family $\mathcal{E}(b)$ of functions ${}_be_n : \mathfrak{D} \to \mathbb{R}$ (or $\mathbb{C}$) is defined by:*

$$ {}_be_n(x) = \bar{G}_n(x|\{-bi,\, i \in \mathbb{N}_0\}), \quad n \geq 0. \qquad [4.1]$$

Note that the ${}_be_n$ are linearly independent and standard at 0. Using [3.81] for $a = 0$ and b with modified sign, we also see that [4.1] is equivalent to:

$$ {}_be_n(x) = \bar{G}_n(x|\tilde{\mathcal{E}}, \{0, 0, \ldots\}), \qquad [4.2]$$

where $\tilde{\mathcal{E}}$ is the family of functions $\tilde{e}_n(x)$ with:

$$\tilde{e}_n(x) \equiv e_n(x + bn), \quad n \geq 0. \qquad [4.3]$$

Moreover, if $\mathcal{E}$ is of generating function $\exp(xg(s))$, [3.83] provides a third possible definition:

$$ {}_be_n(x) = \frac{x}{x + bn}\, e_n(x + bn), \quad n \geq 0. \qquad [4.4]$$

Denote $\tilde{\Delta}$ the operator acting on $\tilde{\mathcal{E}}$ as in [2.1], that is:

$$\tilde{\Delta}\, \tilde{e}_{n+1} = \tilde{e}_n, \quad n \geq 0. \tag{4.5}$$

PROPOSITION 4.1.– *The operator $\tilde{\Delta}$ is given by:*

$$\tilde{\Delta} = S^{-b}\Delta = \Delta S^{-b}, \tag{4.6}$$

and it acts on $\mathcal{E}(b)$ also as in [2.1], that is:

$$\tilde{\Delta}\; {}_be_{n+1} = {}_be_n, \quad n \geq 0. \tag{4.7}$$

Proof. By virtue of [4.5]:

$$\begin{aligned} \tilde{\Delta}\, \tilde{e}_{n+1}(x) &= \tilde{\Delta}\, e_{n+1}(x + b(n+1)) \\ &= \tilde{e}_n(x) = e_n(x + bn), \end{aligned}$$

that is, $\tilde{\Delta}\, S^{b(n+1)}\, e_{n+1}(x) = S^{bn}\, e_n(x)$. This implies $\tilde{\Delta}\, S^b\, e_{n+1}(x) = e_n(x)$, hence $\tilde{\Delta}\, S^b = \Delta$ as stated in [4.6].

In addition, from [4.2] and [4.5], the family $\mathcal{E}(b)$ has an Appell structure with respect to $\tilde{\Delta}$ (it is a copy of $\tilde{\mathcal{E}}$). So, $\tilde{\Delta}$ acts on $\mathcal{E}(b)$ exactly as on $\tilde{\mathcal{E}}$, hence the result [4.7]. □

Let us derive some useful expansions related to $\mathcal{E}(b)$.

PROPOSITION 4.2.– *Let $a \in \mathcal{D}$ and f_n be a pseudopolynomial of degree n. Then:*

$${}_be_n(x + a) = \sum_{i=0}^{n} {}_be_{n-i}(a)\, {}_be_i(x), \tag{4.8}$$

$$f_n(x) = \sum_{i=0}^{n} \tilde{\Delta}^i f_n(0)\, {}_be_i(x), \tag{4.9}$$

$$e_n(x) = \sum_{i=0}^{n} e_{n-i}(-bi)\, {}_be_i(x), \tag{4.10}$$

and when $\mathcal{E}$ is standard at 0*:*

$${}_be_n(x) = \sum_{i=0}^{n} {}_be_{n-i}(bi)\, e_i(x). \tag{4.11}$$

Proof. The expansion [4.8] is a binomial identity like [2.78]. Formula [4.9] comes from an A–G expansion of f_n like [2.29], but with respect to $\tilde{\Delta}$ and $\mathcal{E}(b)$ (using [4.2]). In the special case $f_n = e_n$, it gives [4.10] since by [4.6]:

$$\tilde{\Delta}^i e_n(0) = S^{-bi} \Delta^i \, e_n(0) = e_{n-i}(-bi).$$

Finally, consider the Taylor-type expansion [2.79] of ${}_be_n(x)$ with respect to Δ around $a = 0$, when $\mathcal{E}$ is standard at 0. This yields:

$${}_be_n(x) = \sum_{i=0}^{n} (\Delta^i \, {}_be_n)(0) \, e_i(x),$$

and because of [4.7]:

$$(\Delta^i \, {}_be_n)(0) = S^{bi} (\tilde{\Delta}^i \, {}_be_n)(0) = S^{bi} \, {}_be_{n-i}(0) = {}_be_{n-i}(bi),$$

which leads to the desired formula [4.11]. □

Suppose again that $\mathcal{E}$ is obtained from the formal generating function [3.21]. In this case, a similar result also applies for the new family $\mathcal{E}(b)$.

PROPOSITION 4.3.– *If $\mathcal{E}$ is of generating function $\theta(s)\exp(xg(s))$ with $g(0) = 0$, $g'(0) \neq 0$, then $\mathcal{E}(b)$ admits the formal generating function $\exp(xh(s))$ with $h(0) = 0$, $h'(0) \neq 0$, and the functions g and h are connected by:*

$$g(s) = h\big(se^{-bg(s)}\big). \qquad [4.12]$$

In addition, for $\mathcal{E}$ standard at 0 (i.e. when $\theta(s) = 1$):

$$h(s) = g\big(se^{bh(s)}\big). \qquad [4.13]$$

Proof. Using [4.10], we can write the generating function of $\mathcal{E}$ as:

$$\begin{aligned}
\theta(s)e^{xg(s)} &= \sum_{n=0}^{\infty} e_n(x) \, s^n = \sum_{n=0}^{\infty} s^n \Big(\sum_{i=0}^{n} e_{n-i}(-bi) \, {}_be_i(x) \Big) \\
&= \sum_{i=0}^{\infty} s^i \, {}_be_i(x) \Big(\sum_{n-i=0}^{\infty} e_{n-i}(-bi) \, s^{n-i} \Big) \\
&= \sum_{i=0}^{\infty} s^i \, {}_be_i(x) \, \theta(s) e^{-big(s)},
\end{aligned}$$

which leads to:

$$e^{xg(s)} = \sum_{i=0}^{\infty} {}_be_i(x)\left(se^{-bg(s)}\right)^i \equiv \sum_{i=0}^{\infty} {}_be_i(x)\,\sigma^i, \qquad [4.14]$$

after defining:

$$\sigma = se^{-bg(s)} \equiv \rho(s).$$

Obviously, $\rho(0) = 0$, $\rho'(0) \neq 0$ and when g is analytic for $|s|$ small enough, one can find in the s-complex plane (respectively, σ-complex plane) an open disk $D(0,\epsilon)$ (an open set $\tilde{D}$ with $0 \in \tilde{D}$) such that the restriction of ρ on D is a conformal transformation $\rho_D : D(0,\epsilon) \to \tilde{D}$. By virtue of the local inversion theorem, $\rho_D^{-1} : \tilde{D} \to D(0,\epsilon)$ also exists as an analytic function and:

$$\rho_D^{-1}(\sigma) = \sum_{k=1}^{\infty} c_k\,\sigma^k, \qquad [4.15]$$

for appropriate real numbers c_k that are obtained explicitly from the α_i defining g. Furthermore, getting rid of the analyticity assumption does not affect the evaluation of c_k, and [4.15] still remains valid as a formal series. Consequently, $\mathcal{E}(b)$ admits the formal generating function:

$$\sum_{n=0}^{\infty} {}_be_n(x)\,\sigma^n = e^{xg(\rho_D^{-1}(\sigma))} \equiv e^{xh(\sigma)}, \qquad [4.16]$$

where $h(\sigma) = g(\rho_D^{-1}(\sigma))$ implies $h(\rho_D(s)) = g(s)$, which provides the relation [4.12]. Moreover, we see that as announced:

$$h(0) = g\left(\rho_D^{-1}(0)\right) = g(0) = 0,$$
$$\text{and } h'(0) = g'\left(\rho_D^{-1}(0)\right)/\rho_D'(0) = g'(0) \neq 0.$$

In addition, if $\mathcal{E}$ is standard at 0, we may apply a similar proof based this time on formula [4.11]. Indeed, we express the generating function of $\mathcal{E}(b)$ as:

$$e^{xh(s)} = \sum_{n=0}^{\infty} {}_b e_n(x)\, s^n = \sum_{n=0}^{\infty} s^n \Big(\sum_{i=0}^{n} {}_b e_{n-i}(bi)\, e_i(x) \Big)$$

$$= \sum_{i=0}^{\infty} s^i\, e_i(x) \Big(\sum_{n-i=0}^{\infty} {}_b e_{n-i}(bi)\, s^{n-i} \Big)$$

$$= \sum_{i=0}^{\infty} s^i\, e_i(x)\, e^{bih(s)} = \sum_{i=0}^{\infty} e_i(x) \left(s e^{bh(s)}\right)^i,$$

then using the generating function of $\mathcal{E}$ where $\theta(s) = 1$ by hypothesis:

$$\exp\left(xh(s)\right) = \exp\left\{xg(se^{bh(s)})\right\}. \qquad [4.17]$$

Therefore, [4.17] implies the identity [4.13]. □

In simple terms, the generating functions of $\mathcal{E}$ and $\mathcal{E}(b)$ are connected as follows:

PROPOSITION 4.4.–

$$\sum_{n=0}^{\infty} {}_b e_n(x)\, s^n = \frac{\sum_{n=0}^{\infty} e_n(x+bn)\, s^n}{\sum_{n=0}^{\infty} e_n(bn)\, s^n}, \qquad [4.18]$$

and if $\mathcal{E}$ is standard at 0*:*

$$\sum_{n=0}^{\infty} e_n(x)\, s^n = \frac{\sum_{n=0}^{\infty} {}_b e_n(x-bn)\, s^n}{\sum_{n=0}^{\infty} {}_b e_n(-bn)\, s^n}. \qquad [4.19]$$

Proof. From [4.9] for $f_n(x) = e_n(x+bn)$, we have:

$$e_n(x+bn) = \sum_{i=0}^{n} e_{n-i}(-bi+bn)\, {}_b e_i(x),$$

so that for their generating function:

$$\sum_{n=0}^{\infty} e_n(x+bn)\, s^n = \sum_{i=0}^{\infty} {}_b e_i(x)\, s^i \left(\sum_{n=i}^{\infty} e_{n-i}\big(b(n-i)\big)\, s^{n-i} \right)$$

$$= \sum_{i=0}^{\infty} {}_b e_i(x)\, s^i \left(\sum_{n=0}^{\infty} e_n(bn)\, s^n \right),$$

hence [4.18]. Note that the result could also be derived using [3.68].

When $\mathcal{E}$ is standard at 0, we first apply Taylor's formula [2.79] to ${}_be_n(x - bn)$, and then use [4.6] and [4.7] to obtain:

$$\begin{aligned} {}_be_n(x - bn) &= \sum_{i=0}^{n} \left[\Delta^i\, {}_be_n(0 - bn)\right] e_i(x) \\ &= \sum_{i=0}^{n} \left[S^{bi}\, {}_be_{n-i}(-bn)\right] e_i(x) \\ &= \sum_{i=0}^{n} {}_be_{n-i}(-b(n-i))\, e_i(x). \end{aligned}$$

Therefore:

$$\sum_{n=0}^{\infty} {}_be_n(x - bn)\, s^n = \sum_{i=0}^{\infty} e_i(x)\, s^i \left(\sum_{n=i}^{\infty} {}_be_{n-i}(-b(n-i))\, s^{n-i}\right),$$

which leads to [4.19]. □

4.2. Upper and lower bounds for $\bar{G}_n(\ |U)$

Bounding the A–G pseudopolynomials requires to make specific assumptions. Below we consider one or more of the following hypotheses:

– H_1: there exists some constant $c > 0$ such that $|e_n(x)| \leq c^n$, $n \geq 1$, where x is either fixed or an element of a fixed subset of $\mathcal{D}$.

– H_2: $e_n(x) \geq 0$, $n \geq 1$, when $\mathcal{D} = \mathbb{R}$ and $x \geq 0$ (for $\mathcal{D} = \mathbb{R}^k$, the latter condition is $\mathbf{x} \geq \mathbf{0}$, i.e. $x_j \geq 0$, $1 \leq j \leq k$).

– H_3: $\mathcal{E}$ is built from a formal generating function:

$$\psi(x, s) \;=\; \theta(s)\, e^{xg(s)}, \quad \text{where } \; g(s) = \sum_{i=1}^{\infty} \alpha_i\, s^i$$

with all $\alpha_i \geq 0$ and $\alpha_1 \neq 0$.

– H_4: $x \geq u_0 \geq u_1 \geq \ldots \geq u_n \geq \ldots$ with $\mathcal{D} = \mathbb{R}$ (or $\mathbb{R}^k$).

Note that H_1, H_2, H_3 concern $\mathcal{E}$, while H_4 concerns U.

4.2.1. *Upper bound for* $|\bar{G}_n|$

If H_1 is satisfied, we can easily find an upper bound for the absolute value of the A–G pseudopolynomials.

PROPOSITION 4.5.– *Under* H_1, *then for any* x:

$$|\bar{G}_n(x|U)| \leq (2c)^n, \quad n \geq 1. \qquad [4.20]$$

Proof. As shown in section 3.3 (see [3.57] and [3.58]), $\bar{G}_n$ is the sum of 2^n terms of the form:

$$(-1)^s \, e_{r_1} \, e_{r_2-r_1}(u_{r_1}) \ldots e_{n-r_s}(u_{r_s}).$$

where $r_1 \leq \ldots \leq r_s \leq n-1$. When $|e_i(x)| \leq c^i$ for all i, this implies that:

$$|\bar{G}_n(x|U)| \leq 2^n \left(c^{r_1} \, c^{r_2-r_1} \ldots c^{n-r_s}\right) = (2c)^n,$$

as announced in [4.20]. □

The hypothesis H_1 is quite often verified in practice. This is illustrated below.

i) Consider $\mathcal{D} = \mathbb{R}$ and $\mathcal{E}$ generated by the function $\exp(xg(s))$ where $g(s)$ is the sum of the series defined before but which is assumed here to be convergent for $|s| \leq \zeta$.

Using [3.12], we get:

$$|e_n(x)| = \left| \sum_{k=0}^{n} \frac{x^k}{k!} \, \alpha_n^{*k} \right| \leq \sum_{k=0}^{n} \frac{|x|^k}{k!} \, \frac{1}{\zeta^n} |\alpha_n^{*k} \, \zeta^n|$$

$$\leq \frac{1}{\zeta^n} \sum_{k=0}^{n} \frac{|x|^k}{k!} \left(\sum_{i=1}^{\infty} |\alpha_i| \, \zeta^i \right)^k \equiv \frac{1}{\zeta^n} \sum_{k=0}^{n} \frac{(|x|m)^k}{k!},$$

where m denotes $\sum_{i=1}^{\infty} |\alpha_i| \, \zeta^i$. Therefore:

$$|e_n(x)| < \frac{1}{\zeta^n} \, e^{|x|m} \leq c^n,$$

by choosing $c \geq (1/\zeta) \exp(|x|m/n)$.

ii) As in section 3.5, let $\mathcal{D}=\mathbb{R}^{\mathbb{N}}$ and define $e_n(\mathbf{x})=(1/n!)x_1\ldots x_n$, $n\geq 1$. In addition, we now suppose that $|x_j|\leq cj$, $j\geq 1$, and $|u_{i,j}|\leq cj$, $j\geq 1$. We then have the desired upper bound since obviously:

$$|e_n(\mathbf{x})|\leq c^n,\ \textit{and}\ |e_n(\mathbf{u}_i)|\leq c^n.$$

iii) Take $\mathcal{D}=\mathbb{R}^{\mathbb{N}}$, and suppose $|x_j|\leq m$, $j\geq 1$. We saw in [3.42] of section 3.2 that:

$$e_n(\mathbf{x})=\sum_{\substack{i_1,i_2\ldots,i_n\geq 0:\\ i_1+2i_2+\ldots+ni_n=n}}\frac{x_1^{i_1}}{i_1!}\frac{x_2^{i_2}}{i_2!}\cdots\frac{x_n^{i_n}}{i_n!}.$$

So, if $|x_j|\leq m$ for all j, we get:

$$|e_n(\mathbf{x})|\leq\sum_{\substack{i_1,i_2\ldots,i_n\geq 0:\\ i_1+2i_2+\ldots+ni_n=n}}\frac{m^{i_1}}{i_1!}\frac{m^{i_2}}{i_2!}\cdots\frac{m^{i_n}}{i_n!}$$

$$\leq\sum_{i_1,i_2\ldots,i_n\geq 0}\frac{m^{i_1}}{i_1!}\frac{m^{i_2}}{i_2!}\cdots\frac{m^{i_n}}{i_n!}=(e^m)^n,$$

that is, again the bound c^n where $c=\exp(m)$.

4.2.2. Non-negativity of $\bar{G}_n$

Let $\mathcal{D}=\mathbb{R}$ and assume the e_n are differentiable. We will show that if H_3 and H_4 are satisfied, the A–G pseudoplynomials are non-negative and their partial derivatives with respect to each parameter are non-positive.

PROPOSITION 4.6.– *Under H_3+H_4, then for $x\geq u_0$:*

$$\bar{G}_n(x|U)\geq 0,\quad n\geq 1,\ \textit{and}\qquad [4.21]$$

$$\partial_{u_i}\left[\bar{G}_n(x|U)\right]\leq 0,\quad i=0,1,\ldots,n-1.\qquad [4.22]$$

Proof. By the hypothesis H_3, we know from [3.22] that D, the derivative operator in x, can be expressed as:

$$D = \sum_{i=1}^{\infty} \alpha_i \, \Delta^i.$$

Applied to $\bar{G}_n(x|U)$, we thus get:

$$D\bar{G}_n(x|U) = \sum_{i=1}^{n} \alpha_i \, \bar{G}_{n-i}(x|E^i U). \qquad [4.23]$$

Now, observe that if in the r.h.s. of [4.23], all $\bar{G}_j(x|E^i U)$, $0 \leq j \leq n-1$, are non-negative, then $D\bar{G}_n(x|U)$ is also, that is, $\bar{G}_n(x|U)$ is non-decreasing. In other words, for $x \geq u_0$:

$$\bar{G}_n(x|U) \geq \bar{G}_n(u_0|U) = \delta_{n,0} \geq 0.$$

Since the indices j in $\bar{G}_j$ vary from 0 until $n-1$, it then suffices to establish [4.21] to reason by recursion from $\bar{G}_0 = 1$ and using H_4.

For [4.22], we will essentially rely on the previous result [4.21] and on the formula [3.61] in the remark after proposition 3.8. For that, we first introduce a bivariate function $F_n(\xi, \eta)$, $n \geq 1$, as:

$$F_n(\xi, \eta) = \bar{G}_n(\xi|V_\eta) \quad \text{where} \quad V_\eta = \{\eta, u_1, u_2, \ldots\},$$

and we denote $\partial_1 = \partial/\partial\xi$, $\partial_2 = \partial/\partial\eta$. Since $F_n(\gamma, \gamma) \equiv 0$ for any γ, then:

$$(d/d\gamma)F_n(\gamma, \gamma) = (\partial_1 + \partial_2)F_n(\gamma, \gamma) \equiv 0. \qquad [4.24]$$

Using [3.22] and noting that $E^i V_\eta = E^i U$ when $i \geq 1$, we get:

$$\begin{aligned} \partial_1 \left[F_n(\gamma, \gamma)\right] &= \partial_1 \left[\bar{G}_n(\gamma|V_\gamma)\right] = \sum_{i=1}^{\infty} \alpha_i \, \Delta^i \left[\bar{G}_n(\gamma|V_\gamma)\right] \\ &= \sum_{i=1}^{\infty} \alpha_i \, \bar{G}_{n-i}(\gamma|E^i U)], \end{aligned}$$

which is non-negative for $\gamma \geq u_1$ by virtue of [4.21]. Thus, from [4.24]:

$$\partial_2\big[F_n(\gamma,\gamma)\big] = -\partial_1\big[F_n(\gamma,\gamma)\big] \leq 0 \;\; \text{for } \gamma \geq u_1. \tag{4.25}$$

Consider the partial derivative $\partial_{u_0}[\bar{G}_n(x|U)]$, that is $\partial_2[\bar{G}_n(x|U)]$ in the present notation. As pointed out in the remark mentioned above, this partial derivative is independent of x. Taking $\gamma = u_0$ in [4.25] (where $u_0 \geq u_1$ by H_4), we obtain for any x:

$$\partial_{u_0}\big[\bar{G}_n(x|U)\big] = \partial_2\big[\bar{G}_n(u_0|U)\big] = \partial_2\big[F_n(u_0,u_0)\big] \leq 0, \tag{4.26}$$

hence the result [4.22] when $i = 0$. Now, for the other values of i, we refer to [3.61] that specifies that for all x:

$$\partial_{u_i}\big[\bar{G}_n(x|U)\big] = \bar{G}_i(x|U)\;\partial_{u_i}\big[\bar{G}_{n-i}(a|E^iU)\big],$$

a being chosen arbitrarily. Due to [4.21] and [4.26], we then deduce that [4.22] is also valid when $i \geq 1$. □

COROLLARY 4.1.– *If a family $U^{(s)}$ is smaller than U (i.e. $u_i^{(s)} \leq u_i$ for all i), and U satisfies $H_3 + H_4$, then for $x \geq u_0$:*

$$\bar{G}_n(x|U) \leq \bar{G}_n(x|U^{(s)}), \quad n \geq 1. \tag{4.27}$$

Analogously, if a family $U^{(g)}$ is greater than U (i.e. $u_i^{(g)} \geq u_i$ for all i), and $U^{(g)}$ satisfies $H_3 + H_4$, then for $x \geq u_0^{(g)}$:

$$\bar{G}_n(x|U^{(g)}) \leq \bar{G}_n(x|U), \quad n \geq 1. \tag{4.28}$$

In particular, if U satisfies H_3 and $\mathcal{E}$ is standard at 0:

$$e_n(x-u_0) \leq \bar{G}_n(x|U) \leq e_n(x-u_{n-1}), \quad n \geq 1. \tag{4.29}$$

Proof. By [4.22], we know that for fixed $x \geq u_0$, $\bar{G}_n(x|U)$ is a non-increasing function of each u_i. Thus, the larger U, the smaller $\bar{G}_n(x|U)$, which gives the inequality [4.27]. A similar reasoning leads to [4.28]. For the inequality [4.29], we first get from [4.27] and [4.28]:

$$\bar{G}_n(x|\{u_0,u_0,\ldots\}) \leq \bar{G}_n(x|U) \leq \bar{G}_n(x|\{u_{n-1},u_{n-1},\ldots\}).$$

It then suffices to use the S.I. property implied by H_3 with $\mathcal{E}$ standard at 0 to be able to apply [2.83]. □

EXAMPLE 4.1. (back to u_i geometric).– We consider the model examined in section 3.4, part 2, but without the factor a, $U = \{u_i = q^i, i \in \mathbb{N}_0\}$ with $0 < q < 1$ to guarantee H_4. We want to derive simple bounds for $\bar{G}_n(x|U)$.

For an upper bound, we take any $i_0 \in \mathbb{N}_0$ and define:

$$u_i^{(s)} = q^{i_0}[1 - (1-q)(i - i_0)] \equiv a + bi, \quad i \in \mathbb{N}_0,$$

after putting $a = q^{i_0}[1 + (1-q)i_0]$ and $b = -q^{i_0}(1-q)$. Note that the family $U^{(s)}$ is linear, non-increasing, in i. Moreover:

$$u_{i_0}^{(s)} = q^{i_0} = u_{i_0}, \text{ and } u_{i_0+1}^{(s)} = q^{i_0+1} = u_{i_0+1},$$

thus by the convexity of the function q^i, we have $u_i^{(s)} < u_i$ for $i < i_0$ or $i > i_0 + 1$. From [4.27], we then deduce that:

$$\bar{G}_n(x|U) \leq \bar{G}_n(x|\{a + bi, i \in \mathbb{N}_0\}), \quad \mathrm{n} \geq 1.$$

For a lower bound, we proceed in a similar way and define:

$$u_i^{(g)} = 1 - (1 - q^{i_0})i/i_0 \equiv c + di, \quad 0 \leq i \leq i_0,$$

after putting $c = 1$ and $d = (1 - q^{i_0})/i_0$. The family $U^{(g)}$ is linear, non-increasing, for $i \leq i_0$. Moreover:

$$u_0^{(g)} = 1 = u_0, \text{ and } u_{i_0}^{(g)} = q^{i_0} = u_{i_0},$$

therefore $u_i^{(g)} > u_i$ for $0 < i < i_0$ again by convexity. We thus obtain from [4.28]:

$$\bar{G}_n(x|U) \geq \bar{G}_n(x|\{c + di, i \in \mathbb{N}_0\}), \quad 1 \leq \mathrm{n} \leq \mathrm{i}_0.$$

4.2.3. *Lower bound for* $\bar{G}_n$

Our goal is to obtain a lower bound for the A–G pseudopolynomials if H_3 and H_4 are satisfied and $\mathcal{E}$ is standard at 0. To begin, let us focus on bounding the functions $e_n(x)$.

PROPOSITION 4.7.– *Under $H_3 + \mathcal{E}$ standard at* 0*, then for* $x \geq 0$*:*

$$n_{[k]}\, e_n(x) \geq k!\, e_k(x)\, e_{n-k}(x), \quad 1 \leq k \leq n. \qquad [4.30]$$

Proof. We fix $x \geq 0$ (viewed as a parameter) and denote $f(s) = \exp(xg(s))$ (function of s only). Then, let us introduce a related function $l_k(s)$, for $k \geq 1$, given by:

$$l_k(s) = f^{(k)}(s) - f^{(k)}(0) f(s), \qquad [4.31]$$

where $f^{(k)}(s)$ is the kth derivative of $f(s)$. By definition of $f(s)$, we have $f^{(k)}(s)$ of the form $f(s)h_k(s)$, the function $h_k(s)$ being a power series that has non-negative coefficients (like f and g). From [4.31], it is clear that $l_k(0) = 0$ and:

$$l_k(s) = f(s)h_k(s) - f(0)h_k(0)f(s) = f(s)[h_k(s) - h_k(0)],$$

so $l_k(s)$ is also a power series with non-negative coefficients. It then remains to note that $l_k(s)$ is explicitly given by:

$$l_k(s) = \sum_{n=1}^{\infty} n_{[k]}\, e_n(x)\, s^{n-k} - \sum_{n=1}^{\infty} n_{[k]}\, e_n(x)\, 0^{n-k} \sum_{i=0}^{\infty} e_i(x)\, s^i,$$

which yields:

$$n_{[k]}\, e_n(x) - k_{[k]}\, e_k(x)\, e_{n-k}(x) \geq 0, \quad 1 \leq k \leq n,$$

hence the inequality [4.30]. □

EXAMPLE 4.2.– Taking $k = 1$ in [4.30], we have:

$$n\, e_n(x) \geq e_1(x)\, e_{n-1}(x).$$

Now, suppose in addition that $e_1(x) \geq 0$. Then, we get by recursion:

$$e_n(x) \geq (1/n!) \left[e_1(x)\right]^n.$$

For $k = 2$, [4.30] becomes:

$$n(n-1)\, e_n(x) \geq 2\, e_2(x)\, e_{n-2}(x),$$

so that if $e_1(x)$, $e_2(x) \geq 0$:

$$\begin{cases} e_{2n}(x) \geq (2^n/(2n)!) \, [e_2(x)]^n, \\ e_{2n+1}(x) \geq (2^n/(2n+1)!) \, [e_2(x)]^n \, e_1(x). \end{cases}$$

REMARK.– Other bounding relations can be obtained by a similar argument. Instead of [4.31], let us define, for instance, another function $b(s)$ as:

$$b(s) = f^{(2)}(s) - f^{(2)}(0)f(s) - \left[f^{(1)}(s) - f^{(1)}(0)f(s)\right].$$

It can be shown that $b(s)$ is a power series with non-negative coefficients when we choose $x \geq 1/2\alpha_1$. Consequently, we obtain for such values of x:

$$\begin{aligned} &n(n-1)\, e_n(x) - 2\, e_2(x)\, e_{n-2}(x) \\ &\geq (n-1)\, e_{n-1}(x) - e_1(x)\, e_{n-2}(x), \quad n \geq 2. \end{aligned}$$

Using [4.30], we are able to lower bound $\bar{G}_n(x|U)$ under appropriate assumptions.

PROPOSITION 4.8.– *Under $H_3 + H_4 + \mathcal{E}$ standard at 0, then for $x \geq u_0$:*

$$n_{[k]}\, \bar{G}_n(x|U) \geq k!\, e_k(x-u_0)\, \bar{G}_{n-k}(x|E^k U), \quad 1 \leq k \leq n. \qquad [4.32]$$

Proof. Since the S.I. is implied by H_3 and $\mathcal{E}$ is standard at 0, any Λ G pseudopolynomial admits a Taylor-type expansion given by [2.79]. So, we can write that:

$$\bar{G}_n(x|U) = \sum_{i=0}^{n} \bar{G}_{n-i}(u_0|E^i U)\, e_i(x-u_0), \qquad [4.33]$$

$$\begin{aligned} \bar{G}_{n-k}(x|E^k U) &= \sum_{i=0}^{n-k} \bar{G}_{n-k-i}(u_0|E^{k+i} U)\, e_i(x-u_0) \\ &= \sum_{i=k}^{n} \bar{G}_{n-i}(u_0|E^i U)\, e_{i-k}(x-u_0). \end{aligned} \qquad [4.34]$$

Let us multiply both sides of [4.33] by $n_{[k]}$ and [4.34] by $k!\, e_k(x-u_0)$, then consider their sum. After arrangement of the two terms, we obtain:

$$n_{[k]}\,\bar{G}_n(x|U) - k!\,e_k(x-u_0)\,\bar{G}_{n-k}(x|E^kU)$$
$$= \sum_{i=0}^{k-1} \bar{G}_{n-i}(u_0|E^iU)\,n_{[k]}\,e_i(x-u_0)$$
$$+ \sum_{i=k}^{n} \bar{G}_{n-i}(u_0|E^iU)\,\big[n_{[k]}\,e_i(x-u_0)$$
$$- k!\,e_k(x-u_0)\,e_{i-k}(x-u_0)\big]. \qquad [4.35]$$

From [4.30] and $n_{[k]} \geq i_{[k]}$, the factor $[\ldots]$ in [4.35] is non-negative (note that by H_4, $x - u_0 \geq 0$ as required for [4.30]). Moreover, the $\bar{G}_{n-i}$ are ≥ 0 by virtue of [4.21]. Finally, $e_1(x-u_0) = \alpha_1(x-u_0) \geq 0$, so that by recursion of [4.30], we have $e_n(x-u_0) \geq 0$ for any n. Therefore, the r.h.s. of [4.35] is non-negative, which gives the bound [4.32]. □

COROLLARY 4.2.– *Under $H_3 + H_4 + \mathcal{E}$ standard at 0, then for $x \geq u_0$:*

$$\bar{G}_n(x|U) \geq \frac{1}{n!}\prod_{i=0}^{n-1} e_1(x-u_i), \qquad [4.36]$$

$$\bar{G}_{2n}(x|U) \geq \frac{2^n}{(2n)!}\prod_{i=0}^{n-1} e_2(x-u_{2i}), \qquad [4.37]$$

$$\bar{G}_{2n+1}(x|U) \geq \frac{2^n}{(2n+1)!}\Big(\prod_{i=0}^{n-1} e_2(x-u_{2i})\Big)\,e_1(x-u_{2n}). \qquad [4.38]$$

Proof. Taking $k = 1$ in [4.32], we have:

$$n\,\bar{G}_n(x|U) \geq e_1(x-u_0)\,\bar{G}_{n-1}(x|EU),$$

which yields [4.36] by recursion (and using H_4). For $k = 2$, [4.32] gives:

$$n(n-1)\,\bar{G}_n(x|U) \geq 2\,e_2(x-u_0)\,\bar{G}_{n-2}(x|E^2U),$$

hence directly [4.37], while for [4.38]:

$$\bar{G}_1(x|E^{2n}U) = \bar{G}_1(x|\{u_{2n}\}) = \bar{G}_1(x-u_{2n}|\{0\}) = e_1(x-u_{2n}).$$

Note that for $k = n$, [4.32] becomes $\bar{G}_n(x|U) \geq e_n(x-u_0)$, in agreement with [4.29]. □

4.3. Short visit to the A–G type series

Given a family $\mathcal{G}(U)$ of A–G pseudopolynomials, we introduce a set of A–G type series, denoted $\mathfrak{F}_{\mathcal{G}(U)}$, as follows.

DEFINITION 4.2.– *An A–G type series is any formal series $h(\ )$ of the form:*

$$h(\ ) = \sum_{n=0}^{\infty} a_n \, \bar{G}_n(\ |U). \qquad [4.39]$$

We note that by [2.24], $\Delta^n h(u_n) = a_n$, thus [4.39] is equivalent to:

$$h(\ ) = \sum_{n=0}^{\infty} \Delta^n h(u_n) \, \bar{G}_n(\ |U), \qquad [4.40]$$

which means that h is the formal sum of its series expansion of A–G type.

Recall that $\mathfrak{F}_{\mathcal{E}}$ denotes the vector space of formal series $\sum_{i=0}^{\infty} a_i e_i$ (section 2.1). So, for any $f \in \mathfrak{F}_{\mathcal{E}}$, the $\Delta^n f$ do exist as formal series, but the $\Delta^n f(u_n)$ are not necessarily finite. In that case:

$$\sum_{n=0}^{\infty} \Delta^n f(u_n) \, \bar{G}_n(\ |U), \qquad [4.41]$$

has no meaning, of course. When all $\Delta^n f(u_n)$ are finite, [4.41] may be denoted as $h \in \mathfrak{F}_{\mathcal{G}(U)}$. These objects f and h are clearly distinct ($\mathfrak{F}_{\mathcal{E}} \neq \mathfrak{F}_{\mathcal{G}(U)}$), but both values $f(x)$ and $h(x)$ are in $\bar{\mathbb{R}}$ and can be compared.

For simplicity, we will consider $\mathcal{D} = \mathbb{R}$ in the following (it could be $\mathbb{R}^k$).

DEFINITION 4.3.– *Let $f \in \mathfrak{F}_{\mathcal{E}}$, and suppose that $\Delta^r f(u_r)$ is finite for $0 \leq r \leq k-1$. The expansion:*

$$f(\ ) = \sum_{r=0}^{k-1} \Delta^r f(u_r) \, \bar{G}_r(\ |U) + R_k(\ ) \qquad [4.42]$$

is called the finite A–G expansion of f (of order k).

The remainder R_k in [4.42] has the following representation:

PROPOSITION 4.9.–

$$R_k(\ ) = I_{u_0} I_{u_1} \dots I_{u_{k-1}} \Delta^k f(\ ), \quad k \geq 1. \qquad [4.43]$$

Proof. Denote by r_k the r.h.s. of [4.43]. We will show that $R_k - r_k$ is identically null. For that, we examine its differences $\Delta^j(R_k - r_k)$ for all $j \geq 0$.

On the one hand, we have from the integral definition of r_k and using the identity [2.19]:

$$\Delta^j r_k(u_j) = 0, \quad j \leq k-1, \qquad [4.44]$$

$$\Delta^j r_k(\ ) = \Delta^j f(\ ), \quad j \geq k. \qquad [4.45]$$

On the other hand, the same properties [4.44] and [4.45] are also true when R_k is substituted for r_k. Indeed, from [4.42], we get for $j \leq k-1$:

$$\begin{aligned}\Delta^j f(u_j) &= \sum_{r=j}^{k-1} \Delta^r f(u_r)\, \bar{G}_{r-j}(u_j|E^j U) + \Delta^j R_k(u_j) \\ &= \Delta^j f(u_j) + \Delta^j R_k(u_j),\end{aligned}$$

hence $\Delta^j R_k(u_j) = 0$ as in [4.44]. Moreover, for $j \geq k$, $\Delta^j R_k(\ ) = \Delta^j f(\ )$ as in [4.45]. We can therefore say that:

$$\Delta^j (R_k - r_k)(u_j) = 0, \quad j \leq k-1, \qquad [4.46]$$

$$\Delta^j (R_k - r_k)(\ ) = 0, \quad j \geq k. \qquad [4.47]$$

Now, writing $R_k - r_k = \sum_{i=0}^{\infty} b_i\, e_i$, we see from [4.47] that $b_i = 0$ for $i \geq k$. So, $R_k - r_k$ is a pseudopolynomial of degree at most $k-1$. Thus, it can be expressed in the form $R_k - r_k = \sum_{i=0}^{k-1} c_i\, \bar{G}_i(\ |U)$. Using [4.46], we finally deduce that $R_k - r_k = 0$ as announced. □

The notion of an absolutely monotonic function is well known in mathematical analysis. It is adapted directly to the current context as follows.

DEFINITION 4.4.– *A function $f \in \mathfrak{F}_{\mathcal{E}}$ is absolutely monotone on $\mathbb{R}_+$ if $\Delta^n f$ is finite and non-negative for any $n \geq 0$ and all $x \geq 0$.*

By means of some of the hypotheses considered in section 4.2, we will be able to derive a Bernstein-type inequality for such a function.

PROPOSITION 4.10.– *Let $f \in \mathfrak{F}_{\mathcal{E}}$ be absolutely monotone on $\mathbb{R}_+$. Under $H_3 + H_4$ with all $u_i \geq 0 + \mathcal{E}$ standard at 0, then for $x \geq u_0$:*

$$f(x) \geq \sum_{n=0}^{\infty} \Delta^n f(u_n)\, \bar{G}_n(x|U). \tag{4.48}$$

Proof. We start by showing that the remainder R_k in [4.42] is non-negative for any k. Let us write $f = \sum_{n=0}^{\infty} a_n\, e_n$. From [4.43]:

$$R_k = I_{u_0} \dots I_{u_{k-1}}\, \Delta^k \Big(\sum_{n=0}^{\infty} a_n\, e_n \Big) = \sum_{n=k}^{\infty} a_n\, I_{u_0} \dots I_{u_{k-1}}\, e_{n-k}. \tag{4.49}$$

First, $\mathcal{E}$ being standard at 0, we see that $a_n = \Delta^n f(0)$, which is ≥ 0 since f is absolutely continuous on $\mathbb{R}_+$. Moreover, from [2.20]:

$$I_{u_0} \dots I_{u_{k-1}}\, e_j - \bar{G}_{j+k}(x|\{u_0, \dots u_k\ \ 1, 0, 0, \dots\}), \tag{4.50}$$

which is also ≥ 0 by virtue of [4.21] (valid under $H_3 + H_4$ and for $x \geq u_0$). From [4.49], we thus obtain $R_k \geq 0$, so that by [4.42]:

$$f(\ ) \geq \sum_{r=0}^{k-1} \Delta^r f(u_r)\, \bar{G}_r(\ |U), \quad k \geq 1. \tag{4.51}$$

f being absolutely monotone on $\mathbb{R}_+$, the terms in the r.h.s. of [4.51] are all non-negative, and we can make k tend to ∞, hence the inequality [4.48]. □

COROLLARY 4.3.– *Let U and $U^{(s)}$ be two families satisfying the assumptions of proposition 4.10 and such that $u_i \geq u_i^{(s)}$, $i \in \mathbb{N}_0$. Then, for $x \geq u_0$:*

$$f(x) \geq \sum_{n=0}^{\infty} \Delta^n f(u_n)\, \bar{G}_n(x|U) \geq \sum_{n=0}^{\infty} \Delta^n f(u_n^{(s)})\, \bar{G}_n(x|U^{(s)}). \tag{4.52}$$

Proof. By assumption, we have:

$$U_k^{(s)*} = \{u_0^{(s)}, \dots, u_{k-1}^{(s)}, 0, 0, \dots\} \leq U_k^* = \{u_0, \dots, u_{k-1}, 0, 0, \dots\}.$$

From [4.27], we then know that $\bar{G}_{j+k}(x|U_k^{(s)*}) \geq \bar{G}_{j+k}(x|U_k^*)$ for $x \geq u_0$. So, using [4.50], we get for all j:

$$I_{u_0} \dots I_{u_{k-1}} e_j \leq I_{u_0^{(s)}} \dots I_{u_{k-1}^{(s)}} e_j,$$

and following the previous argument, we deduce that $R_k(x|U) \leq R_k(x|U^{(s)})$. Therefore:

$$f(x) - R_k(x|U) \geq f(x) - R_k(x|U^{(s)}),$$

and it remains to take $k \to \infty$ to obtain [4.52]. □

In other words, the smaller $U^{(s)}$ is, the more difficult it is for Bernstein's inequality to be transformed into equality. Let us try to obtain such an equality.

PROPOSITION 4.11.– *In addition to the assumptions of proposition 4.10, suppose that $\Delta^j f$ for all j is the sum of its Taylor-type expansion around the point $a = \lim_i u_i \geq 0$ in an open interval J with $U \subset J$. Then, the formula [4.48] is true with the equality sign, that is, for $x \geq u_0$:*

$$f(x) = \sum_{n=0}^{\infty} \Delta^n f(u_n)\, \bar{G}_n(x|U). \qquad [4.53]$$

Proof. From [2.79], the Taylor-type expansion of f gives:

$$f(x) = \sum_{j=0}^{\infty} (\Delta^j f)(a)\, e_j(x-a) = \sum_{j=0}^{\infty} (\Delta^j f)(a) \Big(\sum_{i=0}^{j} e_{j-i}(u_i - a)\, \bar{G}_i(x|U) \Big).$$

The two sums of this formula can be permuted because by [4.21], $\bar{G}_i(x|U) \geq 0$ for $x \geq u_0$, and $e_{j-i}(u_i - a) = \bar{G}_{j-i}(u_i - a|\{0, 0, \dots\}) \geq 0$ for $u_i \geq a$, which is true by H_4. By doing the permutation, we obtain:

$$f(x) = \sum_{i=0}^{\infty} \bar{G}_i(x|U) \Big(\sum_{j=0}^{\infty} [\Delta^j(\Delta^i f)](a)\, e_j(u_i - a) \Big). \qquad [4.54]$$

The inner sum $(\dots)$ in [4.54] corresponds to the Taylor expansion [2.79] of $\Delta^i f(u_i)$ around a. Therefore, [4.54] becomes equality [4.53]. □

Note that the assumption made on the Taylor formula for all $\Delta^j f$ is satisfied in the historical case since f expandable around a implies the same for $\Delta^j f$, $j \geq 0$.

A–G type exponential in $\mathfrak{F}_{\mathcal{G}(U)}$

When $\mathcal{E}$ is standard at 0, the series defined by:

$$f(\ ) \equiv \sum_{n=0}^{\infty} \beta^n e_n(\ ), \qquad [4.55]$$

satisfies $\Delta f = \beta f$ with $f(0) = 1$, so that f may be considered as a formal A–G exponential.

Due to $\Delta^n f = \beta^n f$, [4.55] can be rewritten as:

$$f(\ ) = \sum_{n=0}^{\infty} \Delta^n f(0)\, \bar{G}_n(\ |\{0, 0, \ldots\}). \qquad [4.56]$$

When the binomial identity applies to $\mathcal{E}$, it directly extends to f, which implies that $f(x) = f(a) f(x - a)$ for all a. Thus:

$$\begin{aligned} f(x) &= \sum_{n=0}^{\infty} f(a)\, \beta^n e_n(x - a) \\ &= \sum_{n=0}^{\infty} (\Delta^n f)(a)\, \bar{G}_n(x|\{a, a, \ldots\}). \end{aligned} \qquad [4.57]$$

A natural question is what about a more general A–G expansion like:

$$\sum_{n=0}^{\infty} \Delta^n f(u_n)\, \bar{G}_n(x|U). \qquad [4.58]$$

As noticed before, when all $(\Delta^n f)(u_n)$ are finite, the series [4.58] has a formal sum h, and we are interested in having the equality $f(x) = h(x)$.

PROPOSITION 4.12.– *If for all $n \geq 1$, $\bar{G}_n(x|U) \geq 0$ and $e_n(u_i) \geq 0$, $i \in \mathbb{N}_0$, then an A–G exponential $f(x)$, when finite, is the sum of its A–G expansion [4.58].*

Proof. By [4.55] and using [2.32], we have:

$$f(x) = \sum_{n=0}^{\infty} \beta^n e_n(x) = \sum_{n=0}^{\infty} \beta^n \Big(\sum_{i=0}^{n} e_{n-i}(u_i)\, G_i(x|U) \Big).$$

According to the assumptions made, all terms are non-negative, so the two sums can be permuted. This gives:

$$f(x) = \sum_{i=0}^{\infty} \beta^i \bar{G}_i(x|U)\Big(\sum_{j=0}^{\infty} \beta^j e_j(u_i)\Big)$$

$$= \sum_{i=0}^{\infty} \beta^i \bar{G}_i(x|U) f(u_i) = \sum_{i=0}^{\infty} (\Delta^i f)(u_i) \bar{G}_i(x|U),$$

because $\Delta f = \beta f$, hence the equality with [4.58]. □

Note that the above non-negativity conditions are satisfied under $H_3 + H_4$ with all $u_i \geq 0 + \mathcal{E}$ standard at 0. Moreover, when [4.57] applies, we get after multiplication by β^j:

$$\Delta^j f = \sum_{n=0}^{\infty} \Delta^n (\Delta^j f)(a)\, e_n(x-a),$$

so that the condition on the Taylor-type formula put in proposition 4.11 is indeed verified.

4.4. Bilinear forms and biorthogonality

Below we introduce the notion of bilinear forms into our framework. This will also make the link with the well-known property of biorthogonality.

DEFINITION 4.5.– *Let U and V be any two given families in $\mathcal{D}$. The associated bilinear form $<\ ,\ >_{U,V}$ is the map defined on $\mathcal{E} \times \mathcal{E}$ by:*

$$< e_i(\ ), e_j(\ ) >_{U,V} = \sum_{k=0}^{i \wedge j} e_{i-k}(u_k)\, e_{j-k}(v_k), \quad i, j \in \mathbb{N}_0. \qquad [4.59]$$

PROPOSITION 4.13.– *The bilinear form $<\ ,\ >_{U,V}$ can be reexpressed equivalently on $\mathfrak{F} \times \mathfrak{F}$ by each of the following three conditions:*

$$< Q(\ ), R(\ ) >_{U,V} = \sum_{k=0}^{\infty} \Delta^k Q(u_k)\, \Delta^k R(v_k), \quad Q, R \in \mathfrak{F}, \; [4.60]$$

$$< Q(\ ), \bar{G}_n(\ |V) >_{U,V} = \Delta^n Q(u_n), \quad n \in \mathbb{N}_0, \qquad [4.61]$$

$$< \bar{G}_{n_1}(\ |U), \bar{G}_{n_2}(\ |V) >_{U,V} = \delta_{n_1,n_2}, \quad n_1, n_2 \in \mathbb{N}_0. \qquad [4.62]$$

Proof. We first show that [4.59] implies [4.60]. Denoting $Q = \sum_{i=0}^{\infty} \alpha_i e_i$ and $R = \sum_{i=0}^{\infty} \beta_i e_i$, we have:

$$< Q, R >_{U,V} = \sum_{i,j \geq 0} \alpha_i \beta_j < e_i, e_j >_{U,V},$$

and using [4.59], we get successively:

$$\begin{aligned} < Q, R >_{U,V} &= \sum_{i,j \geq 0} \alpha_i \beta_j \Big(\sum_{k,l \geq 0} \delta_{k,l}\, e_{i-k}(u_k)\, e_{j-l}(v_l) \Big) \\ &= \sum_{k,i \geq 0} \alpha_i e_{i-k}(u_k) \Big(\sum_{l,j \geq 0} \delta_{k,l}\, \beta_j e_{j-l}(v_l) \Big) \\ &= \sum_{k,i \geq 0} \alpha_i e_{i-k}(u_k) \Big(\sum_{l \geq 0} \delta_{k,l}\, \Delta^l R(v_l) \Big) \\ &= \sum_{k,i \geq 0} \alpha_i e_{i-k}(u_k)\, \Delta^k R(v_k) \; = \sum_{k \geq 0} \Delta^k Q(u_k)\, \Delta^k R(v_k), \end{aligned}$$

which is [4.60]. Then, we obtain from [2.24] that [4.60] implies [4.61], hence [4.62]. So, it remains to show that [4.62] implies [4.59]. This is true since from [2.32]:

$$\begin{aligned} < e_i, e_j >_{U,V} &= < \sum_{k \geq 0} e_{i-k}(u_k) \bar{G}_k(\ |U), \sum_{l \geq 0} e_{j-l}(v_l) \bar{G}_l(\ |V) >_{U,V} \\ &= \sum_{k,l \geq 0} e_{i-k}(u_k)\, e_{j-l}(v_l) < \bar{G}_k(\ |U), \bar{G}_l(\ |V) >_{U,V} \\ &= \sum_{k,i \geq 0} e_{i-k}(u_k)\, e_{j-l}(v_l)\, \delta_{k,l}, \end{aligned}$$

by using [4.62], hence the desired formula [4.59]. □

Note that by [4.60], $< Q, R >_{U,V} = < R, Q >_{V,U}$, but $< Q, R >_{U,V} \neq < R, Q >_{U,V}$. Moreover, the forms $< \, , \, >_{U,V}$ and $< \, , \, >_{V,U}$ are not identical: for example, by [4.62]:

$$< \bar{G}_{n_1}(\ |U), \bar{G}_{n_2}(\ |V) >_{U,V} = \delta_{n_1,n_2},$$

while from [4.60]:

$$< \bar{G}_{n_1}(\ |U), \bar{G}_{n_2}(\ |V) >_{V,U} = \sum_{k\geq 0} \bar{G}_{n_1-k}(v_k|E^kU)\, \bar{G}_{n_2-k}(u_k|E^kV).$$

Recall that a bilinear form is degenerate if it has a non-trivial kernel (left or right). This is not the case here.

PROPOSITION 4.14.– *The bilinear form* $< , >_{U,V}$ *is not degenerate on the space* $\mathfrak{F} \times \mathfrak{F}$.

Proof. To build the left kernel, we consider the condition:

$$< Q, R >_{U,V} = 0 \quad \text{for any } R.$$

In particular, we must have:

$$0 =< Q, \bar{G}_i(\ |V) >_{U,V} = \Delta^i Q(u_i), \quad i \geq 0,$$

the last equality coming from [4.61]. Using [2.29], we then get:

$$Q(\ ) = \sum_{i\geq 0} \Delta^i Q(u_i)\, \bar{G}_i(\ |U) = 0.$$

The same conclusion holds for the right kernel. Since both kernels reduce to 0, the bilinear form is not degenerate. □

Two cases of special interest are when V is identical to U or is the family $V_0 \equiv \{0, 0, \ldots\}$.

COROLLARY 4.4.– *i) When* $V = U$*, the form* $< , >_{U,U}$ *reduces to a simple scalar product,* $< , >_U$ *say, which makes* $\mathcal{G}(U)$ *orthogonal to itself.*

ii) When $V = V_0$ *and* $\mathcal{E}$ *is standard at* 0*, then:*

$$< e_i, e_j >_{U,V_0} = e_{i-j}(u_j),$$

$$< Q, e_j >_{U,V_0} = \Delta^j Q(u_j),$$

$$< \bar{G}_n(\ |U), e_j(\ ) >_{U,V_0} = \delta_{n,j},$$

$$< e_j(\ ), \bar{G}_n(\ |U) >_{U,V_0} = \sum_{k\geq 0} e_{j-k}(u_k)\, \bar{G}_{n-k}(0|E^kU).$$

Proof. For i), it suffices to observe that the bilinear form then becomes symmetric. For ii), this is immediate from [4.59] to [4.62]. □

4.5. An alternative generalization

The method that we have proposed for generalizing the A–G polynomials is certainly not the only possible one. Below we describe another related approach but which will not be explored here in detail. This section therefore constitutes a temporary excursion before returning to the central theme of the monograph.

Let us take up the expression of the historical polynomials $G_n(x|U)$, which is given in the form [1.8] of a multiple integral. An alternative feasible extension consists of adding functions inside the successive nested integrals. More precisely, let $U = \{u_i,\ i \in \mathbb{N}_0\}$ a family of real numbers u_i, and introduce now a set $H = \{h_i,\ i \in \mathbb{N}_0\}$ of real functions h_i, continuous for simplicity. We then build an associated family $\mathcal{G}(U, H)$ of *generalized A–G pseudopolynomials* $\hat{G}_n(x|U, H)$, $n \in \mathbb{N}_0$, as follows.

DEFINITION 4.6. *Starting with* $\hat{G}_0(x|U, H) = 1$, *then:*

$$\hat{G}_n(x|U, H) = \int_{u_0}^{x} h_0(x_1)\, dx_1 \int_{u_1}^{x_1} h_1(x_2)\, dx_2 \ldots \int_{u_{n-1}}^{x_{n-1}} h_{n-1}(x_n)\, dx_n, \quad n \geq 1. \qquad [4.63]$$

Equivalently, a recursive construction of [4.63] is therefore:

$$\hat{G}_n(x|U, H) = \int_{u_0}^{x} h_0(x_1)\, \hat{G}_{n-1}(x_1|EU, EH)\, dx_1, \quad n \geq 1, \qquad [4.64]$$

where $EU = \{u_{1+i},\ i \in \mathbb{N}_0\}$ as before, and similarly EH is the shifted set $\{h_{1+i},\ i \in \mathbb{N}_0\}$. By introducing an integration operator $I_u^{(h)}$ defined by:

$$(I_u^{(h)} f)(x) = \int_u^x h(\xi)\, f(\xi)\, d\xi,$$

for any continuous real function f, we can rewrite [4.63] and [4.64] as:

$$\hat{G}_n(\ |U,H) = I_{u_0}^{(h_0)} I_{u_1}^{(h_1)} \dots I_{u_{n-1}}^{(h_{n-1})} \hat{G}_0, \qquad [4.65]$$

$$= I_{u_0}^{(h_0)} \hat{G}_{n-1}(\ |EU,EH), \qquad [4.66]$$

which generalize the formulas [2.20] and [2.21] defining the A–G pseudopolynomials. Observe that as in [2.24] for $m = 0$:

$$\hat{G}_n(u_0|U,H) = \delta_{0,n}, \quad n \in \mathbb{N}_0. \qquad [4.67]$$

We will see in the following that the main results obtained on $\bar{G}_n$ easily extend to $\hat{G}_n$. All proofs will be based on the integral representations [4.63] and [4.64].

PROPOSITION 4.15.– *Given the sets U,H and $r,s \in \mathbb{N}_0$, the family $\{\hat{G}_i(\ |E^sU,E^rH),\ 0 \le i \le n\}$ is linearly independent whatever $n \in \mathbb{N}_0$.*

Proof. Since this is true for $n = 0$, we proceed by induction and assume that the property is also true for $n = k-1 \ge 0$. Consider the situation where there exist real coefficients a_i not all zero such that:

$$\sum_{i=0}^{k} a_i \hat{G}_i(x|E^sU,E^rH) = 0, \quad \text{for all } x, \qquad [4.68]$$

and let us show that this is not feasible. Indeed, taking $x = u_s$, we have from [4.67] that $\hat{G}_i(u_s|E^sU,E^rH) = 0$ for $1 \le i \le k$, so [4.68] implies $a_0 = 0$. Furthermore, by differentiating [4.68] with respect to x and using [4.64], we obtain:

$$h_0(x) \sum_{i=1}^{k} a_i \hat{G}_{i-1}(x|E^{1+s}U,E^{1+r}H) = 0, \quad \text{for all } x. \qquad [4.69]$$

By the linear independence assumed for $n = k-1$, we deduce from [4.69] that $a_1 = \dots = a_k = 0$, hence the desired property. □

Let us now introduce some concepts that will be used later. First, a family of basic functions is defined as follows.

DEFINITION 4.7.– *For any* $r \in \mathbb{N}_0$*:*

$$\hat{e}_n(\ |E^r H) = \hat{G}_n(\ |U_0, E^r H), \quad \textit{where } U_0 \equiv \{0, 0, \ldots\}, \qquad [4.70]$$

and let $E^r\hat{\mathcal{E}} = \{\hat{e}_n(\ |E^r H),\ n \in \mathbb{N}_0\}$.

The real vectorial space generated by $E^r\hat{\mathcal{E}}$ is denoted $E^r\hat{\mathfrak{F}}$. This space is written $E^r\hat{\mathfrak{F}}_n$ if it is generated only by $\{\hat{e}_i(\ |E^r H),\ i = 0, \ldots, n\}$. So, an element f_n of $E^r\hat{\mathfrak{F}}_n$ is a finite sum of the form $\sum_{i=0}^{n} a_i \hat{e}_i(\ |E^r H)$. When $a_n \neq 0$, f_n is called a *generalized pseudopolynomial* of degree n.

Then, to move to "differentiation" operators analogous to Δ (defined in [2.1]), some precautions must be taken. While Δ is an operator from $\mathcal{E}$ to $\mathcal{E}$, the corresponding operators will be here of the form $E^r\mathcal{E}$ to $E^{1+r}\mathcal{E}$, for different values of r. This requires us to explicitly add a reference to r, hence the notation $\hat{\Delta}^{(r)}$.

DEFINITION 4.8.– *For any* $r \in \mathbb{N}_0$, *the operator* $\hat{\Delta}^{(r)}$*:* $E^r\mathcal{E} \to E^{1+r}\mathcal{E}$ *is defined as:*

$$\hat{\Delta}^{(r)} = \frac{1}{h_r} D, \qquad [4.71]$$

where D *is the standard differentiation operator. So, from [4.70],* $\hat{\Delta}^{(r)}\hat{e}_0 = 0$ *and:*

$$\hat{\Delta}^{(r)}\hat{e}_{n+1}(\ |E^r H) = \frac{1}{h_r} D\hat{e}_{n+1}(\ |E^r H) = \hat{e}_n(\ |E^{r+1} H), \quad n \in \mathbb{N}_0. \quad [4.72]$$

Note that [4.71] also yields:

$$\hat{\Delta}^{(r)}\hat{G}_{n+1}(\ |E^s U, E^r H) = \hat{G}_n(\ |E^{1+s} U, E^{1+r} H), \quad n \in \mathbb{N}_0. \quad [4.73]$$

Now, iteration of operators like $\hat{\Delta}^{(r)}$ will always be of the form:

$$\hat{\Delta}^{(r+i-1)}\hat{\Delta}^{(r+i-2)} \ldots \hat{\Delta}^{(r)} \equiv \hat{\Delta}^{(r),i} \text{ say}, \quad i \geq 1. \qquad [4.74]$$

Using [4.74], the formula [4.73] directly generalizes as follows, similarly to [2.23].

PROPOSITION 4.16.–

$$\hat{\Delta}^{(r),i}\hat{G}_n(\ |E^s U, E^r H) = \hat{G}_{n-i}(\ |E^{i+s}U, E^{i+r}H), \quad n \geq i. \qquad [4.75]$$

We can then adapt the formula [2.29] of the A–G expansion, hence the Maclaurin formula type [2.35] with $\omega = 0$.

PROPOSITION 4.17.– *For* $f_n \equiv f_n(\ |E^r H) \in E^r\hat{\mathfrak{F}}_n$*:*

$$f_n = \sum_{i=0}^{n} \hat{\Delta}^{(r),i} f_n(u_i)\, \hat{G}_i(\ |U, E^r H), \qquad [4.76]$$

and in particular, if $U = U_0$*:*

$$f_n = \sum_{i=0}^{n} \hat{\Delta}^{(r),i} f_n(0)\, \hat{e}_i(\ |E^r H). \qquad [4.77]$$

Proof. By proposition 4.15, f_n can be expressed as:

$$f_n(x) = \sum_{k=0}^{n} b_k\, \hat{G}_k(x|U, E^r H). \qquad [4.78]$$

Applying $\hat{\Delta}^{(r),i}$ (defined in [4.74]), then taking $x = u_i$, we get from [4.75] and [4.67]:

$$\hat{\Delta}^{(r),i} f_n(u_i) = \sum_{k=0}^{n} b_k\, \hat{G}_{k-i}(u_i|E^i U, E^{i+r}H) = \sum_{k=0}^{n} b_k\, \delta_{i,k} = b_i,$$

so that [4.78] becomes [4.76]. The formula [4.77] follows directly using [4.70]. □

Let us introduce on H a shift operator S^a, with a real, by:

$$S^a H = \{S^a h_n,\ n \in \mathbb{N}_0\}, \quad \text{where } S^a h_n(x) = h_n(x+a),$$

as well as a multiplication operator M^ρ, with ρ real, by:

$$M^\rho H = \{M^\rho h_n,\ n \in \mathbb{N}_0\}, \quad \text{where } M^\rho h_n(x) = h_n(\rho x).$$

PROPOSITION 4.18.–

$$\hat{G}_n(x+a|U+a,H) = \hat{G}_n(x|U,S^a H), \quad n \in \mathbb{N}_0, \tag{4.79}$$

$$\hat{G}_n(\rho x|\rho U,H) = \rho^n\, \hat{G}_n(x|U,M^\rho H), \quad n \in \mathbb{N}_0. \tag{4.80}$$

Proof. We will reason by recurrence on n for the two formulas. First, [4.64] gives:

$$\begin{aligned}\hat{G}_n(x+a|U+a,H) &= \int_{u_0+a}^{x+a} h_0(x_1)\, \hat{G}_{n-1}(x_1|EU+a,EH)\, dx_1 \\ &= \int_{u_0}^{x} h_0(\xi+a)\, \hat{G}_{n-1}(\xi+a|EU+a,EH)\, d\xi.\end{aligned}$$

Using the recurrence hypothesis for $\hat{G}_{n-1}(\xi+a|EU+a,EH)$, we then obtain:

$$\hat{G}_n(x+a|U+a,H) = I_{u_0}^{(S^a h_0)}\, \hat{G}_{n-1}(x|EU,S^a EH),$$

which is $\hat{G}_n(x|U,S^a H)$ from [4.66], so [4.79] is true.

Second, we have from [4.64]:

$$\begin{aligned}\hat{G}_n(\rho x|\rho U,H) &= \int_{\rho u_0}^{\rho x} h_0(x_1)\, \hat{G}_{n-1}(x_1|\rho EU,EH)\, dx_1 \\ &= \rho \int_{u_0}^{x} h_0(\rho\xi)\, \hat{G}_{n-1}(\rho\xi|\rho EU,EH)\, d\xi,\end{aligned}$$

and by the recurrence hypothesis for $\hat{G}_{n-1}(\rho\xi|\rho EU,EH)$:

$$\hat{G}_n(\rho x|\rho U,H) = \rho \int_{u_0}^{x} M^\rho h_0(\xi)\, \rho^{n-1}\, \hat{G}_{n-1}(\xi|EU,M^\rho EH)\, d\xi,$$

which is indeed $\rho^n \hat{G}_n(x|U,M^\rho H)$, hence [4.80]. □

We are able to derive a new binomial identity which extends the formula [2.78].

PROPOSITION 4.19.–

$$\hat{e}_n(x+a|H) = \sum_{i=0}^{n} \hat{e}_{n-i}(a|E^i H)\, \hat{e}_i(x|S^a H), \quad n \in \mathbb{N}_0. \qquad [4.81]$$

Proof. Applying the A–G type expansion [4.76] with $r = 0$ to $f_n(x+a) = \hat{e}_n(x+a|H)$, we have:

$$\hat{e}_n(x+a|H) = \sum_{i=0}^{n} \hat{e}_{n-i}(u_i|E^i H)\, \hat{G}_i(x+a|U, H).$$

Now, take all the u_i equal to a, so $U = U_0 + a$. From [4.79], we then obtain:

$$\hat{G}_i(x+a|U, H) = \hat{G}_i(x+a|U_0 + a, H) = \hat{G}_i(x|U_0, S^a H),$$

which is $\hat{e}_i(x|S^a H)$ by [4.70]. The identity [4.81] is thus proven. □

We can also establish a new Taylor-type formula which extends [2.79].

PROPOSITION 4.20.– *For all* $f_n \equiv f_n(\ |H) \in \hat{\mathfrak{F}}_n$*:*

$$f_n(x+a) = \sum_{i=0}^{n} \Delta^{(0),i} f_n(a)\, \hat{e}_i(x|S^a H), \quad n \in \mathbb{N}_0. \qquad [4.82]$$

Proof. The argument is similar. Expand $f_n(x+a)$ using [4.76] with $r = 0$ and choose $u_i = a$ for all i. We then get:

$$f_n(x+a) = \sum_{i=0}^{n} \Delta^{(0),i} f_n(a)\, \hat{G}_i(x+a|U_0 + a, H),$$

where $\Delta^{(0),i} = \hat{\Delta}^{(i-1)} \ldots \hat{\Delta}^{(0)}$ by [4.74]. This leads to [4.82] as desired. □

For illustration, we close the section by examining the particular case where H is a family of exponential functions.

When H is the set $\{h_i(x) \equiv \exp[(\lambda_i - \lambda_{i+1})x],\ i \in \mathbb{N}_0\}$

The particular exponential form chosen for these h_i comes from that encountered in the context of a nonlinear birth process (see section 7.7). Our goal is to obtain a simplified recursion that explicitly provides the generalized A–G pseudopolynomials.

For the sake of brevity, we will assume that for all $n \geq 1$, the exponent λ_n is distinct from the previous ones $\lambda_0, \ldots, \lambda_{n-1}$. This assumption can however be relaxed with some adaptations in the final results.

LEMMA 4.1.– *Under the above condition on the λ_n, each function* $\exp[(\lambda_0 - \lambda_n)x]$ *is an element of the space* $\hat{\mathfrak{F}}_n$.

Proof. First, we will show by recurrence that $\hat{e}_n(x|H)$ can be expressed as:

$$\hat{e}_n(x|H) = \sum_{i=0}^{n} a_{n,i}(H)\, e^{(\lambda_0 - \lambda_i)x}, \quad \text{with } a_{n,n}(H) \neq 0, \tag{4.83}$$

for well-chosen real coefficients $a_{n,i}(H)$. Since this is true for $n = 0$, let us assume that it is also the case for $n = k - 1$. By the definitions [4.70] with $r = 0$ and [4.64], we have when $n = k$:

$$\hat{e}_k(x|H) = \int_{\xi=0}^{x} e^{(\lambda_0 - \lambda_1)\xi}\, \hat{e}_{k-1}(\xi|EH)\, d\xi,$$

and using [4.83] for $n = k - 1$:

$$\begin{aligned}
\hat{e}_k(x|H) &= \int_{\xi=0}^{x} e^{(\lambda_0 - \lambda_1)\xi} \left(\sum_{i=0}^{k-1} a_{k-1,i}(EH)\, e^{(\lambda_1 - \lambda_{i+1})\xi}\, d\xi \right) \\
&= \sum_{i=0}^{k-1} a_{k-1,i}(EH) \int_{\xi=0}^{x} e^{(\lambda_0 - \lambda_{i+1})\xi}\, d\xi \\
&= \sum_{i=0}^{k-1} a_{k-1,i}(EH)\, \frac{e^{(\lambda_0 - \lambda_{i+1})x} - 1}{\lambda_0 - \lambda_{i+1}}.
\end{aligned} \tag{4.84}$$

From [4.84], we therefore see that $\hat{e}_k(x|H)$ is indeed of the form [4.83] when $n = k$, with:

$$a_{k,k}(H) = a_{k-1,k-1}(EH)/(\lambda_0 - \lambda_k) \neq 0.$$

Now, we just need to write from [4.83] that:

$$e^{(\lambda_0-\lambda_n)x} = \frac{1}{a_{n,n}(H)}\left[\hat{e}_n(x|H) - \sum_{i=0}^{n-1} a_{n,i}(H)\, e^{(\lambda_0-\lambda_i)x}\right].$$

By recursion, we then deduce that the function $\exp[(\lambda_0-\lambda_n)x]$ is in $\hat{\mathfrak{F}}_n$ as announced. □

Thanks to this preliminary result, we are able to derive a prettier recursion for $\hat{G}_n(\ |U,H)$. For this, we introduce a polynomial $p_k(x)$ of degree k defined by $p_0(x)=1$ and:

$$p_k(x) = (\lambda_0-x)(\lambda_1-x)\dots(\lambda_{k-1}-x), \quad k\geq 1. \qquad [4.85]$$

PROPOSITION 4.21.–

$$p_n(\lambda_n)\,\hat{G}_n(x|U,H) = e^{(\lambda_0-\lambda_n)x} - \sum_{k=0}^{n-1} p_k(\lambda_n)\, e^{(\lambda_k-\lambda_n)u_k}\,\hat{G}_k(x|U,H). \qquad [4.86]$$

Proof. We know from lemma 4.1 that:

$$e^{(\lambda_i-\lambda_n)x} = E^i e^{(\lambda_0-\lambda_{n-i})x} \in E^i\hat{\mathfrak{F}}_{n-i}.$$

By definition 4.8, we can thus apply the operator $\hat{\Delta}^{(i)}$ to this function, from which [4.71] gives:

$$\begin{aligned}\hat{\Delta}^{(i)} e^{(\lambda_i-\lambda_n)x} &= e^{-(\lambda_i-\lambda_{i+1})x}\, D\, e^{(\lambda_i-\lambda_n)x}\\ &= (\lambda_i-\lambda_n)\, e^{(\lambda_{i+1}-\lambda_n)x}. \qquad [4.87]\end{aligned}$$

Now, let us take $i=0$ above, so $\hat{\Delta}^{(0)}=\hat{\Delta}$, and consider its kth iterate which was defined in [4.74]. From [4.87], we obtain by recursion:

$$\begin{aligned}\hat{\Delta}^k e^{(\lambda_0-\lambda_n)x} &= \hat{\Delta}^{(k-1)}\dots\hat{\Delta}^{(1)}\hat{\Delta}^{(0)} e^{(\lambda_0-\lambda_n)x}\\ &= (\lambda_0-\lambda_n)(\lambda_1-\lambda_n)\dots(\lambda_{k-1}-\lambda_n)\, e^{(\lambda_k-\lambda_n)x}\\ &= p_k(\lambda_n)\, e^{(\lambda_k-\lambda_n)x}. \qquad [4.88]\end{aligned}$$

Using [4.88], the A–G type expansion [4.78] with $r = 0$ then leads to:

$$e^{(\lambda_0-\lambda_n)x} = \sum_{k=0}^{n} p_k(\lambda_n)\, e^{(\lambda_k-\lambda_n)u_k}\, \hat{G}_k(x|U,H) \qquad [4.89]$$

which directly provides the formula [4.86]. □

COROLLARY 4.5.–

$$\hat{G}_n(x|U,H) = \sum_{i=0}^{n} c_{n,i}\, e^{(\lambda_0-\lambda_i)x}, \qquad [4.90]$$

where the constants $c_{n,i}$ are obtained recursively from the relations:

$$\sum_{k=i}^{n} c_{k,i}\, p_k(\lambda_n)\, e^{(\lambda_k-\lambda_n)u_k} = \delta_{i,n}. \qquad [4.91]$$

Proof. By inserting [4.90] for $n = k$ into [4.89], we get:

$$e^{(\lambda_0-\lambda_n)x} = \sum_{k=0}^{n} p_k(\lambda_n)\, e^{(\lambda_k-\lambda_n)u_k} \left(\sum_{i=0}^{k} c_{k,i}\, e^{(\lambda_0-\lambda_i)x}\right)$$

$$= \sum_{i=0}^{n} e^{(\lambda_0-\lambda_i)x} \left(\sum_{k=i}^{n} c_{k,i}\, p_k(\lambda_n)\, e^{(\lambda_k-\lambda_n)u_k}\right).$$

It remains to observe that this identity is indeed verified under the condition [4.91]. □

5

Multi-index A–G Pseudopolynomials

5.1. Key definitions and expansions

For better clarity, we will only examine the case of two indices, but the methods used have completely general validity. As before, let $\mathcal{D}$ be a non-empty set. We define on $\mathcal{D}$ a family of real functions with two indices, $\mathcal{E} = \{e_{n_1,n_2}, (n_1, n_2) \in \mathbb{N}_0^2\}$ where $e_{0,0} = 1$, which are linearly independent. The vector space generated by $\mathcal{E}$ is always denoted $\mathfrak{F}$. An element of $\mathfrak{F}$ is a finite sum of the form:

$$f_{n_1,n_2}(x) = \sum_{i_1=0}^{n_1} \sum_{i_2=0}^{n_2} a_{i_1,i_2}\, e_{i_1,i_2}(x), \quad x \in \mathcal{D}.$$

Such f_{n_1,n_2} is called an A–G pseudopolynomial with two indices of degree (n_1, n_2) if $a_{n_1,n_2} \neq 0$.

Let us introduce two marginal "differentiation" operators, Δ_j for $j = 1, 2$, as follows.

DEFINITION 5.1.– *On $\mathcal{E}$, then for $(n_1, n_2) \in \mathbb{N}_0^2$:*

$$\begin{cases} \Delta_1 e_{n_1,n_2} = e_{n_1-1,n_2}, & \textit{with } e_{-1,n_2} = 0, \\ \Delta_2 e_{n_1,n_2} = e_{n_1,n_2-1}, & \textit{with } e_{n_1,-1} = 0. \end{cases} \qquad [5.1]$$

These operators extend linearly over $\mathfrak{F}$ and their iterates $\Delta_j^{k_j}$, $k_j \geq 0$, are defined in the traditional way. Now, to construct A–G pseudopolynomials with two indices, we need to introduce a family U of elements of $\mathcal{D}$, also with two indices, namely:

$$U = \{u_{i_1,i_2},\ (i_1,i_2) \in \mathbb{N}_0^2\}.$$

We denote by $E^{k_1,k_2}U$, for $(k_1,k_2) \in \mathbb{N}_0^2$, the shifted family $\{u_{k_1+i_1,k_2+i_2},\ (i_1,i_2) \in \mathbb{N}_0^2\}$.

DEFINITION 5.2.– *The associated family* $\mathcal{G}(U) = \{\bar{G}_{n_1,n_2}(\ |U),$ $(n_1,n_2) \in \mathbb{N}_0^2\}$ *of pseudopolynomials of degree* (n_1,n_2) *is of A–G type if for each* (n_1,n_2)*:*

$$\Delta_1^{k_1}\Delta_2^{k_2}\bar{G}_{n_1,n_2}(u_{k_1,k_2}|U) = \delta_{n_1,k_1}\delta_{n_2,k_2},\ 0 \le k_1 \le n_1, 0 \le k_2 \le n_2. \quad [5.2]$$

Note that $\mathcal{G}(U)$ is a basic family for $\mathfrak{F}$ as before (see proposition 2.9). The condition [5.2] recalls the characterization [2.31] derived in the case of a single index. This is confirmed in the property below where we also obtain a recursive method, extending [2.33], to determine the multi-index A–G polynomials.

PROPOSITION 5.1.– *The A–G pseudopolynomials are univocally defined by [5.2]. In addition, they satisfy the recursive relations:*

$$\bar{G}_{n_1,n_2}(\ |U) = e_{n_1,n_2}(\ ) - \sum_{\substack{(k_1,k_2)=(0,0):\\ k_1+k_2<n_1+n_2}}^{(n_1,n_2)} e_{n_1-k_1,n_2-k_2}(u_{k_1,k_2})\,\bar{G}_{k_1,k_2}(\ |U), \quad (n_1,n_2) \in \mathbb{N}_0^2. \quad [5.3]$$

Proof. For the uniqueness part, let us show that [5.2] implies [5.3]. By definition, $\bar{G}_{0,0}(x|U) = c_{0,0}$ say, and by [5.2], $\bar{G}_{0,0}(u_{0,0}|U) = 1$, thus we have:

$$\bar{G}_{0,0}(x|U) = 1 = e_{0,0}(x),$$

as desired. Next, using the family $\mathcal{G}(U)$ as a basis in $\mathfrak{F}$, we can write:

$$e_{n_1,n_2}(x) = \sum_{(k_1,k_2)=(0,0)}^{(n_1,n_2)} a_{k_1,k_2}(n_1,n_2)\,\bar{G}_{k_1,k_2}(x|U), \quad n_1+n_2 \ge 1. \quad [5.4]$$

Apply $\Delta_1^{r_1}\Delta_2^{r_2}$ to both sides of [5.4] and choose $x = u_{r_1,r_2}$. From [5.1] and [5.2], we obtain:

$$e_{n_1-r_1,n_2-r_2}(u_{r_1,r_2}) = \sum_{(k_1,k_2)=(0,0)}^{(n_1,n_2)} a_{k_1,k_2}(n_1,n_2)\,\delta_{r_1,k_1}\delta_{r_2,k_2} = a_{r_1,r_2}(n_1,n_2),$$

where $0 \leq r_1 \leq n_1, 0 \leq r_2 \leq n_2$. After substitution in [5.4], we then deduce [5.3].

For the existence part, let us show that [5.3] implies [5.2]. This is obvious for $(k_1,k_2) = (n_1,n_2)$ since by [5.3], $\Delta_1^{n_1}\Delta_2^{n_2}\bar{G}_{n_1,n_2}(x|U) = \Delta_1^{n_1}\Delta_2^{n_2}e_{n_1,n_2}(x) = e_{0,0}(x) = 1$. For $0 \leq k_1 \leq n_1, 0 \leq k_2 \leq n_2$, $k_1 + k_2 < n_1 + n_2$, we get from [5.1] and [5.3]:

$$\Delta_1^{k_1}\Delta_2^{k_2}\bar{G}_{n_1,n_2}(u_{k_1,k_2}|U) = e_{n_1-k_1,n_2-k_2}(u_{k_1,k_2}) - \sum_{\substack{(r_1,r_2)=(0,0):\\ r_1+r_2<n_1+n_2}}^{(n_1,n_2)} e_{n_1-r_1,n_2-r_2}(u_{r_1,r_2})\,\Delta_1^{k_1}\Delta_2^{k_2}\bar{G}_{r_1,r_2}(u_{k_1,k_2}|U). \quad [5.5]$$

Arguing by induction, we obtain from [5.5]:

$$\Delta_1^{k_1}\Delta_2^{k_2}\bar{G}_{n_1,n_2}(u_{k_1,k_2}|U) = e_{n_1-k_1,n_2-k_2}(u_{k_1,k_2}) - e_{n_1-k_1,n_2-k_2}(u_{k_1,k_2}) = 0.$$

It remains to prove [5.2] for $k_1 = k_2 = 0$, that is, $\bar{G}_{n_1,n_2}(u_{0,0}|U) = \delta_{n_1,0}\delta_{n_2,0}$, which follows from [5.3] also by induction. □

The A–G type expansion [5.6] below generalizes the formula [2.29] established in the case of a single index. Its proof is obvious when $f_{n_1,n_2} = e_{n_1,n_2}$ and extends directly by linearity.

PROPOSITION 5.2.– *Given a family of A–G pseudopolynomials $\mathcal{G}(U)$, any pseudopolynomial $f_{n_1,n_2} \in \mathfrak{F}$ admits an A–G type expansion given by:*

$$f_{n_1,n_2}(\;) = \sum_{(k_1,k_2)=(0,0)}^{(n_1,n_2)} \Delta_1^{k_1}\Delta_2^{k_2} f_{n_1,n_2}(u_{k_1,k_2})\,\bar{G}_{k_1,k_2}(\;|U). \quad [5.6]$$

Furthermore, by reasoning by induction from [5.3], we obtain the identities [5.7] that generalize the previous ones [2.23].

PROPOSITION 5.3.– *For* $0 \leq k_1 \leq n_1, 0 \leq k_2 \leq n_2$*:*

$$\Delta_1^{k_1} \Delta_2^{k_2} \bar{G}_{n_1,n_2}(\ |U) = \bar{G}_{n_1-k_1,n_2-k_2}(\ |E^{k_1,k_2}U). \qquad [5.7]$$

5.2. Explicit formulas for $\bar{G}_{n_1,n_2}(\ |U)$

The pseudopolynomials can be calculated from the recursion [5.3]. A simpler method, however, is to use the A–G expansion [5.6] applied to e_{n_1,n_2}, that is,

$$e_{n_1,n_2}(\ ) = \sum_{(k_1,k_2)=(0,0)}^{(n_1,n_2)} e_{n_1-k_1,n_2-k_2}(u_{k_1,k_2})\, \bar{G}_{k_1,k_2}(\ |U). \qquad [5.8]$$

This expansion yields:

$$\bar{G}_{n_1,n_2}(\ |U) = [e_{n_1,n_2}(\ ) - e_{n_1,n_2}(u_{0,0})] - \sum_{\substack{(k_1,k_2)=(0,0): \\ 1 \leq k_1+k_2 < n_1+n_2}}^{(n_1,n_2)} e_{n_1-k_1,n_2-k_2}(u_{k_1,k_2})\, \bar{G}_{k_1,k_2}(\ |U), \quad [5.9]$$

which extends [3.54] derived in the single index case. By recursion, [5.9] then leads to an explicit formula for $\bar{G}_{n_1,n_2}(\ |U)$ whose form is similar to [3.53]. The details are however omitted here for the sake of brevity.

First bi-index A–G pseudopolynomials. For any family $\mathcal{E}$,

$$\bar{G}_{1,0}(\ |U) = e_{1,0} - e_{1,0}(u_{0,0}),$$

$$\bar{G}_{0,1}(\ |U) = e_{0,1} - e_{0,1}(u_{0,0}),$$

$$
\begin{aligned}
\bar{G}_{2,0}(\ |U) &= [e_{2,0} - e_{2,0}(u_{0,0})] - e_{1,0}(u_{1,0})\,\bar{G}_{1,0}(\ |U) \\
&= e_{2,0} - e_{1,0}(u_{1,0})e_{1,0} + [e_{1,0}(u_{1,0})e_{1,0}(u_{0,0}) - e_{2,0}(u_{0,0})], \\
\bar{G}_{0,2}(\ |U) &= [e_{0,2} - e_{0,2}(u_{0,0})] - e_{0,1}(u_{0,1})\,\bar{G}_{0,1}(\ |U) \\
&= e_{0,2} - e_{0,1}(u_{0,1})e_{0,1} + [e_{0,1}(u_{0,1})e_{0,1}(u_{0,0}) - e_{0,2}(u_{0,0})], \\
\bar{G}_{1,1}(\ |U) &= [e_{1,1} - e_{1,1}(u_{0,0})] - e_{0,1}(u_{1,0})\,\bar{G}_{1,0}(\ |U) - e_{1,0}(u_{0,1})\,\bar{G}_{0,1}(\ |U) \\
&= e_{1,1} - e_{0,1}(u_{1,0})e_{1,0} - e_{1,0}(u_{0,1})e_{0,1} \\
&\quad + [e_{0,1}(u_{1,0})e_{1,0}(u_{0,0}) + e_{1,0}(u_{0,1})e_{0,1}(u_{0,0}) - e_{1,1}(u_{0,0})], \\
\bar{G}_{2,1}(\ |U) &= [e_{2,1} - e_{2,1}(u_{0,0})] - e_{2,0}(u_{0,1})\,\bar{G}_{0,1}(\ |U) - e_{1,1}(u_{1,0})\,\bar{G}_{1,0}(\ |U) \\
&\quad - e_{1,0}(u_{1,1})\,\bar{G}_{1,1}(\ |U) - e_{0,1}(u_{2,0})\,\bar{G}_{2,0}(\ |U).
\end{aligned}
$$

We can also generalize the explicit results [3.68], [3.76], [3.81] and [3.83] obtained for the particular cases discussed in section 3.4.

PROPOSITION 5.4.– *When $U = \{u_{i_1,i_2} \equiv a,\ (i_1,i_2) \in \mathbb{N}_0^2\}$, then for the joint generating function of arguments s_1, s_2:*

$$
\sum_{(n_1,n_2)\geq(0,0)} \bar{G}_{n_1,n_2}(x|U)\, s_1^{n_1} s_2^{n_2} = \frac{\sum_{(n_1,n_2)\geq(0,0)} e_{n_1,n_2}(x)\, s_1^{n_1} s_2^{n_2}}{\sum_{(n_1,n_2)\geq(0,0)} e_{n_1,n_2}(a)\, s_1^{n_1} s_2^{n_2}}. \qquad [5.10]
$$

When $U = \{u_{i_1,i_2} \equiv aq^{i_1+i_2},\ (i_1,i_2) \in \mathbb{N}_0^2\}$, then if $\mathcal{E}$ satisfies the multiplicative property, that is, $e_{n_1,n_2}(\lambda x) = \lambda^{n_1+n_2} e_{n_1,n_2}(x)$ for any $\lambda \in \mathbb{R}$:

$$
\bar{G}_{n_1,n_2}(x|\mathcal{E},U) = q^{(n_1+n_2)^2/2}\,\bar{G}_{n_1,n_2}(x|\hat{\mathcal{E}},\{a,a,\ldots\}), \qquad [5.11]
$$

where $\hat{\mathcal{E}}$ is the family of functions $\hat{e}_{n_1,n_2}$ given by:

$$
\hat{e}_{n_1,n_2}(x) = q^{-(n_1+n_2)^2/2}\, e_{n_1,n_2}(x), \quad (n_1,n_2) \in \mathbb{N}_0^2. \qquad [5.12]
$$

When $U = \{u_{i_1,i_2} \equiv a + b_1 i_1 + b_2 i_2,\ (i_1,i_2) \in \mathbb{N}_0^2\}$, then if the S.I. property is satisfied, that is, Δ_1, Δ_2 commute with the shift operator:

$$
\bar{G}_{n_1,n_2}(x|\mathcal{E},U) = \bar{G}_{n_1,n_2}(x|\tilde{\mathcal{E}},\{a,a,\ldots\}), \qquad [5.13]
$$

where $\tilde{\mathcal{E}}$ is the family of functions $\tilde{e}_{n_1,n_2}$ given by:

$$\tilde{e}_{n_1,n_2}(x) = e_{n_1,n_2}(x - u_{n_1,n_2}), \quad (n_1, n_2) \in \mathbb{N}_0^2. \qquad [5.14]$$

Take $\mathcal{D} = \mathbb{R}$ or $\mathbb{Z}$, and suppose that $\mathcal{E}$ is constructed from the generating function $\exp(xg(s_1, s_2))$ *where $g(0,0) = 0$ and $g^{(1,0)}(0,0), g^{(0,1)}(0,0) \neq 0$. Given the previous affine family U, then:*

$$\bar{G}_{n_1,n_2}(x|U) = \frac{x - u_{0,0}}{x - u_{n_1,n_2}} e_{n_1,n_2}(x - u_{n_1,n_2}), \quad (n_1, n_2) \in \mathbb{N}_0^2. \qquad [5.15]$$

Proof. Concerning [5.10], we observe that if all the u_{i_1,i_2} are equal to a, [5.8] reduces to:

$$e_{n_1,n_2}(\ ) = \sum_{(k_1,k_2)=(0,0)}^{(n_1,n_2)} e_{n_1-k_1,n_2-k_2}(a)\, \bar{G}_{k_1,k_2}(\ |U),$$

hence the announced result in terms of generating functions.

For [5.11] and [5.12], a proof is postponed, for ease, to proposition 5.7 in the next section.

Concerning [5.13], we start with the A–G expansion of $e_{n_1,n_2}(x - u_{n_1,n_2})$ that gives, thanks to the S.I. hypothesis:

$$e_{n_1,n_2}(x - u_{n_1,n_2}) = \sum_{(k_1,k_2)=(0,0)}^{(n_1,n_2)} e_{n_1-k_1,n_2-k_2}(u_{k_1,k_2} - u_{n_1,n_2})\, \bar{G}_{k_1,k_2}(x|U),$$

or through the substitution [5.14]:

$$\tilde{e}_{n_1,n_2}(x) = \sum_{(k_1,k_2)=(0,0)}^{(n_1,n_2)} \tilde{e}_{n_1-k_1,n_2-k_2}(a)\, \bar{G}_{k_1,k_2}(x|U), \qquad [5.16]$$

because all the u_{i_1,i_2} are equal to $a + b_1 i_1 + b_2 i_2$. On the other hand, by using the operators $\tilde{\Delta}_j$, $j = 1, 2$, associated with $\tilde{\mathcal{E}}$, we also have the expansion:

$$\tilde{e}_{n_1,n_2}(x) = \sum_{(k_1,k_2)=(0,0)}^{(n_1,n_2)} \tilde{e}_{n_1-k_1,n_2-k_2}(a)\, \bar{G}_{k_1,k_2}(x|\tilde{\mathcal{E}}, \{a, a, \ldots\}).$$

Compared to [5.16], this leads to the desired identity.

For [5.15], we proceed as for proposition 3.10. First, writing $g(s_1, s_2) = \sum_{(i_1,i_2)\geq(0,0)} \alpha_{i_1,i_2}\, s_1^{i_1} s_2^{i_2}$, we get when $\mathcal{E}$ is standard at 0

$$\frac{n_1 e_{n_1,n_2}(x)}{x} = \sum_{(i_1,i_2)\geq(0,0)} i_1 \alpha_{i_1-1,i_2}\, e_{n_1-i_1,n_2-i_2}(x),$$

and consequently:

$$\Delta_1\Big(\frac{n_1 e_{n_1,n_2}(x)}{x}\Big) = \frac{(n_1-1)e_{n_1-1,n_2}(x)}{x},$$

with a similar formula valid for Δ_2 (compare to [3.17] and [3.15]). Now, denoting the r.h.s. of [5.15] by $K_{n_1,n_2}(x|U)$, that is,

$$K_{n_1,n_2}(x|U) = \left[1 + \frac{u_{n_1,n_2} - u_{0,0}}{x - u_{n_1,n_2}}\right] e_{n_1,n_2}(x - u_{n_1,n_2}),$$

we then manage to show that for an affine family U:

$$\begin{cases} \Delta_1 K_{n_1,n_2}(x|U) = K_{n_1-1,n_2}(x|E^{1,0}U), \\ \Delta_2 K_{n_1,n_2}(x|U) = K_{n_1,n_2-1}(x|E^{0,1}U). \end{cases}$$

Since $K_{n_1,n_2}(u_{0,0}|U) = \delta_{n_1,0}\delta_{n_2,0}$, [5.2] with [5.7] imply that $K_{n_1,n_2}(x|U) = G_{n_1,n_2}(x|U)$. □

REMARK.– It is possible to define and use on the same $\mathcal{D}$ two basic families, one with a single index, the other with two indices. For instance, let $\hat{\mathcal{E}} = \{\hat{e}_i,\ i \in \mathbb{N}_0\}$, and on $\hat{\mathcal{E}}$ define $\mathcal{E} = \{e_{i_1,i_2},\ (i_1,i_2) \in \mathbb{N}_0^2\}$ by:

$$e_{0,0} = \hat{e}_0 = 1;\quad e_{1,0} = \hat{e}_1, e_{0,1} = \hat{e}_2;\quad e_{2,0} = \hat{e}_3, e_{1,1} = \hat{e}_4, e_{0,2}$$
$$= \hat{e}_5;\ ,\ e_{3,0} = \hat{e}_6, \ldots, e_{0,3} = \hat{e}_9,$$

more generally:

$$e_{n-i,i} = \hat{e}_{i+n(n+1)/2},\quad 0 \leq i \leq n. \qquad [5.17]$$

We can then define the operators $\hat{\Delta}$ on $\hat{\mathcal{E}}$ and Δ_1, Δ_2 on $\mathcal{E}$. Of course, these operators are linked. So, for $0 \leq i \leq n-1$, $\Delta_1\, e_{n-i,i} = e_{n-1-i,i}$ by [5.1], which, using [5.17], gives:

$$\Delta_1 \hat{e}_{i+n(n+1)/2} = \hat{e}_{i+(n-1)n/2}$$
$$= \hat{e}_{i+n(n+1)/2-n} = \hat{\Delta}^n \hat{e}_{i+n(n+1)/2}.$$

Similarly, for $0 < i \leq n$, $\Delta_2 e_{n-i,i} = e_{n-i,i-1}$ leads to:

$$\Delta_2 \hat{e}_{i+n(n+1)/2} = \hat{\Delta}^{n+1} \hat{e}_{i+n(n+1)/2}.$$

5.3. Multivariate case $\bar{G}_{n_1,n_2}(\ |U^{(1)}, U^{(2)})$

We will particularize the previous A–G pseudopolynomials with two indices $\bar{G}_{n_1,n_2}(x|U)$ to the bivariate case where $\mathcal{D} = \mathcal{D}_1 \times \mathcal{D}_2$. Thus, any element $x \in \mathcal{D}$ becomes a bivariate element $(x_1, x_2) \in \mathcal{D}_1 \times \mathcal{D}_2$. The same change occurs with the family U, i.e. $U = (U^{(1)}, U^{(2)})$ where:

$$U^{(1)} = \{u^{(1)}_{i_1,i_2},\ (i_1, i_2) \in \mathbb{N}_0^2\}, \quad \text{and} \quad U^{(2)} = \{u^{(2)}_{i_1,i_2},\ (i_1, i_2) \in \mathbb{N}_0^2\}.$$

For $(k_1, k_2) \in \mathbb{N}_0^2$, denote $E^{k_1,k_2}U^{(j)} = \{u^{(j)}_{k_1+i_1,k_2+i_2},\ (i_1, i_2) \in \mathbb{N}_0^2\}$, where $j = 1, 2$.

Here as previously, $\mathcal{E}$ is the family $\{e_{n_1,n_2}(x_1, x_2),\ (n_1, n_2) \in \mathbb{N}_0^2\}$. It is standard at (ω_1, ω_2) if:

$$e_{n_1,n_2}(\omega_1, \omega_2) = \delta_{n_1+n_2,0}, \quad (n_1, n_2) \in \mathbb{N}_0^2. \qquad [5.18]$$

When $\mathcal{E}$ is standard at (ω_1, ω_2), a copy of $\mathcal{E}$ is a family $\hat{\mathcal{E}}$ defined by:

$$\hat{e}_{n_1,n_2} = \sum_{(k_1,k_2)=(0,0)}^{(n_1,n_2)} \hat{e}_{n_1-k_1,n_2-k_2}(\omega_1, \omega_2)\, e_{k_1,k_2}, \qquad [5.19]$$

which extends [2.13].

Concerning the operators Δ_j defined via [5.1], they are adjusted in a simple way. We use a mimicry of basic properties of the operators $\partial/\partial x_j$ in standard differential calculus when $\mathcal{D} = \mathbb{R} \times \mathbb{R}$ and $e_{n_1,n_2}(x_1, x_2) = x_1^{n_1} x_2^{n_2}/n_1!n_2!$. The commutativity property $\Delta_1\Delta_2 = \Delta_2\Delta_1$ is obvious from [5.1], it suffices to retain the following two axioms.

a) For any function $h : \mathcal{D} \to \mathrm{I\!R}$ on which Δ_j is applied, $\Delta_j h = 0$ iff h is independent of x_j. From [5.1], we thus see that e_{0,n_2} does not depend on x_1, and $e_{n_1,0}$ does not depend on x_2. This suggests to adopt the notations $e_{\diamond,n_2}$ and $e_{n_1,\diamond}$, respectively, the sign $\diamond$ indicating that this component in question does not matter.

b) Moreover, $\Delta_j h = 0$ implies $\Delta_j(hf) = h\Delta_j f$ for any functions $h, f : \mathcal{D} \to \mathrm{I\!R}$. In other words, when h is independent of x_j, h acts as a constant with respect to Δ_j and the classical linearity of Δ_j applies. Note that this formula implicitly says that $\Delta_j(hf)$ exists iff $\Delta_j f$ exists.

PROPOSITION 5.5.– *When $\mathcal{E}$ is standard at some element (ω_1, ω_2):*

$$e_{n_1,n_2}(x_1, x_2) = e_{n_1,0}(x_1, \diamond)\, e_{0,n_2}(\diamond, x_2), \quad (n_1, n_2) \in \mathrm{I\!N}_0^2. \qquad [5.20]$$

Proof. Let us introduce a new family $\bar{\mathcal{E}}$ through:

$$\bar{e}_{n_1,n_2}(x_1, x_2) = e_{n_1,0}(x_1, \omega_2)\, e_{0,n_2}(\omega_1, x_2), \quad (n_1, n_2) \in \mathrm{I\!N}_0^2, \quad [5.21]$$

and on which, due to axiom (b), Δ_1, Δ_2 act as indicated by [5.1]. For example:

$$\begin{aligned}\Delta_2[e_{n_1,n_2}(x_1, x_2)] &= \Delta_2[e_{n_1,0}(x_1, \omega_2)\, e_{0,n_2}(\omega_1, x_2)] \\ &= e_{n_1,0}(x_1, \omega_2)\, \Delta_2[e_{0,n_2}(\omega_1, x_2)] \\ &= e_{n_1,0}(x_1, \omega_2)\, e_{0,n_2-1}(\omega_1, x_2) = e_{n_1,n_2-1}(x_1, x_2).\end{aligned}$$

Next, define the differences:

$$d_{n_1,n_2} \equiv e_{n_1,n_2} - \bar{e}_{n_1,n_2}.$$

We want to show that these differences are all zero, which implies the identity [5.20].

Note that $d_{0,0} = 1 - 1 = 0$, and for $n_1 + n_2 > 0$, $d_{n_1,n_2}(\omega_1, \omega_2) = 0$. Now, we argue by recursion to establish that $d_{n_1,n_2} = 0$ for all (n_1, n_2). Thus, supposing that $d_{k_1,k_2} = 0$ for all $k_1 \le n_1, k_2 \le n_2, k_1 + k_2 < n_1 + n_2$, we find that:

$$\Delta_1 d_{n_1,n_2} = d_{n_1-1,n_2} = 0, \quad \text{and} \quad \Delta_2 d_{n_1,n_2} = d_{n_1,n_2-1} = 0,$$

so that d_{n_1,n_2} is independent of x_1 and x_2. Being a constant, it is 0 since $d_{n_1,n_2}(\omega_1, \omega_2) = 0$. □

Let $\mathcal{G}(U^{(1)}, U^{(2)})$ the family of bivariate A–G pseudoplynomials $\bar{G}_{n_1,n_2}(x_1, x_2|U^{(1)}, U^{(2)})$. For each $(n_1, n_2) \in \mathbb{N}_0^2$, they are defined from [5.2] by:

$$\Delta_1^{k_1} \Delta_2^{k_2} \bar{G}_{n_1,n_2}\left(u_{k_1,k_2}^{(1)}, u_{k_1,k_2}^{(2)} | U^{(1)}, U^{(2)}\right) = \delta_{n_1,k_1} \delta_{n_2,k_2},$$
$$0 \le k_1 \le n_1, 0 \le k_2 \le n_2. \qquad [5.22]$$

In fact, these A–G pseudopolynomials satisfy the identities [5.7], which are written here:

$$\Delta_1^{k_1} \Delta_2^{k_2} \bar{G}_{n_1,n_2}\left(x_1, x_2 | U^{(1)}, U^{(2)}\right)$$
$$= \bar{G}_{n_1-k_1,n_2-k_2}\left(x_1, x_2 | E^{k_1,k_2}(U^{(1)}, U^{(2)})\right). \qquad [5.23]$$

Moreover, they can be determined recursively through the relations [5.3], that is,

$$\bar{G}_{n_1,n_2}\left(x_1, x_2 | U^{(1)}, U^{(2)}\right) = e_{n_1,n_2}(x_1, x_2)$$
$$- \sum_{\substack{(k_1,k_2)=(0,0): \\ k_1+k_2<n_1+n_2}}^{(n_1,n_2)} e_{n_1-k_1,n_2-k_2}\left(u_{k_1,k_2}^{(1)}, u_{k_1,k_2}^{(2)}\right) \bar{G}_{k_1,k_2}\left(x_1, x_2 | U^{(1)}, U^{(2)}\right).$$
$$[5.24]$$

First bivariate A–G pseudopolynomials. For any family $\mathcal{E}$:

$$\bar{G}_{1,0}(x_1, x_2 | U^{(1)}, U^{(2)}) = e_{1,0}(x_1, x_2) - e_{1,0}(u_{0,0}^{(1)}, x_2),$$

$$\bar{G}_{0,1}(x_1, x_2 | U^{(1)}, U^{(2)}) = e_{0,1}(x_1, x_2) - e_{0,1}(x_1, u_{0,0}^{(2)}),$$

$$\bar{G}_{2,0}(x_1, x_2 | U^{(1)}, U^{(2)}) = [e_{2,0}(x_1, x_2) - e_{2,0}(u_{0,0}^{(1)}, x_2)]$$
$$- e_{1,0}(u_{1,0}^{(1)}, x_2)\, \bar{G}_{1,0}(x_1, x_2 | U^{(1)}, U^{(2)})$$
$$= e_{2,0}(x_1, x_2) - e_{1,0}(u_{1,0}^{(1)}, x_2)\, e_{1,0}(x_1, x_2)$$
$$+ [e_{1,0}(u_{1,0}^{(1)}, x_2)\, e_{1,0}(u_{0,0}^{(1)}, x_2) - e_{2,0}(u_{0,0}^{(1)}), x_2)],$$

$$\bar{G}_{0,2}(x_1, x_2|U^{(1)}, U^{(2)}) = [e_{0,2}(x_1, x_2) - e_{0,2}(x_1, u^{(2)}_{0,0})]$$

$$- e_{0,1}(x_1, u^{(2)}_{0,1})\, \bar{G}_{0,1}(x_1, x_2|U^{(1)}, U^{(2)})$$

$$= e_{0,2}(x_1, x_2) - e_{0,1}(x_1, u^{(2)}_{0,1})\, e_{0,1}(x_1, x_2)$$

$$+ [e_{0,1}(x_1, u^{(2)}_{0,1})\, e_{0,1}(x_1, u^{(2)}_{0,0}) - e_{0,2}(x_1, u^{(2)}_{0,0})],$$

$$\bar{G}_{1,1}(x_1, x_2|U^{(1)}, U^{(2)}) = [e_{1,1}(x_1, x_2) - e_{1,1}(u^{(1)}_{0,0}, u^{(2)}_{0,0})]$$

$$- e_{0,1}(x_1, u^{(2)}_{1,0})\, \bar{G}_{1,0}(x_1, x_2|U^{(1)}, U^{(2)})$$

$$- e_{1,0}(u^{(1)}_{0,1}, x_2)\, \bar{G}_{0,1}(x_1, x_2|U^{(1)}, U^{(2)}),$$

$$\bar{G}_{2,1}(x_1, x_2|U^{(1)}, U^{(2)}) = [e_{2,1}(x_1, x_2) - e_{2,1}(u^{(1)}_{0,0}, u^{(2)}_{0,0})]$$

$$- e_{2,0}(u^{(1)}_{0,1}, x_2)\, \bar{G}_{0,1}(x_1, x_2|U^{(1)}, U^{(2)})$$

$$- e_{1,1}(u^{(1)}_{1,0}, u^{(2)}_{1,0})\, \bar{G}_{1,0}(x_1, x_2|U^{(1)}, U^{(2)})$$

$$- e_{1,0}(u^{(1)}_{1,1}, x_2)\, \bar{G}_{1,1}(x_1, x_2|U^{(1)}, U^{(2)})$$

$$- e_{0,1}(x_1, u^{(2)}_{2,0})\, \bar{G}_{2,0}(x_1, x_2|U^{(1)}, U^{(2)}).$$

Of course, a binomial-type identity holds in the bivariate case if it is valid on each component $\mathcal{D}_1$, $\mathcal{D}_2$. A somewhat more surprising result occurs in the following case. Note that a sign $\diamond$ is also used below for an index which may be unspecified.

PROPOSITION 5.6.– *If, in addition, the family $U^{(2)}$ is independent of i_1, that is:*

$$u^{(2)}_{i_1,i_2} = u^{(2)}_{\diamond,i_2}, \quad (i_1, i_2) \in \mathbb{N}_0^2, \tag{5.25}$$

then any bivariate A–G pseudopolynomial is the product of two univariate A–G pseudopolynomials, namely:

$$\bar{G}_{n_1,n_2}(x_1, x_2|\mathcal{E}, U^{(1)}, U^{(2)}) = \bar{G}_{n_1}(x_1|\mathcal{E}_1, \{u^{(1)}_{i_1,n_2}, i_1 \in \mathbb{N}_0\})$$

$$\times \bar{G}_{n_2}(x_2|\mathcal{E}_2, \{u^{(2)}_{\diamond,i_2}, i_2 \in \mathbb{N}_0\}), \tag{5.26}$$

where

$$\mathcal{E}_1 = \{e_{n_1,0},\, n_1 \in \mathbb{N}_0\}, \quad \textit{and} \quad \mathcal{E}_2 = \{\mathrm{e}_{0,\mathrm{n}_2},\, \mathrm{n}_2 \in \mathbb{N}_0\}.$$

Proof. We first note that like the left side, the right side of [5.26] is also an element of $\mathfrak{F}$. In fact, if $\mathfrak{F}_j$ is the family generated by $\mathcal{E}_j$, the right side will be an element of $\mathfrak{F}_1 \times \mathfrak{F}_2$, that is, a linear combination of terms $e_{n_1,0}\, e_{0,n_2} = e_{n_1,n_2}$ that belong to $\mathfrak{F}$.

We must then verify that the right side of [5.26] satisfies property [5.22]. Inserting [5.25], this would mean that:

$$\begin{aligned}
&\bar{G}_{n_1-k_1}\big(u^{(1)}_{k_1,n_2}|\mathcal{E}_1, \{u^{(1)}_{k_1+i_1,n_2}, i_1 \in \mathbb{N}_0\}\big) \\
&\times \bar{G}_{n_2-k_2}\big(u^{(2)}_{\diamond,k_2}|\mathcal{E}_2, \{u^{(2)}_{\diamond,k_2+i_2}, i_2 \in \mathbb{N}_0\}\big) \\
&= \delta_{n_1,k_1}\delta_{n_2,k_2}, \quad 0 \le k_1 \le n_1, 0 \le k_2 \le n_2.
\end{aligned} \qquad [5.27]$$

Now, consider the term $\bar{G}_{n_2-k_2}(\ldots)$. When $k_2 = n_2$, its value is 1, which corresponds to δ_{n_2,k_2}, and at the same time $\bar{G}_{n_1-k_1}(\ldots) = \delta_{n_1,k_1}$. When $k_2 < n_2$, its value is 0, so [5.27] is true. □

As in section 5.1, an A–G type expansion applies to any pseudopolynomial $f_{n_1,n_2}(x_1,x_2) \in \mathfrak{F}$. More precisely, given a family of bivariate A–G pseudopolynomials $\mathcal{G}(U^{(1)}, U^{(2)})$, we have:

$$\begin{aligned}
f_{n_1,n_2}(x_1,x_2) = \sum_{(k_1,k_2)=(0,0)}^{(n_1,n_2)} \Delta_1^{k_1}\Delta_2^{k_2} f_{n_1,n_2}(u^{(1)}_{k_1,k_2}, u^{(2)}_{k_1,k_2}) \\
\bar{G}_{k_1,k_2}(x_1,x_2|U^{(1)},U^{(2)}).
\end{aligned} \qquad [5.28]$$

Several formulas in section 5.2 are also easy to adapt. Below we focus on two less obvious results. The first deals with the case of geometric parameters.

REMARK.– We emphasize that the proof of proposition 5.7 will not require the particularization made in section 5.3 concerning the operators Δ_j, $j = 1,2$. Therefore, taking $q_1 = q = q_2$, we have $(u^{(1)}_{i_1,i_2}, u^{(2)}_{i_1,i_2}) = (a_1,a_2)q^{i_1+i_2}$, and the same proof also allows us to obtain [5.11] and [5.12] as announced.

PROPOSITION 5.7.– *When* $(U^{(1)}, U^{(2)}) = \{(u^{(1)}_{i_1,i_2}, u^{(2)}_{i_1,i_2}) \equiv (a_1 q_1^{i_1} q^{i_2}, a_2 q^{i_1} q_2^{i_2}), (i_1, i_2) \in \mathbb{N}_0^2\}$, *then if* $\mathcal{E}$ *satisfies the multiplicative property, that is,* $e_{n_1,n_2}(\lambda_1 x_1, \lambda_2 x_2) = \lambda_1^{n_1} \lambda_2^{n_2} e_{n_1,n_2}(x_1, x_2)$ *for any* $\lambda_1, \lambda_2 \in \mathbb{R}$*:*

$$\bar{G}_{n_1,n_2}(x_1, x_2|\mathcal{E}, U^{(1)}, U^{(2)}) = q_1^{n_1^2/2} q^{n_1 n_2} q_2^{n_2^2/2}\, \bar{G}_{n_1,n_2}(x_1, x_2|\hat{\mathcal{E}}, (a_1, a_2), (a_1, a_2), \ldots\}), \qquad [5.29]$$

where $\hat{\mathcal{E}}$ *is the family of functions* $\hat{e}_{n_1,n_2}$ *given by:*

$$\hat{e}_{n_1,n_2}(x_1, x_2) = q_1^{-n_1^2/2} q^{-n_1 n_2} q_2^{-n_2^2/2}\, e_{n_1,n_2}(x_1, x_2), \quad (n_1, n_2) \in \mathbb{N}_0^2. \qquad [5.30]$$

Proof. We start from the A–G expansion [5.28] applied to $e_{n_1,n_2}(x_1, x_2)$, that is:

$$e_{n_1,n_2}(x_1, x_2) = \sum_{(k_1,k_2)=(0,0)}^{(n_1,n_2)} e_{n_1-k_1,n_2-k_2}(u^{(1)}_{k_1,k_2}, u^{(2)}_{k_1,k_2}) \times \bar{G}_{k_1,k_2}(x_1, x_2|U^{(1)}, U^{(2)}),$$

which becomes from the definition of $(u^{(1)}_{k_1,k_2}, u^{(2)}_{k_1,k_2})$, then applying the multiplication property:

$$\begin{aligned} e_{n_1,n_2}(x_1, x_2) &= \sum_{(k_1,k_2)=(0,0)}^{(n_1,n_2)} e_{n_1-k_1,n_2-k_2}(a_1 q_1^{k_1} q^{k_2}, a_2 q^{k_1} q_2^{k_2}) \\ &\times \bar{G}_{k_1,k_2}(x_1, x_2|U^{(1)}, U^{(2)}) \\ &= \sum_{(k_1,k_2)=(0,0)}^{(n_1,n_2)} (q_1^{k_1} q^{k_2})^{n_1-k_1} (q^{k_1} q_2^{k_2})^{n_2-k_2}\, e_{n_1-k_1,n_2-k_2}(a_1, a_2) \\ &\times \bar{G}_{k_1,k_2}(x_1, x_2|U^{(1)}, U^{(2)}). \end{aligned} \qquad [5.31]$$

Notice that:

$$k_1(n_1-k_1)=[n_1^2-(n_1-k_1)^2-k_1^2]/2,$$
$$k_2(n_2-k_2)=[n_2^2-(n_2-k_2)^2-k_2^2]/2,$$
$$\text{and } k_2(n_1-k_1)+k_1(n_2-k_2)=n_1n_2-(n_1-k_1)(n_2-k_2)-k_1k_2.$$

Therefore, [5.31] can be rewritten as:

$$e_{n_1,n_2}(x_1,x_2)=q_1^{n_1^2/2}q^{n_1n_2}q_2^{n_2^2/2}\sum_{(k_1,k_2)=(0,0)}^{(n_1,n_2)}q_1^{-(n_1-k_1)^2/2}$$
$$\times\, q^{-(n_1-k_1)(n_2-k_2)}q_2^{-(n_2-k_2)^2/2}$$
$$e_{n_1-k_1,n_2-k_2}(a_1,a_2)\,q_1^{-k_1^2/2}q^{-k_1k_2}q_2^{-k_2^2/2}\,\bar{G}_{k_1,k_2}(x_1,x_2|U^{(1)},U^{(2)}),$$

or equivalently, by the definition [5.30]:

$$\hat{e}_{n_1,n_2}(x_1,x_2)=\sum_{(k_1,k_2)=(0,0)}^{(n_1,n_2)}\hat{e}_{n_1-k_1,n_2-k_2}(a_1,a_2)\,q_1^{-k_1^2/2}q^{-k_1k_2}q_2^{-k_2^2/2}$$
$$\times\,\bar{G}_{k_1,k_2}(x_1,x_2|U^{(1)},U^{(2)}). \qquad [5.32]$$

In parallel, using the operators $\hat{\Delta}_j$, $j=1,2$, associated with $\hat{\mathcal{E}}$, we have the expansion:

$$\hat{e}_{n_1,n_2}(x_1,x_2)=\sum_{(k_1,k_2)=(0,0)}^{(n_1,n_2)}\hat{e}_{n_1-k_1,n_2-k_2}(a_1,a_2)$$
$$\times\,\bar{G}_{k_1,k_2}(x_1,x_2|\hat{\mathcal{E}},\{(a_1,a_2),(a_1,a_2),\ldots\}),$$

hence [5.29] after comparison with [5.32]. □

The second result concerns the case of affine parameters, our objective this time being to extend the formula [5.15].

PROPOSITION 5.8.– *Take* $\mathcal{D}_1, \mathcal{D}_2 = \mathbb{R}$ *or* $\mathbb{Z}$, *and suppose that* $\mathcal{E}$ *is constructed from the generating function* $\exp[x_1 g_1(s_1)+x_2 g_2(s_2)]$ *where* $g_1(0) = g_2(0) = 0$ *and* $g_1^{(1)}(0), g_2^{(1)}(0) \neq 0$. *When* $(U^{(1)}, U^{(2)}) = \{(u^{(1)}_{i_1,i_2}, u^{(2)}_{i_1,i_2}) \equiv (b_1^{(1)} i_1 + b_2^{(1)} i_2, b_1^{(2)} i_1 + b_2^{(2)} i_2),\ (i_1, i_2) \in \mathbb{N}_0^2\}$, *then:*

$$\bar{G}_{n_1,n_2}(x_1, x_2|U^{(1)}, U^{(2)}) = \begin{vmatrix} e_{1,0}(x_1 - u^{(1)}_{0,n_2}, \diamond) & e_{0,1}(\diamond, u^{(2)}_{n_1,0}) \\ e_{1,0}(u^{(1)}_{0,n_2}, \diamond) & e_{0,1}(\diamond, x_2 - u^{(2)}_{n_1,0}) \end{vmatrix}$$

$$\frac{e_{n_1,n_2}(x_1 - u^{(1)}_{n_1,n_2}, x_2 - u^{(2)}_{n_1,n_2})}{e_{1,1}(x_1 - u^{(1)}_{n_1,n_2}, x_2 - u^{(2)}_{n_1,n_2})}, \quad (n_1, n_2) \in \mathbb{N}_0^2. \qquad [5.33]$$

Proof. Note that by construction, $\mathcal{E}$ is standard at $(0, 0)$, thus from [5.20]:

$$e_{1,1}(x_1, x_2) = e_{1,0}(x_1, \diamond)\, e_{0,1}(\diamond, x_2). \qquad [5.34]$$

In addition, for $g_j(s_j)$, $j = 1, 2$, of the form $\sum_{i=1}^{\infty} \alpha_{j,i}(s_j)^i$, we have inter alia (see [3.12]):

$$e_{1,0}(x_1, \diamond) = \alpha_{1,1} x_1, \quad \text{and} \quad e_{0,1}(\diamond, x_2) = \alpha_{2,1} x_2. \qquad [5.35]$$

Now, let $K_{n_1,n_2}(x_1, x_2|U^{(1)}, U^{(2)})$ be the r.h.s. of [5.33]. We will prove below that $K_{n_1,n_2}(\ldots)$ enjoys properties that allow it to be identified with the l.h.s. of [5.33]. Note that this identification is only possible because on both sides, we are working within the same $\mathfrak{F}$, which is a family of bivariate polynomials in x_1, x_2. Therefore, it suffices to verify that $K_{n_1,n_2}(x_1, x_2|U^{(1)}, U^{(2)})$ is indeed such a polynomial. By [5.35], this is obvious for the determinant. It is also true for the fraction since by [5.20], this fraction factors into the product of a fraction in x_1 by a fraction in x_2, and these two fractions belong to $\mathfrak{F}_1$ (respectively, $\mathfrak{F}_2$), thus reducing to polynomials in x_1 (respectively, x_2). After multiplying the two parts, the r.h.s. corresponds to products of polynomials that are still polynomials, so in $\mathfrak{F}$ as desired.

We reason in a similar way to deriving [5.15]. First, we show that $K_{n_1,n_2}(u^{(1)}_{0,0}, u^{(2)}_{0,0}|U^{(1)}, U^{(2)}) = \delta_{n_1,0}\,\delta_{n_2,0}$. Indeed, for $n_1 = n_2 = 0$, the present affine case implies $u^{(1)}_{0,\diamond} = u^{(2)}_{\diamond,0} = 0$, so $K_{0,0}(x_1, x_2|U^{(1)}, U^{(2)})$ becomes:

$$\begin{vmatrix} e_{1,0}(x_1, \diamond) & e_{0,1}(\diamond, 0) \\ e_{1,0}(0, \diamond) & e_{0,1}(\diamond, x_2) \end{vmatrix} \frac{e_{0,0}(x_1, x_2)}{e_{1,1}(x_1, x_2)} = \frac{e_{1,0}(x_1, \diamond)\, e_{0,1}(\diamond, x_2)}{e_{1,1}(x_1, x_2)} = 1.$$

by virtue of [5.34]. For $n_1 + n_2 > 0$, since $u_{0,0}^{(1)} = u_{0,0}^{(2)} = 0$, $K_{n_1,n_2}(u_{0,0}^{(1)}, u_{0,0}^{(2)}|U^{(1)}, U^{(2)})$ reduces to:

$$\begin{vmatrix} e_{1,0}(-b_2^{(1)} n_2, \diamond) & e_{0,1}(\diamond, b_1^{(2)} n_1) \\ e_{1,0}(b_2^{(1)} n_2, \diamond) & e_{0,1}(\diamond, -b_1^{(2)} n_1) \end{vmatrix} \frac{e_{n_1,n_2}(-u_{n_1,n_2}^{(1)}, -u_{n_1,n_2}^{(2)})}{e_{1,1}(-u_{n_1,n_2}^{(1)}, -u_{n_1,n_2}^{(2)})} = 0,$$

because the determinant is zero by virtue of [5.35].

Thus, to prove [5.33], it is sufficient to show that $K_{n_1,n_2}(x_1, x_2|U^{(1)}, U^{(2)})$ satisfies the property [5.23], that is:

$$\Delta_1 K_{n_1,n_2}(x_1, x_2|U^{(1)}, U^{(2)}) = K_{n_1-1,n_2}(x_1, x_2|E^{1,0}(U^{(1)}, U^{(2)})), \quad [5.36]$$

$$\Delta_2 K_{n_1,n_2}(x_1, x_2|U^{(1)}, U^{(2)}) = K_{n_1,n_2-1}(x_1, x_2|E^{0,1}(U^{(1)}, U^{(2)})).$$

For this, we consider the shifted r.h.s.:

$$S^{u_{n_1,n_2}^{(1)}, u_{n_1,n_2}^{(2)}} K_{n_1,n_2}(x_1, x_2|U^{(1)}, U^{(2)}),$$

where $S^{u_{n_1,n_2}^{(1)}, u_{n_1,n_2}^{(2)}} \equiv S_1^{u_{n_1,n_2}^{(1)}} S_2^{u_{n_1,n_2}^{(2)}}$ with $S_1^{u_{n_1,n_2}^{(1)}}$ acting on x_1 and $S_2^{u_{n_1,n_2}^{(2)}}$ on x_2. Given the affine assumption, it is given by:

$$S^{u_{n_1,n_2}^{(1)}, u_{n_1,n_2}^{(2)}} K_{n_1,n_2}(\ldots) = \begin{vmatrix} e_{1,0}(x_1 + u_{n_1,0}^{(1)}, \diamond) & e_{0,1}(\diamond, u_{n_1,0}^{(2)}) \\ e_{1,0}(u_{0,n_2}^{(1)}, \diamond) & e_{0,1}(\diamond, x_2 + u_{0,n_2}^{(2)}) \end{vmatrix}$$

$$\frac{e_{n_1,n_2}(x_1, x_2)}{e_{1,1}(x_1, x_2)}. \quad [5.37]$$

When applying Δ_1, only the terms in x_1 and n_1 are relevant, and the other terms can be considered as fixed. Also using [5.20], the r.h.s. of [5.37] can then be rewritten equivalently as $r_{n_1}(x_1)$ where:

$$r_{n_1}(x_1) = \begin{vmatrix} \alpha_{1,1}(x_1 + u_{n_1,0}^{(1)}) & \alpha_{2,1} u_{n_1,0}^{(2)} \\ c_1 & c_2 \end{vmatrix} c_3 \frac{e_{n_1,0}(x_1, \diamond)}{e_{1,0}(x_1, \diamond)}, \quad [5.38]$$

c_1, c_2, c_3 denoting some constants. Again by assumption, [5.38] reduces to:

$$r_{n_1}(x_1) = \left[c_2\, \alpha_{1,1}(x_1 + u_{n_1,0}^{(1)}) - c_1\, \alpha_{2,1} u_{n_1,0}^{(2)}\right] c_3 \frac{e_{n_1,0}(x_1, \diamond)}{e_{1,0}(x_1, \diamond)}$$

$$= (c_4 + c_5 n_1/x_1)\, e_{n_1,0}(x_1, \diamond),$$

for appropriate constants c_4, c_5. Using the S.I. property inherited by $\mathcal{E}$, we thus obtain:

$$\Delta_1 \left[S_2^{u^{(2)}_{n_1,n_2}} K_{n_1,n_2}(x_1, x_2|U^{(1)}, U^{(2)})\right]$$
$$= S_1^{-u^{(1)}_{n_1,n_2}} \Delta_1 \left[(c_4 + c_5 n_1/x_1) e_{n_1,0}(x_1, \diamond)\right], \qquad [5.39]$$

where the r.h.s. has exactly the same structure as [3.85] in the univariate case, except for some constants. Because of the proof leading to [3.86], we obtain here from [5.39] that:

$$S_1^{-u^{(1)}_{n_1,n_2}} \Delta_1 \left[(c_4 + c_5 n_1/x_1) e_{n_1,0}(x_1, \diamond)\right]$$
$$= E^{1,0} \left[S_2^{u^{(2)}_{n_1-1,n_2}} K_{n_1-1,n_2}(x_1, x_2|U^{(1)}, U^{(2)})\right],$$

and therefore:

$$\Delta_1 \left[S_2^{u^{(2)}_{n_1,n_2}} K_{n_1,n_2}(x_1, x_2|U^{(1)}, U^{(2)})\right]$$
$$= S_2^{u^{(2)}_{n_1,n_2}} K_{n_1-1,n_2}(x_1, x_2|E^{1,0}(U^{(1)}, U^{(2)})).$$

After simplification by $S_2^{u^{(2)}_{n_1,n_2}}$, this directly gives the desired relation [5.36]. Obviously, the result for Δ_2 and $E^{0,1}$ follows similarly. □

REMARK.– The explicit expression of the pseudopolynomials A–G is intuitively guessed for a multivariate affine case. In fact, by adapting the previous proof, we can show that with 3 variables for example, when $(U^{(1)}, U^{(2)}, U^{(3)}) = \{(u^{(1)}_{i_1,i_2,i_3}, u^{(2)}_{i_1,i_2,i_3}, u^{(3)}_{i_1,i_2,i_3}),\ (i_1, i_2, i_3) \in \mathbb{N}_0^3\}$ where:

$$u^{(j)}_{i_1,i_2,i_3} \equiv b_1^{(j)} i_1 + b_2^{(j)} i_2 + b_3^{(j)} i_3, \quad j = 1, 2, 3,$$

then $\bar{G}_{n_1,n_2,n_3}(x_1, x_2, x_3|U^{(1)}, U^{(2)}, U^{(3)})$ is given by:

$$\frac{\begin{vmatrix} e_{1,0,0}(x_1 - u^{(1)}_{0,n_2,n_3}, \diamond, \diamond) & e_{0,1,0}(\diamond, u^{(2)}_{n_1,0,0}, \diamond) & e_{0,0,1}(\diamond, \diamond, u^{(3)}_{n_1,0,0}) \\ e_{1,0,0}(u^{(1)}_{0,n_2,0}, \diamond, \diamond) & e_{0,1,0}(\diamond, x_2 - u^{(2)}_{n_1,0,n_3}, \diamond) & e_{0,0,1}(\diamond, \diamond, u^{(3)}_{0,n_2,0}) \\ e_{1,0,0}(u^{(1)}_{0,0,n_3}, \diamond, \diamond) & e_{0,1,0}(\diamond, u^{(2)}_{0,0,n_3}, \diamond) & e_{0,0,1}(\diamond, \diamond, x_3 - u^{(3)}_{n_1,n_2,0}) \end{vmatrix}}{}$$

$$\frac{e_{n_1,n_2,n_3}(x_1 - u^{(1)}_{n_1,n_2,n_3}, x_2 - u^{(2)}_{n_1,n_2,n_3}, x_3 - u^{(3)}_{n_1,n_2,n_3})}{e_{1,1,1}(x_1 - u^{(1)}_{n_1,n_2,n_3}, x_2 - u^{(2)}_{n_1,n_2,n_3}, x_3 - u^{(3)}_{n_1,n_2,n_3})}, \quad (n_1, n_2, n_3) \in \mathbb{N}_0^3.$$

Many other results obtained for the univariate case can be extended without real difficulty. This is particularly true for the Taylor-type formula [2.79], which easily adapts to the bivariate case. Further analysis is left to readers so as not to be too long or tedious. However, we will return below to the integral definition of the A–G pseudopolynomials, the generalization of which seems less direct.

5.4. Integral multivariate representation

Recall that indeed, the A–G pseudopolynomials in the one-dimensional case can be defined via an integration operator I_u (see Chapter 2, sections 2.3 and 2.4). Our objective here is to generalize this approach to the two-dimensional case.

5.4.1. *Single-index* $\bar{G}_n$

We showed in [2.23] that $\Delta\bar{G}_n(\ |U) = \bar{G}_{n-1}(\ |EU)$ for any $n \geq 1$. In this part, we use a compact notation that omits U and rewrite this relation:

$$\Delta\bar{G}_n = E\bar{G}_{n-1}, \quad n \geq 1, \qquad [5.40]$$

in order to simplify the presentation of certain important formulas.

Recall that by [2.14], the operator I_u, $u \in \mathcal{D}$, is defined by:

$$I_u e_n = e_{n+1} - e_{n+1}(u)\, e_0, \quad n \in \mathbb{N}_0, \qquad [5.41]$$

so that

$$\Delta I_u = \Delta^0, \quad \text{and} \quad I_u\Delta f = f - f(u), \quad f \in \mathfrak{F}.$$

Moreover, by definition 2.6:

$$\delta_u \equiv \Delta^0 - I_u\Delta \ \text{ is such that } \ \delta_u f = f(u). \qquad [5.42]$$

Let us move on to the A–G pseudopolynomials $\bar{G}_n$. Of course, we get from [2.24] that:

$$I_{u_0}\Delta\bar{G}_n = \bar{G}_n.$$

PROPOSITION 5.9.– *With the notation of [5.40]:*

$$\bar{G}_n = I_{u_0} E \bar{G}_{n-1}, \quad \textit{and} \quad \bar{G}_n = (I_{u_0} E)^n e_0. \qquad [5.43]$$

Proof. For the first result, we obtain from the definition [2.20]:

$$\begin{aligned}\bar{G}_n(\ |U) &= I_{u_0} I_{u_1} \dots I_{u_{n-1}} e_0 \\ &= I_{u_0} E(I_{u_0} \dots I_{u_{n-2}} e_0) = I_{u_0} E \bar{G}_{n-1}, \quad n \geq 1.\end{aligned}$$

For the next result, taking $n = 3$ for example, we have:

$$\begin{aligned}(I_{u_0} E)^3 e_0 &= (I_{u_0} E)(I_{u_0} E)(I_{u_0} E) e_0 = I_{u_0} I_{u_1} E^2 (I_{u_0} E) e_0 \\ &= I_{u_0} I_{u_1} I_{u_2} E^3 e_0 = I_{u_0} I_{u_1} I_{u_2} e_0 = \bar{G}_3,\end{aligned}$$

as stated in [5.43]. □

5.4.2. *Return to bivariate $\bar{G}_{n_1,n_2}$*

This time, $u = (u^{(1)}, u^{(2)}) \in \mathcal{D}_1 \times \mathcal{D}_2$, and Δ_j, $j = 1, 2$, generates an integration operator $I^{(j)}_{u^{(j)}}$ and an associated operator $\delta^{(j)}_{u^{(j)}}$ defined by [5.41], [5.42].

It is clear that:

$$\Delta_1 \Delta_2 = \Delta_2 \Delta_1, \quad \text{and} \quad \delta^{(1)}_{u^{(1)}} \delta^{(2)}_{u^{(2)}} = \delta^{(2)}_{u^{(2)}} \delta^{(1)}_{u^{(1)}}.$$

Consider now the case of integration operators.

PROPOSITION 5.10.– *$I^{(1)}_{u^{(1)}}$ and $I^{(2)}_{u^{(2)}}$ permute, as well as Δ_1 and $I^{(2)}_{u^{(2)}}$, or Δ_2 and $I^{(1)}_{u^{(1)}}$.*

Proof. From [5.41], we get:

$$I^{(1)}_{u^{(1)}} I^{(2)}_{u^{(2)}} e_{n_1,n_2}(x_1, x_2) = I^{(1)}_{u^{(1)}} \left[e_{n_1,n_2+1}(x_1, x_2) - e_{n_1,n_2+1}(x_1, u^{(2)})\right]$$

$$= e_{n_1+1,n_2+1}(x_1, x_2) - e_{n_1+1,n_2+1}(u^{(1)}, x_2)$$

$$- \left[e_{n_1+1,n_2+1}(x_1, u^{(2)}) - e_{n_1+1,n_2+1}(u^{(1)}, u^{(2)})\right],$$

which is the same as $I^{(2)}_{u^{(2)}} I^{(1)}_{u^{(1)}} e_{n_1,n_2}(x_1, x_2)$ by symmetry. On the other hand:

$$\begin{aligned}\Delta_1 I^{(2)}_{u^{(2)}} e_{n_1,n_2}(x_1, x_2) &= \Delta_1 \left[e_{n_1,n_2+1}(x_1, x_2) - e_{n_1,n_2+1}(x_1, u^{(2)})\right] \\ &= e_{n_1-1,n_2+1}(x_1, x_2) - e_{n_1-1,n_2+1}(x_1, u^{(2)}) \\ &= I^{(2)}_{u^{(2)}} e_{n_1-1,n_2}(x_1, x_2) = I^{(2)}_{u^{(2)}} \Delta_1 e_{n_1,n_2}(x_1, x_2),\end{aligned}$$

with a similar conclusion for $\Delta_2 I^{(1)}_{u^{(1)}}$. □

We note, however, that $\Delta_j I^{(j)}_{u^{(j)}} \neq I^{(j)}_{u^{(j)}} \Delta_j$. The A–G pseudopolynomials then have the following integral representation.

PROPOSITION 5.11.– *For* $(n_1, n_2) \in \mathbb{N}_0^2$*:*

$$\bar{G}_{n_1,n_2} = \left(I^{(1)}_{u^{(1)}_{0,0}} \Delta_1 + I^{(2)}_{u^{(2)}_{0,0}} \Delta_2 \, \delta^{(1)}_{u^{(1)}_{0,0}}\right) \bar{G}_{n_1,n_2}, \qquad [5.44]$$

or equivalently,

$$\bar{G}_{n_1,n_2} = I^{(1)}_{u^{(1)}_{0,0}} E^{1,0} \bar{G}_{n_1-1,n_2} + I^{(2)}_{u^{(2)}_{0,0}} \delta^{(1)}_{u^{(1)}_{0,0}} E^{0,1} \bar{G}_{n_1,n_2-1}. \qquad [5.45]$$

Proof. We can obviously express $\bar{G}_{n_1,n_2}$ as:

$$\bar{G}_{n_1,n_2} = [\bar{G}_{n_1,n_2} - \bar{G}_{n_1,n_2}(u^{(1)}_{0,0}, \,)] + [\bar{G}_{n_1,n_2}(u^{(1)}_{0,0}, \,) - \bar{G}_{n_1,n_2}(u^{(1)}_{0,0}, u^{(2)}_{0,0})].$$

Observe from [5.41] that the first bracket is $I^{(1)}_{u^{(1)}_{0,0}} \Delta_1 \bar{G}_{n_1,n_2}$, and the second is $I^{(2)}_{u^{(2)}_{0,0}} \Delta_2 \, \delta^{(1)}_{u^{(1)}_{0,0}} \bar{G}_{n_1,n_2}$. This gives the formula [5.44], and the equivalence [5.45] follows from [5.43]. □

The expression [5.45] obtained for $\bar{G}_{n_1,n_2}$ clearly provides us with a recursive way to calculate the A–G pseudopolynomials, starting from $G_{0,0} = 1$. Of course, swapping the roles played by Δ_1 and Δ_2 in [5.44] leads to an alternative expression:

$$\bar{G}_{n_1,n_2} = \left(I^{(1)}_{u^{(1)}_{0,0}} \Delta_1 \, \delta^{(2)}_{u^{(2)}_{0,0}} + I^{(2)}_{u^{(2)}_{0,0}} \Delta_2\right) \bar{G}_{n_1,n_2}, \quad (n_1, n_2) \in \mathbb{N}_0^2.$$

To obtain better symmetry in [5.44], we introduce a double integration operator $I_{u^{(1)},u^{(2)}}$.

DEFINITION 5.3.– *For* $(u^{(1)}, u^{(2)}) \in \mathcal{D}_1 \times \mathcal{D}_2$*:*

$$I_{u^{(1)},u^{(2)}} = I^{(1)}_{u^{(1)}} \, I^{(2)}_{u^{(2)}}, \qquad [5.46]$$

which implies:

$$\Delta_1 \, I_{u^{(1)},u^{(2)}} = I^{(2)}_{u^{(2)}}, \quad \text{and} \quad \Delta_2 \, I_{u^{(1)},u^{(2)}} = I^{(1)}_{u^{(1)}}. \qquad [5.47]$$

PROPOSITION 5.12.– *For* $(n_1, n_2) \in \mathbb{N}_0^2$*:*

$$\bar{G}_{n_1,n_2} = \left(\Delta_2 \, I_{u^{(1)}_{0,0},u^{(2)}_{0,0}} \, \Delta_1 + \Delta_1 \, I_{u^{(1)}_{0,0},u^{(2)}_{0,0}} \, \Delta_2 - I_{u^{(1)}_{0,0},u^{(2)}_{0,0}} \, \Delta_1 \, \Delta_2\right) \bar{G}_{n_1,n_2}, \qquad [5.48]$$

or equivalently,

$$\bar{G}_{n_1,n_2} = \Delta_2 \, I_{u^{(1)}_{0,0},u^{(2)}_{0,0}} \, E^{1,0} \bar{G}_{n_1-1,n_2} + \Delta_1 \, I_{u^{(1)}_{0,0},u^{(2)}_{0,0}} \times E^{0,1} \bar{G}_{n_1,n_2-1} - I_{u^{(1)}_{0,0},u^{(2)}_{0,0}} \, E^{1,1} \bar{G}_{n_1-1,n_2-1}. \qquad [5.49]$$

Proof. By writing explicitly in [5.44] $\delta^{(1)}_{u^{(1)}_{0,0}}$ given in [5.42], we see that:

$$\bar{G}_{n_1,n_2} = \left(I^{(1)}_{u^{(1)}_{0,0}} \, \Delta_1 + I^{(2)}_{u^{(2)}_{0,0}} \, \Delta_2 - I^{(1)}_{u^{(1)}_{0,0}} \, I^{(2)}_{u^{(2)}_{0,0}} \, \Delta_1 \Delta_2\right) \bar{G}_{n_1,n_2},$$

and using [5.46] and [5.47], we deduce [5.48] and [5.49]. □

5.5. Special case of A–G polynomials

Let $\mathcal{D} = \mathbb{R} \times \mathbb{R}$ and choose:

$$e_{n_1,n_2}(x_1, x_2) = \frac{x_1^{n_1}}{n_1!} \frac{x_2^{n_2}}{n_2!}, \quad (n_1, n_2) \in \mathbb{N}_0^2. \qquad [5.50]$$

By [5.1], this implies that:

Δ_j corresponds to the partial derivative operator $\partial/\partial x_j, \quad j = 1, 2.$ [5.51]

Moreover, we notice from (5.50) that the family $\mathcal{E}$ is standard at $(0, 0)$, that it satisfies the multiplicative property, and that the associated generating function is simply $\exp(x_1 s_1 + x_2 s_2)$.

In this particular case, the bivariate pseudopolynomials A–G become bivariate polynomials of type A–G, denoted $G_{n_1,n_2}(x_1, x_2|U^{(1)}, U^{(2)})$. Originally, multivariate A–G polynomials were introduced by Lefèvre and Picard (1990), with epidemic theory as motivation. These polynomials have also been studied independently by Kung and Yan (2003), notably for applications in combinatorics. The reader is referred to the review made in Lefèvre and Picard (2023).

Very briefly, $G_{0,0} = 1$ as usual, and for $n_1 + n_2 \geq 1$, the $G_{n_1,n_2}(x_1, x_2|U^{(1)}, U^{(2)})$ are derived recursively from the partial differential equations:

$$\begin{aligned}
&\frac{\partial}{\partial x_1} G_{n_1,n_2}(x_1, x_2|U^{(1)}, U^{(2)}) = G_{n_1-1,n_2}(x_1, x_2|E^{1,0}(U^{(1)}, U^{(2)})),\\
&\quad n_1 \geq 1, n_2 \geq 0,\\
&\frac{\partial}{\partial x_2} G_{n_1,n_2}(x_1, x_2|U^{(1)}, U^{(2)}) = G_{n_1,n_2-1}(x_1, x_2|E^{0,1}(U^{(1)}, U^{(2)})),\\
&\quad n_1 \geq 0, n_2 \geq 1,
\end{aligned} \qquad [5.52]$$

with the boundary conditions:

$$G_{n_1,n_2}(u_{0,0}^{(1)}, u_{0,0}^{(2)}|U^{(1)}, U^{(2)}) = 0. \qquad [5.53]$$

Obviously, when $n_1 = 0$ for example, $G_{0,n_2}(x_1, x_2|U^{(1)}, U^{(2)})$ reduces to the univariate polynomial $G_{n_2}(x_2|U^{(2)})$ (see [1.13] and [1.14]). In general, to determine the $G_{n_1,n_2}(x_1, x_2|U^{(1)}, U^{(2)})$, it will suffice to use the recursion relations [5.24] (with [5.50]), which is often simpler than [5.52] and [5.53].

First bivariate A–G polynomials.

$$\begin{aligned}
&G_{1,0}(x_1, x_2|U^{(1)}, U^{(2)}) = x_1 - u_{0,0}^{(1)},\\
&G_{0,1}(x_1, x_2|U^{(1)}, U^{(2)}) = x_2 - u_{0,0}^{(2)},\\
&G_{2,0}(x_1, x_2|U^{(1)}, U^{(2)}) = x_1^2/2 - u_{1,0}^{(1)} x_1 + u_{1,0}^{(1)} u_{0,0}^{(1)} - (u_{0,0}^{(1)})^2/2,\\
&G_{0,2}(x_1, x_2|U^{(1)}, U^{(2)}) = x_2^2/2 - u_{0,1}^{(2)} x_2 + u_{0,1}^{(2)} u_{0,0}^{(2)} - (u_{0,0}^{(2)})^2/2,
\end{aligned}$$

$$G_{1,1}(x_1, x_2|U^{(1)}, U^{(2)}) = x_1x_2 - u_{1,0}^{(2)}x_1 - u_{0,1}^{(1)}x_2 + u_{0,1}^{(1)}u_{0,0}^{(2)}$$

$$+ u_{0,0}^{(1)}u_{1,0}^{(2)} - u_{0,0}^{(1)}u_{0,0}^{(2)},$$

$$G_{2,1}(x_1, x_2|U^{(1)}, U^{(2)}) = x_1^2x_2/2 - u_{2,0}^{(2)}\, G_{2,0}(x_1, x_2|U^{(1)}, U^{(2)})$$

$$- u_{1,1}^{(1)}\, G_{1,1}(x_1, x_2|U^{(1)}, U^{(2)})$$

$$- u_{1,0}^{(1)}u_{1,0}^{(2)}\, G_{1,0}(x_1, x_2|U^{(1)}, U^{(2)})$$

$$- [(u_{0,1}^{(1)})^2/2]\, G_{0,1}(x_1, x_2|U^{(1)}, U^{(2)}) - (u_{0,0}^{(1)})^2u_{0,0}^{(2)}/2.$$

The other results of section 5.3 are easily rewritten for the current situation. In particular, the A–G polynomials are explicitly known in the affine case studied in proposition 5.8. Indeed, from [5.33], we obtain the formula [5.54].

PROPOSITION 5.13.– *When* $(U^{(1)}, U^{(2)}) = \{(u_{i_1,i_2}^{(1)}, u_{i_1,i_2}^{(2)}) \equiv (b_1^{(1)}i_1 + b_2^{(1)}i_2, b_1^{(2)}i_1 + b_2^{(2)}i_2), (i_1, i_2) \in \mathbb{N}_0^2\}$, *then:*

$$G_{n_1,n_2}\big(x_1, x_2|U^{(1)}, U^{(2)}\big) = \frac{(x_1 - u_{n_1,n_2}^{(1)})^{n_1-1}}{n_1!}\frac{(x_2 - u_{n_1,n_2}^{(2)})^{n_2-1}}{n_2!}$$

$$\big(x_1x_2 - x_1u_{n_1,0}^{(2)} - x_2u_{0,n_2}^{(1)}\big), \quad (n_1, n_2) \in \mathbb{N}_0^2, \qquad [5.54]$$

which correspond to the bivariate Abel polynomials (compare with [1.5]).

The integral representation of section 5.4 is also directly adapted. Indeed, from [5.41] and [5.51], we see that:

$$I_{u^{(1)}}^{(1)} f(x_1, x_2) \text{ corresponds to the integral } \int_{u^{(1)}}^{x_1} f(\xi_1, x_2)\, d\xi_1, \qquad [5.55]$$

and similarly for $I_{u^{(2)}}^{(2)} f(x_1, x_2)$. So, using [5.55], the formulas [5.44] and [5.49] give:

$$G_{n_1,n_2}\big(x_1, x_2|U^{(1)}, U^{(2)}\big) = \int_{u_{0,0}^{(1)}}^{x_1} G_{n_1-1,n_2}\big(\xi_1, x_2|E^{1,0}(U^{(1)}, U^{(2)})\big)\, d\xi_1$$

$$+ \int_{u_{0,0}^{(2)}}^{x_2} G_{n_1,n_2-1}\big(u_{0,0}^1, \xi_2|E^{0,1}(U^{(1)}, U^{(2)})\big)\, d\xi_2,$$

and

$$G_{n_1,n_2}\big(x_1,x_2|U^{(1)},U^{(2)}\big) = \int_{u_{0,0}^{(1)}}^{x_1} G_{n_1-1,n_2}\big(\xi_1,x_2|E^{1,0}(U^{(1)},U^{(2)})\big)\,d\xi_1$$

$$+\int_{u_{0,0}^{(2)}}^{x_2} G_{n_1,n_2-1}\big(x_1,\xi_2|E^{0,1}(U^{(1)},U^{(2)})\big)\,d\xi_2$$

$$-\int_{u_{0,0}^{(1)}}^{x_1}\int_{u_{0,0}^{(2)}}^{x_2} G_{n_1-1,n_2-1}\big(\xi_1,\xi_2|E^{1,1}(U^{(1)},U^{(2)})\big)\,d\xi_1 d\xi_2.$$

6

Randomizing A-G Pseudopolynomials

6.1. How to integrate stochasticity?

So far, all components incorporated in the A-G theory are fixed, that is, they take deterministic values. It is natural at this point, and useful for some applications, to introduce a stochastic version of the A–G pseudopolynomials $\bar{G}_n(x|U)$.

A simple way to proceed would be to substitute a random variable X for any element x of $\mathbb{R}$. However, such an approach is not really interesting. Indeed, according to [2.79], the Taylor-type expansion of $\bar{G}_n(x|U)$ around 0 would then give:

$$\bar{G}_n(X|U) = \sum_{i=0}^{n} \bar{G}_{n-i}(0|E^i U)\, e_i(X), \quad n \in \mathbb{N}_0, \qquad [6.1]$$

provided that the S.I. property is true and that $\mathcal{E}$ is standard at 0. We therefore observe that there is no direct link between X and the A–G structure.

A more fruitful approach is to randomize the family of parameters $U = \{u_i,\ i \in \mathbb{N}_0\}$. This will certainly cause additional difficulties but, as shown later, partial results can nevertheless be obtained. Moreover, these results will also prove to be of practical interest. In this section, we will briefly discuss the case where the u_i random variables are independent and equidistributed (i.i.d.). Then, we will examine in detail the cases where the u_i are either partial sums or partial products of i.i.d. random variables. Note that most of the results obtained will concern the mathematical expectation of the

A–G pseudopolynomials, with some possible transformations. To avoid confusion of notation, an expectation is denoted $\mathbb{E}$ (E remaining used for the shift operator).

The case where U is an i.i.d. sequence is reminiscent of the one where the u_i are constant. It leads us to generalize [3.68] as follows.

PROPOSITION 6.1.– *If U is a family of i.i.d. random variables X_i, $i \geq 0$, then the family of expectations $\{\mathbb{E}[\bar{G}_n(x|EU)],\ n \in \mathbb{N}_0\}$ is a copy of $\mathcal{E}$, and its generating function is given by:*

$$\sum_{n=0}^{\infty} \mathbb{E}[\bar{G}_n(x|U)]\, s^n = \frac{\sum_{n=0}^{\infty} e_n(x)\, s^n}{\sum_{n=0}^{\infty} \mathbb{E}[e_n(X_0)]\, s^n}. \qquad [6.2]$$

Proof. By the expansion (2.32):

$$e_n(x) = \sum_{i=0}^{n} e_{n-i}(u_i)\, \bar{G}_i(x|U),$$

where $\bar{G}_i(x|U)$ is independent of u_i from the definition [2.20]. Since the u_i are i.i.d. random variables, taking the expectation yields:

$$e_n(x) = \sum_{i=0}^{n} \mathbb{E}[e_{n-i}(u_i)]\, \mathbb{E}[\bar{G}_i(x|U)], \qquad [6.3]$$

with $\mathbb{E}[e_{n-i}(u_i)] = \mathbb{E}[e_{n-i}(X_0)]$. Therefore, $\mathcal{E}$ is a copy of the family $\{\mathbb{E}[\bar{G}_n(x|U)],\ n \in \mathbb{N}_0\}$, and simultaneously, this family is a copy of $\mathcal{E}$. Moreover, we directly deduce [6.2] from [6.3] in terms of generating functions. □

REMARK.– Note that the case where all the u_i are equal to the same random variable X is of no interest. Indeed, under S.I. and $\mathcal{E}$ standard at 0, [2.83] implies:

$$\bar{G}_n(x|\{u, u, \ldots\}) = e_n(x - X),$$

which is similar to the ruled out situation [6.1].

6.2. With u_i partial sums of i.i.d. variables

We consider here the case where the family U corresponds to the sequence of partial sums of i.i.d. random variables. More precisely:

$$u_i \equiv -S_i = -(X_1 + \ldots + X_i), \quad i \in \mathbb{N}, \quad \text{with } u_0 = 0 \equiv S_0, \quad [6.4]$$

the X_i, $i \geq 1$, being i.i.d. random variables.

REMARK.– The modeling [6.4] will be suitable for various applications. This choice is also natural with reference to the historical polynomial case. Indeed, as recalled in [1.5] and [1.10]:

$$G_n(x|\{-bi,\, i \in \mathbb{N}_0\}) = \frac{x(x+bn)^{n-1}}{n!} = A_n(x|bn),$$

where A_n are the classical Abel polynomials. This result was generalized in [3.83] to the A–G pseudopolynomials by:

$$\bar{G}_n(x|\{-bi,\, i \in \mathbb{N}_0\}) = \frac{x}{x+bn}\, e_n(x+bn) \equiv \bar{A}_n(x|bn), \text{ say}, \quad [6.5]$$

at least when the generating function of $\mathcal{E}$ is exponential. We will continue with the randomized A–G pseudopolynomials $\bar{G}_n(x|\{-S_i,\, i \in \mathbb{N}_0\})$.

Of course, the r.h.s. of [6.5] admits as simple randomization:

$$\bar{A}_n(x|S_n) \equiv \frac{x}{x+S_n}\, e_n(x+S_n). \quad [6.6]$$

These $\bar{G}_n$ and the associated $\bar{A}_n$ are clearly not identical, but both have the same expectation (see proposition 6.4). In the following, we will also discuss these randomized pseudopolynomials $\bar{A}_n(x|S_n)$.

Our first result shows how A–G pseudopolynomials randomized in this way can be evaluated, in expectation, from the family $\mathcal{E}$.

PROPOSITION 6.2.– *For the additive case [6.4], when $\mathcal{E}$ is generated by* $\exp[xg(s)]$, *then:*

$$\mathbb{E}[\bar{G}_r(x|U)] = \sum_{j=1}^{r} (j/r)\, e_j(x)\, \mathbb{E}[e_{r-j}(S_r)], \quad r \geq 1, \quad [6.7]$$

provided that $\mathbb{E}[|X_1|^k] < \infty$ *for* $1 \leq k \leq r-1$.

Proof. The result [6.7] is obvious for $r = 1$, so we consider in what follows $r \geq 2$. Furthermore, for reasons of clarity, we assume for the moment that the variable $|X_1|$ is bounded by M.

We start with the formula [3.65] when for $i \in \mathbb{N}_0$, $a_i = -S_i$ and $n_i = 1$, thus $b_0 = 0$ and for $j \geq 1$, $b_j = a_{j-1} - a_j = -S_{j-1} + S_j = X_j$. Taking the expectation gives:

$$\mathbb{E}[\bar{G}_r(x|U)] = \sum_{i_1,\ldots,i_{r-1} \in \mathbb{N}_0} e_{1+i_1}(x) \Big(\prod_{j=1}^{r-1} \mathbb{E}[e_{w_j}(X_j)] \Big), \quad r \geq 2, \qquad [6.8]$$

where $e_{w_j} = 0$ if $w_j < 0$, with $w_j = 1 + i_{j+1} - i_j$ for $1 \leq j \leq r-2$ and $w_{r-1} = 1 - i_{r-1}$. In a condense way, [6.8] can be written as:

$$\mathbb{E}[\bar{G}_r(x|U)] = \sum_{i_1 \in \mathbb{N}_0} e_{1+i_1}(x)\, B_{1+i_1}, \qquad [6.9]$$

after denoting

$$B_{1+i_1} = \sum_{i_2,\ldots,i_{r-1} \in \mathbb{N}_0} \Big(\prod_{j=1}^{r-1} \mathbb{E}[e_{w_j}(X_j)] \Big). \qquad [6.10]$$

We are going to show how the sums B_{1+i_1} can be evaluated outside of any random context. To this end, we first note that by assumption:

$$e^{X_1 g(s)} = \sum_{n=0}^{\infty} e_n(X_1) s^n, \qquad [6.11]$$

in which $g(0) = 0$ and $g'(0) \neq 0$. Let us therefore take the expectation of [6.11] and consider the corresponding formal series:

$$\mathbb{E}\big[e^{X_1 g(s)}\big] = \sum_{n=0}^{\infty} \mathbb{E}[e_n(X_1)]\, s^n \qquad [6.12]$$

$$= \sum_{k=0}^{\infty} \mathbb{E}(X^k)\, [g(s)]^k/k! \equiv 1 + \sum_{j=1}^{\infty} \gamma_j\, s^j, \text{ say.} \qquad [6.13]$$

There exists a power series $h(s) \equiv \sum_{j=1}^{\infty} \beta_j\, s^j$, such that:

$$\mathbb{E}\big[e^{X_1 g(s)}\big] = e^{h(s)}, \qquad [6.14]$$

where $h(0) = 0$ and $h'(0) \neq 0$. Indeed:

$$e^{h(s)} = 1 + \sum_{k=1}^{\infty} [h(s)]^k / k! = 1 + \sum_{k=1}^{\infty} (\beta_1 s + \beta_2 s^2 + \ldots)^k / k!, \quad [6.15]$$

so that from [6.13], [6.14] and [6.15], we get:

$$\gamma_1 = \beta_1, \ \ \gamma_2 = \beta_2 + \beta_1^2/2!, \ \ \gamma_3 = \beta_3 + 2\beta_1\beta_2/2! + \beta_1^3/3!, \ \ \ldots,$$

which allows us to derive the unknown β_j by recursion.

Having thus determined $h(s)$, we may introduce a new basic family $\mathcal{E}^{(h)} = \{e_n^{(h)}, n \in \mathbb{N}_0\}$ standard at 0 and obtained from the generating function $\exp[xh(s)]$. To $\mathcal{E}^{(h)}$ is attached an operator $\Delta_{(h)}$ defined as before through $\Delta_{(h)} e_{n+1}^{(h)} = e_n^{(h)}$. From [6.14], we then have:

$$\mathbb{E}\big[e^{X_1 g(s)}\big] = \sum_{n=0}^{\infty} e_n^{(h)}(1)\, s^n, \quad [6.16]$$

and comparing [6.16] with [6.12]:

$$e_n^{(h)}(1) = \mathbb{E}[e_n(X_1)] \quad [6.17]$$

so, due to [6.17], B_{1+i_1} given by [6.10] can be expressed as:

$$B_{1+i_1} = \sum_{i_2,\ldots,i_{r-1} \in \mathbb{N}_0} \Big(\prod_{j=1}^{r-1} e_{w_j}^{(h)}(1) \Big). \quad [6.18]$$

Let us now consider the particular family of fixed parameters $U^{(h)} = \{u_i^{(h)} \equiv -i, \ i \in \mathbb{N}_0\}$. The associated A–G pseudopolynomials $\bar{G}_r(x|\mathcal{E}^{(h)}, U^{(h)})$ are again provided by [6.9], which here simplifies as:

$$\bar{G}_r(x|\mathcal{E}^{(h)}, U^{(h)}) = \sum_{i_1 \in \mathbb{N}_0} e_{1+i_1}^{(h)}(x)\, B_{1+i_1}^{(h)} \quad [6.19]$$

moreover, from [6.10] we see that $B_{1+i_1}^{(h)}$ is given precisely by the r.h.s. of [6.18]. It is therefore equal to B_{1+i_1}. In other words, B_{1+i_1} can be determined deterministically as announced.

Let us return to the initial framework. From [6.19], we find that:

$$
\begin{aligned}
B_{1+i_1} = B^{(h)}_{1+i_1} &= \Delta^{1+i_1}_{(h)} \bar{G}_r(0|\mathcal{E}^{(h)}, U^{(h)}) \\
&= \bar{G}_{r-1-i_1}(0|\mathcal{E}^{(h)}, E^{1+i_1} U^{(h)}) \\
&= \bar{G}_{r-1-i_1}(1+i_1|\mathcal{E}^{(h)}, U^{(h)}),
\end{aligned}
$$

the last equality resulting from [2.72]. Since the $u_i^{(h)}$ are of affine form, [3.83] then implies:

$$
\begin{aligned}
B_{1+i_1} &= \frac{e_1^{(h)}(1+i_1)\, e_{r-1-i_1}^{(h)}(1+i_1+r-1-i_1)}{e_1^{(h)}(1+i_1+r-1-i_1)} \\
&= \frac{e_1^{(h)}(1+i_1)\, e_{r-1-i_1}^{(h)}(r)}{e_1^{(h)}(r)} \qquad [6.20]
\end{aligned}
$$

we continue with the $e_n^{(h)}(k)$, $n \geq 1$. When $k = 1$, we have [6.17]. When $k = 2$, the binomial identity [2.78] yields:

$$
\begin{aligned}
e_n^{(h)}(2) = e_n^{(h)}(1+1) &= \sum_{r=0}^{n} e_{n-r}^{(h)}(1)\, e_r^{(h)}(1) \\
&= \sum_{r=0}^{n} \mathbb{E}[e_{n-r}(X_1)]\, \mathbb{E}[e_r(X_2)] = \mathbb{E}[e_n(S_2)],
\end{aligned}
$$

by virtue of [6.17] and the i.i.d. assumption. More generally, we find that:

$$
e_n^{(h)}(k) = \mathbb{E}[e_n(S_k)], \quad k \geq 1 \qquad [6.21]
$$

by inserting [6.21] into [6.20], we deduce that:

$$
\begin{aligned}
B_{1+i_1} &= \frac{\mathbb{E}[e_1(S_{1+i_1})]}{\mathbb{E}[e_1(S_r)]}\, \mathbb{E}[e_{r-1-i_1}(S_r)] \\
&= [(1+i_1)/r]\, \mathbb{E}[e_{r-1-i_1}(S_r)], \qquad [6.22]
\end{aligned}
$$

after simplification by the factor $\mathbb{E}(X_1)$ in the fraction. The desired formula [6.5] follows from [6.9] with [6.22].

Finally, let us recall that by hypothesis, we have considered until now the variable $X_1\, 1_{|X_1| \leq M}$. However, we can easily make M tend toward ∞, hence the condition $\mathbb{E}[|X_1|^k] < \infty$, $1 \leq k \leq r-1$. Moreover, before [6.22], we implicitly assumed that X_1 has a non-zero mean, but again we can make $\mathbb{E}(X_1)$ tend toward 0 without problem. □

REMARK.– The building of $\mathcal{E}^{(h)}$ can be of interest outside the previous proof. In particular, consider again the i.i.d. case of Ssction 6.1, and suppose that, as in [6.12], [6.14] and [6.16]:

$$\sum_{n=0}^{\infty} \mathbb{E}[e_n(X_0)]\, s^n = \mathbb{E}\big[e^{X_0 g(s)}\big] \equiv e^{h(s)} = \sum_{n=0}^{\infty} e_n^{(h)}(1)\, s^n. \qquad [6.23]$$

Then, [6.23] implies that:

$$1/\big(\sum_{n=0}^{\infty} \mathbb{E}[e_n(X_0)]\, s^n\big) = e^{-h(s)} = \sum_{n=0}^{\infty} e_n^{(h)}(-1)\, s^n,$$

so that [6.2] becomes:

$$\sum_{n=0}^{\infty} \mathbb{E}[\bar{G}_n(x|U)]\, s^n = \big(\sum_{n=0}^{\infty} e_n(x)\, s^n\big)\big(\sum_{n=0}^{\infty} e_n^{(h)}(-1)\, s^n\big).$$

Therefore, we deduce the simple formula:

$$\mathbb{E}[\bar{G}_n(x|U)] = \sum_{i=0}^{n} e_{n-i}^{(h)}(-1)\, e_i(x). \qquad [6.24]$$

For example, if the X_i are i.i.d. with a normal law $\mathcal{N}(0, \sigma^2)$, then assuming $g(s)$ converges:

$$\mathbb{E}\big[e^{X_0 g(s)}\big] = e^{(\sigma^2/2)\,(g(s))^2} \equiv e^{h(s)}.$$

Thus, we have

$$\sum_{n=0}^{\infty} e_n^{(h)}(-1)\, s^n = e^{(-\sigma^2/2)\,(g(s))^2} = \sum_{k=0}^{\infty} \frac{(-\sigma^2/2)^k}{k!}\big(\sum_{i=2k}^{\infty} \alpha_i^{*2k}\, s^i\big),$$

which yields

$$e_n^{(h)}(-1) = \sum_{k=0}^{[n/2]} \frac{(-\sigma^2/2)^k}{k!}\, \alpha_n^{*2k}.$$

After insertion in [6.24], we obtain a completely explicit expression for $\mathbb{E}[\bar{G}_n(x|U)]$.

The formula [6.7] for $\mathbb{E}[\bar{G}_r(x|U)]$ can be extended to a more powerful result by introducing an exponential factor of S_r.

PROPOSITION 6.3.– *Under [6.4], when $\mathcal{E}$ is generated by* $\exp[xg(s)]$, *then:*

$$\mathbb{E}[e^{aS_r}\,\bar{G}_r(x|U)] = \sum_{j=1}^{r}(j/r)\,e_j(x)\,\mathbb{E}[e^{aS_r}\,e_{r-j}(S_r)], \quad r \geq 1, \quad [6.25]$$

for any constant a, provided that $\mathbb{E}[|e^{aX_1}X_1|^k] < \infty$ for $1 \leq k \leq r-1$.

Proof. Let us introduce in [6.8], before taking the expectation, the factor e^{aS_r} rewritten as:

$$e^{aS_r} = \mathbb{E}(e^{aS_r})\prod_{j=1}^{r}\frac{e^{aX_j}}{\mathbb{E}(e^{aX_j})}. \quad [6.26]$$

So, the terms $e_{w_j}(X_j)$ in [6.8] are replaced by:

$$e_{w_j}(X_j)\,\frac{e^{aX_j}}{\mathbb{E}(e^{aX_j})}, \quad [6.27]$$

and $h(s)$ is defined through the relation:

$$e^{h(s)} = \sum_{n=0}^{\infty}\mathbb{E}\big[e_n(X_1)\,\frac{e^{aX_1}}{\mathbb{E}(e^{aX_1})}\big]\,s^n. \quad [6.28]$$

By adapting the previous reasoning to [6.28], we find that:

$$e_n^{(h)}(1) = \mathbb{E}\big[e_n(X_1)\,\frac{e^{aX_1}}{\mathbb{E}(e^{aX_1})}\big],$$

thus for $e_n^{(h)}(2)$:

$$\begin{aligned} e_n^{(h)}(2) &= \sum_{r=0}^{n} e_{n-r}^{(h)}(1)\,e_r^{(h)}(1) \\ &= \sum_{r=0}^{n}\mathbb{E}\big[e_{n-r}(X_1)\,\frac{e^{aX_1}}{\mathbb{E}(e^{aX_1})}\big]\,\mathbb{E}\big[e_r(X_2)\,\frac{e^{aX_2}}{\mathbb{E}(e^{aX_2})}\big] \\ &= \mathbb{E}\big[e_n(S_2)\,\frac{e^{aS_2}}{\mathbb{E}(e^{aS_2})}\big], \end{aligned}$$

and more generally:

$$e_n^{(h)}(k) = \mathbb{E}\big[e_n(S_k)\,\frac{e^{aS_k}}{\mathbb{E}(e^{aS_k})}\big], \quad k \geq 1.$$

This implies that the new B_{1+i_1} (i.e. adapted according to [6.27]) is given by:

$$B_{1+i_1} = [(1+i_1)/r]\,\frac{\mathbb{E}[e_{r-1-i_1}(S_r)\,e^{aS_r}]}{\mathbb{E}(e^{aS_r})}.$$

From [6.8] modified by the factor [6.26], we then deduce [6.25] as desired. □

Consider again the expectations evaluated in [6.25], that is,

$$\mathbb{E}[e^{aS_n}\,\bar{G}_n(x|U)], \text{ denoted } H_n(x), \quad n \in \mathbb{N}_0,$$

and in parallel, the similar expectations defined from [6.6], that is,

$$\mathbb{E}[e^{aS_n}\,\bar{A}_n(x|S_n)], \text{ denoted } L_n(x), \quad n \in \mathbb{N}_0.$$

We will start by showing how the operator Δ affects these two types of expectations. Obviously, $\Delta H_0 = \Delta L_0 = 0$, so let $n \geq 1$.

PROPERTY 6.1.– *Under [6.4], when the generating function of $\mathcal{E}$ is $\exp(xg(s))$, then:*

$$\Delta H_n(x) = \mathbb{E}[e^{aX_1}\,H_{n-1}(x+X_1)], \qquad [6.29]$$

$$\Delta L_n(x) = \mathbb{E}[e^{aX_1}\,L_{n-1}(x+X_1)], \quad n \geq 1. \qquad [6.30]$$

Proof. On the one hand, applying Δ to $H_n(x)$ gives:

$$\Delta H_n(x) = \mathbb{E}[e^{aS_n}\,\bar{G}_{n-1}(x|U)] = \mathbb{E}[e^{a(X_1+\tilde{S}_{n-1})}\,\bar{G}_{n-1}(x|EU)],$$

where $\tilde{S}_{n-1} \equiv S_n - X_1$ is distributed as S_{n-1}. Denoting $\tilde{U} = \{-\tilde{S}_i,\, i \in \mathbb{N}_0\}$, we then obtain:

$$\begin{aligned}\Delta H_n(x) &= \mathbb{E}\big[e^{a(X_1+\tilde{S}_{n-1})}\,\bar{G}_{n-1}(x+X_1|\tilde{U})\big]\\ &= \mathbb{E}\big[e^{aX_1}\,\mathbb{E}_{X_1}\{e^{a\tilde{S}_{n-1}}\,\bar{G}_{n-1}(x+X_1|\tilde{U})\}\big], \qquad [6.31]\end{aligned}$$

where $\mathbb{E}_{X_1}$ is the expectation conditionally on X_1. Formula [6.31] then yields [6.29].

On the other hand, $\Delta L_n(x)$ is, by definition, equal to

$$\Delta L_n(x) = \mathbb{E}\big[e^{aS_n}\,\Delta \bar{A}_n(x|S_n)\big] = \mathbb{E}\big[e^{aS_n}\,\Delta\{\frac{x}{x+S_n}\,e_n(x+S_n)\}\big].$$

To avoid any confusion with S_n, we will denote by $\mathcal{S}^a$ the shift operator S^a, thus:

$$\Delta L_n(x) = \mathbb{E}\big[e^{aS_n}\,\Delta\,\mathcal{S}^{S_n}\{(1-\frac{S_n}{x})\,e_n(x)\}\big]. \quad [6.32]$$

By S.I., Δ and $\mathcal{S}$ commute. Moreover, the generating function of $\mathcal{E}$ being $\exp(xg(s))$, we can use [3.15] to transform [6.32] into:

$$\begin{aligned}\Delta L_n(x) &= \mathbb{E}\big[e^{aS_n}\,\mathcal{S}^{S_n}\Delta\,\{e_n(x) - \frac{S_n}{n}\frac{ne_n(x)}{x}\}\big] \\ &= \mathbb{E}\big[e^{aS_n}\,\mathcal{S}^{S_n}\{e_{n-1}(x) - \frac{S_n}{n}\frac{(n-1)e_{n-1}(x)}{x}\}\big] \\ &= \mathbb{E}\big[e^{aS_n}\,\mathcal{S}^{S_n}\{(1-\frac{S_n}{x}\frac{n-1}{n})\,e_{n-1}(x)\}\big]. \quad [6.33]\end{aligned}$$

Now, let us decompose the term $(\ldots)$ in [6.33] into the following sum of two terms:

$$1 - \frac{S_n}{x}\frac{n-1}{n} = (1 - \frac{S_n - X_1}{x}) + (\frac{S_n}{n} - X_1)\,\frac{1}{x},$$

and consider the two corresponding terms in the r.h.s. of [6.33]. The first is:

$$\begin{aligned}&\mathbb{E}\big[e^{aS_n}\,\mathcal{S}^{S_n}\,\{(1-\frac{S_n - X_1}{x})\,e_{n-1}(x)\}\big] \\ &= \mathbb{E}\big[e^{aX_1}\,\mathbb{E}_{X_1}\{e^{a\tilde{S}_{n-1}}\,\frac{x+X_1}{x+X_1+\tilde{S}_{n-1}}\,e_{n-1}(x+X_1+\tilde{S}_{n-1})\}\big], \quad [6.34]\end{aligned}$$

and [6.34] is exactly the r.h.s. of [6.30]. The second is:

$$\begin{aligned}&\mathbb{E}\big[e^{aS_n}\,\mathcal{S}^{S_n}\,\{(\frac{S_n}{n} - X_1)\,\frac{e_{n-1}(x)}{x}\}\big] \\ &= \mathbb{E}\big[e^{aS_n}\,\mathcal{S}^{S_n}\,\{\mathbb{E}_{S_n}(\frac{S_n}{n} - X_1)\,\frac{e_{n-1}(x)}{x}\}\big], \quad [6.35]\end{aligned}$$

after taking the conditional expectation on S_n. Due to exchangeability of the X_i, we have:

$$\mathbb{E}_{S_n}\big(\frac{S_n}{n} - X_1\big) = 0,$$

so that [6.35] reduces to zero. From [6.33], we then obtain [6.30]. □

Consequently, we will deduce that $\bar{G}_n(x|U)$ and $\bar{A}_n(x|S_n)$ have the same expectation.

PROPOSITION 6.4.– *Under the hypotheses of property 6.1:*

$$\mathbb{E}[e^{aS_n}\,\bar{G}_n(x|U)] = \mathbb{E}[e^{aS_n}\,\bar{A}_n(x|S_n)], \quad n \in \mathbb{N}_0. \qquad [6.36]$$

Proof. From [6.29] and [6.30], we get for $n \geq 1$:

$$\Delta(H_n - L_n)(x) = \mathbb{E}[e^{aX_1}\,(H_{n-1} - L_{n-1})(x + X_1)], \qquad [6.37]$$

with $(H_n - L_n)(0) = 0$ from the definition. Furthermore, for $n = 0$, we have $H_0 \quad L_0 = 0$.

To prove [6.36], we will reason by recurrence. Suppose $H_{n-1} = L_{n-1}$ for some $n \geq 1$. From [6.37], we deduce that $H_n - L_n$ is a constant. Since the value of $H_n - L_n$ is 0 at point 0, this constant is necessarily zero, that is, $H_n = L_n$. □

We underline that when $a = 0$, [6.36] gives, as announced:

$$\mathbb{E}[\bar{G}_n(x|U)] = \mathbb{E}[\bar{A}_n(x|S_n)].$$

Moreover, arguing with Laplace transform (if applicable) from [6.36], we easily obtain:

$$\mathbb{E}_{S_n}[\bar{G}_n(x|U)] = \bar{A}_n(x|S_n). \qquad [6.38]$$

Now, let f be a pseudopolynomial in $\mathcal{F}$ (of arbitrary degree).

COROLLARY 6.1.– *Under the hypotheses of property 6.1:*

$$\mathbb{E}[e^{aS_n} f(S_n)\, \bar{G}_n(x|U)] = \mathbb{E}[e^{aS_n} f(S_n)\, \bar{A}_n(x|S_n)], \quad n \in \mathbb{N}_0, \qquad [6.39]$$

provided that $\mathbb{E}[|e^{aX_1} X_1|^k]) < \infty$ *for* $1 \leq k \leq n$.

Proof. Differentiating [6.36] with respect to a, we see that [6.39] is true when $f(S_n) = (S_n)^r$, $r \geq 0$. By summation, the result also holds if $f(x)$ is any polynomial in x, in particular for $f(x) = e_n(x)$. Therefore, it remains valid for all pseudopolynomials of $\mathcal{F}$. □

In addition, any pseudopolynomial f_n admits an Abel-type expansion on expectation.

COROLLARY 6.2.– *Under the hypotheses of property 6.1:*

$$f_n(x) = \mathbb{E}[\sum_{i=0}^{n} \Delta^i f_n(-S_i)\, \bar{A}_i(x|S_i)], \quad n \in \mathbb{N}_0, \qquad [6.40]$$

provided that $\mathbb{E}[|X_1|^k] < \infty$ *for* $1 \leq k \leq n$.

Proof. By [2.29], the A–G expansion of $f_n(x)$ with respect to the $\bar{G}_n(x|U)$ is:

$$f_n(x) = \sum_{i=0}^{n} \Delta^i f_n(-S_i)\, \bar{G}_i(x|U). \qquad [6.41]$$

Let us take the expectation of [6.41]. Obviously, $\mathbb{E}[f_n(x)] = f_n(x)$. Moreover, [6.39] with $a = 0$ implies, for $f(S_i) = \Delta^i f_n(-S_i)$:

$$\mathbb{E}[\Delta^i f_n(-S_i)\, \bar{G}_i(x|U)] = \mathbb{E}[\Delta^i f_n(-S_i)\, \bar{A}_i(x|S_i)],$$

hence the formula [6.40]. □

REMARK.– A family linked to $U = \{-S_i, i \in \mathbb{N}_0\}$ is the finite family $\{S_j, S_{j-1}, S_{j-2}, \ldots, S_{i+1}\}$ where $j \geq i+1$, $i \geq 0$. In fact, we observe that:

$$S_j = S_j - 0, \quad S_{j-1} = S_j - X_j, \quad S_{j-2} = S_j - X_j - X_{j-1},$$
$$\ldots, S_{i+1} = S_j - X_j - X_{j-1} - \ldots - X_{i+2}. \qquad [6.42]$$

So, consider two functions f and g of x (that may also depend on S_i and S_j). We will show that under the previous hypotheses:

$$\mathbb{E}\big[f(x)\,\bar{G}_{j-i}\big(g(x)|\{S_j,S_{j-1},\ldots,S_{i+1}\}\big)\big]$$
$$=\mathbb{E}\Big[f(x)\,\frac{g(x)-S_j}{g(x)-S_i}\,e_{j-i}\big(g(x)-S_i\big)\Big], \qquad [6.43]$$

which implies, when $f(x)=1$, $g(x)=x$ and $i=0$, that:

$$\mathbb{E}\big[\bar{G}_j(x|\{S_j,S_{j-1},\ldots,S_1\})\big]=\frac{x-\mathbb{E}(S_j)}{x}\,e_j(x).$$

Indeed, using [6.42] and [2.72], we can rewrite the l.h.s. of [6.43] as:

$$\mathbb{E}\big[f(x)\,\bar{G}_{j-i}\big(g(x)-S_j|\{0,-X_j,-X_j-X_{j-1},\ldots,$$
$$-X_j-X_{j-1}-\ldots-X_{i+2}\}\big)\big].$$

Take in front of $\bar{G}_{j-i}(\ldots)$ the conditional expectation $\mathbb{E}_{S_j,S_i}$. From [6.38], we then obtain the r.h.s. of [6.43] after a few simple adjustments.

6.3. Multivariate additive extension

We continue by similarly randomizing the bivariate A–G pseudopolynomials $\bar{G}_{n_1,n_2}(x_1,x_2|U^{(1)},U^{(2)})$ discussed in section 5.3. To do this, we first introduce two independent sequences of i.i.d. bidimensional random vectors:

$$\{\mathbf{X}_i^{(1)}=(X_i^{(1,1)},X_i^{(1,2)}),\ i\in\mathbb{N}\},\quad \text{each being distributed as}$$
$$\mathbf{X}_1^{(1)}=(X_1^{(1,1)},X_1^{(1,2)}),$$
$$\{\mathbf{X}_j^{(2)}=(X_j^{(2,1)},X_j^{(2,2)}),\ j\in\mathbb{N}\},\quad \text{each being distributed as}$$
$$\mathbf{X}_1^{(2)}=(X_1^{(2,1)},X_1^{(2,2)}).$$

No hypothesis is made on the dependence between the two random variables composing the vectors $\mathbf{X}_i^{(1)}$ and $\mathbf{X}_j^{(2)}$. Then, we consider the associated partial sums:

$$\{\mathbf{S}_{i_1}^{(1)} = \sum_{i=1}^{i_1} \mathbf{X}_i^{(1)} = \sum_{i=1}^{i_1} (X_i^{(1,1)}, X_i^{(1,2)}) \equiv (S_{i_1}^{(1,1)}, S_{i_1}^{(1,2)}),\ i_1 \in \mathbb{N}_0\},$$

with $S_0^{(1,1)} = S_0^{(1,2)} = 0$,

$$\{\mathbf{S}_{i_2}^{(2)} = \sum_{j=1}^{i_2} \mathbf{X}_j^{(2)} = \sum_{j=1}^{i_2} (X_j^{(2,1)}, X_j^{(2,2)}) \equiv (S_{i_2}^{(2,1)}, S_{i_2}^{(2,2)}),\ i_2 \in \mathbb{N}_0\},$$

with $S_0^{(2,1)} = S_0^{(2,2)} = 0$.

Finally, we define the family of random parameters $U = (U^{(1)}, U^{(2)})$ in the additive case by:

$$\begin{cases} U^{(1)} = \{u_{i_1,i_2}^{(1)} \equiv -S_{i_1}^{(1,1)} - S_{i_2}^{(2,1)},\ (i_1, i_2) \in \mathbb{N}_0^2\}, \\ U^{(2)} = \{u_{i_1,i_2}^{(2)} \equiv -S_{i_1}^{(1,2)} - S_{i_2}^{(2,2)},\ (i_1, i_2) \in \mathbb{N}_0^2\}. \end{cases} \qquad [6.44]$$

We note that $U^{(1)}$ and $U^{(2)}$ can be stochastically dependent, but within each of them, the summed variables are independent (with two possible distributions).

For $\mathcal{D} = \mathbb{R}^2$, consider the A–G pseudopolynomials $\bar{G}_{n_1,n_2}(x_1, x_2|U^{(1)}, U^{(2)})$, $(n_1, n_2) \in \mathbb{N}_0^2$, randomized according to the additive case [6.44]. In parallel, we generalize the random polynomials $\bar{A}_n(x|S_n)$ in [6.6] to the bivariate case by:

$$\bar{A}_{n_1,n_2}(x_1, x_2|\mathbf{S}_{n_1}^{(1)}, \mathbf{S}_{n_2}^{(2)}) = \begin{vmatrix} e_{1,0}(x_1 + S_{n_2}^{(2,1)}, x_2 + S_{n_2}^{(2,2)}) & e_{0,1}(-S_{n_1}^{(1,1)}, -S_{n_1}^{(1,2)}) \\ e_{1,0}(-S_{n_2}^{(2,1)}, -S_{n_2}^{(2,2)}) & e_{0,1}(x_1 + S_{n_1}^{(1,1)}, x_2 + S_{n_1}^{(1,2)}) \end{vmatrix}$$

$$\frac{e_{n_1,n_2}(x_1 + S_{n_1}^{(1,1)} + S_{n_2}^{(2,1)}, x_2 + S_{n_1}^{(1,2)} + S_{n_2}^{(2,2)})}{e_{1,1}(x_1 + S_{n_1}^{(1,1)} + S_{n_2}^{(2,1)}, x_2 + S_{n_1}^{(1,2)} + S_{n_2}^{(2,2)})}, \quad (n_1, n_2) \in \mathbb{N}_0^2. \qquad [6.45]$$

As in [5.37], we can also use the shift operator $\mathcal{S}^{\mathbf{S}_{n_1}^{(1)} + \mathbf{S}_{n_2}^{(2)}} = \mathcal{S}_1^{S_{n_1}^{(1,1)} + S_{n_1}^{(2,1)}} \mathcal{S}_2^{S_{n_2}^{(1,2)} + S_{n_2}^{(2,2)}}$. The r.h.s. of [6.45] is then reexpressed as:

$$\mathcal{S}^{\mathbf{S}_{n_1}^{(1)} + \mathbf{S}_{n_2}^{(2)}} \left\{ \begin{vmatrix} e_{1,0}(x_1 - S_{n_1}^{(1,1)}, x_2 - S_{n_1}^{(1,2)}) & e_{0,1}(-S_{n_1}^{(1,1)}, -S_{n_1}^{(1,2)}) \\ e_{1,0}(-S_{n_2}^{(2,1)}, -S_{n_2}^{(2,2)}) & e_{0,1}(x_1 - S_{n_2}^{(2,1)}, x_2 - S_{n_2}^{(2,2)}) \end{vmatrix} \frac{e_{n_1,n_2}(x_1, x_2)}{e_{1,1}(x_1, x_2)} \right\}.$$

[6.46]

We will now extend property 6.1 to the bivariate case. To do this, we first introduce an exponential factor given by:

$$e^{<\mathbf{a}^{(1)},\mathbf{S}_{n_1}^{(1)}>+<\mathbf{a}^{(2)},\mathbf{S}_{n_2}^{(2)}>}, \qquad [6.47]$$

where $\mathbf{a}^{(1)} = (a^{(1,1)}, a^{(1,2)})$ and $\mathbf{a}^{(2)} = (a^{(2,1)}, a^{(2,2)})$ are in $\mathbb{R}^2$, while $< .,. >$ denotes the canonical scalar product on $\mathbb{R}^2$. So, [6.47] is explicitly:

$$e^{[a^{(1,1)}S_{n_1}^{(1,1)}+a^{(1,2)}S_{n_1}^{(1,2)}]+[a^{(2,1)}S_{n_2}^{(2,1)}+a^{(2,2)}S_{n_2}^{(2,2)}]}.$$

Then, we consider the expectations of $\bar{G}_{n_1,n_2}$ and $\bar{A}_{n_1,n_2}$ multiplied by this factor, that is:

$$H_{n_1,n_2}(x_1,x_2) = \mathbb{E}\big[e^{<\mathbf{a}^{(1)},\mathbf{S}_{n_1}^{(1)}>+<\mathbf{a}^{(2)},\mathbf{S}_{n_2}^{(2)}>}\,\bar{G}_{n_1,n_2}(x_1,x_2|U^{(1)},U^{(2)})\big],$$

$$L_{n_1,n_2}(x_1,x_2) = \mathbb{E}\big[e^{<\mathbf{a}^{(1)},\mathbf{S}_{n_1}^{(1)}>+<\mathbf{a}^{(2)},\mathbf{S}_{n_2}^{(2)}>}\,\bar{A}_{n_1,n_2}(x_1,x_2|\mathbf{S}_{n_1}^{(1)},\mathbf{S}_{n_2}^{(2)})\big].$$

PROPERTY 6.2.– *Under [6.44], when the generating function of $\mathcal{E}$ is $\exp[x_1 g(s_1) + x_2 g(s_2)]$:*

$$\begin{cases}\Delta_1 H_{n_1,n_2}(x_1,x_2) = \mathbb{E}\big[e^{a^{(1,1)}X_1^{(1,1)}+a^{(1,2)}X_1^{(1,2)}}\,H_{n_1-1,n_2}(x_1+X_1^{(1,1)},x_2+X_1^{(1,2)})\big],\\ \Delta_2 H_{n_1,n_2}(x_1,x_2) = \mathbb{E}\big[e^{a^{(2,1)}X_1^{(2,1)}+a^{(2,2)}X_1^{(2,2)}}\,H_{n_1,n_2-1}(x_1+X_1^{(2,1)},x_2+X_1^{(2,2)})\big].\end{cases} \qquad [6.48]$$

Furthermore, identical identities hold for $\Delta_1 L_{n_1,n_2}(x_1,x_2)$ and $\Delta_2 L_{n_1,n_2}(x_1,x_2)$.

Proof. We only deal with the Δ_1 operator, the case of Δ_2 being similar. For the expectation H_{n_1,n_2}, the application of Δ_1 gives:

$$\Delta_1 H_{n_1,n_2}(x_1,x_2) = \mathbb{E}\big[e^{<\mathbf{a}^{(1)},\mathbf{S}_{n_1}^{(1)}>+<\mathbf{a}^{(2)},\mathbf{S}_{n_2}^{(2)}>}$$
$$\times\,\bar{G}_{n_1-1,n_2}(x_1,x_2|E^{1,0}(U^{(1)},U^{(2)}))\big], \qquad [6.49]$$

in which by [6.44], $\bar{G}_{n_1-1,n_2}(x_1,x_2|E^{1,0}(U^{(1)},U^{(2)}))$ is equal to:

$$\bar{G}_{n_1-1,n_2}(x_1,x_2|\{-S_{1+i_1}^{(1,1)}-S_{i_2}^{(2,1)}\},\{-S_{1+i_1}^{(1,2)}-S_{i_2}^{(2,2)}\})$$
$$= \bar{G}_{n_1-1,n_2}(x_1+X_1^{(1,1)},x_2+X_1^{(1,2)}|\{-(S_{1+i_1}^{(1,1)}-X_1^{(1,1)})-S_{i_2}^{(2,1)}\},$$
$$\{-(S_{1+i_1}^{(1,2)}-X_1^{(1,2)})-S_{i_2}^{(2,2)}\}). \qquad [6.50]$$

Note that $\tilde{S}_{i_1}^{(1,1)} \equiv (S_{1+i_1}^{(1,1)} - X_1^{(1,1)})$ and $\tilde{S}_{i_1}^{(1,2)} \equiv (S_{1+i_1}^{(1,2)} - X_1^{(1,2)})$ are independent and distributed as $S_{i_1}^{(1,1)}$ and $S_{i_1}^{(1,2)}$, respectively. Moreover, the exponential [6.47] can be decomposed into:

$$e^{<\mathbf{a}^{(1)}, \mathbf{X}_1^{(1)}> + <\mathbf{a}^{(1)}, \tilde{\mathbf{S}}_{n_1-1}^{(1)}> + <\mathbf{a}^{(2)}, \mathbf{S}_{n_2}^{(2)}>}. \qquad [6.51]$$

Inserting [6.50] and [6.51] in [6.49], we obtain:

$$\Delta_1 H_{n_1,n_2}(x_1, x_2) = \mathbb{E}\big[e^{<\mathbf{a}^{(1)}, \mathbf{X}_1^{(1)}>} \, \mathbb{E}_{\mathbf{X}_1^{(1)}} \{e^{<\mathbf{a}^{(1)}, \tilde{\mathbf{S}}_{n_1-1}^{(1)}> + <\mathbf{a}^{(2)}, \mathbf{S}_{n_2}^{(2)}>}$$

$$\bar{G}_{n_1-1,n_2}(x_1 + X_1^{(1,1)}, x_2 + X_1^{(1,2)} | \{-\tilde{S}_{i_1}^{(1,1)} - S_{i_2}^{(2,1)}\},$$

$$\{-\tilde{S}_{i_1}^{(1,2)} - S_{i_2}^{(2,2)}\})\}\big],$$

which provides the result [6.48].

Let us move on to the expectation L_{n_1,n_2}. We will imitate the proof of proposition 5.8 and for this use the expression [6.45] for $\bar{A}_{n_1,n_2}$. To apply Δ_1, we observe that in the determinant of [6.46], x_1, n_1 (respectively, x_2, n_2) only occur in the first (respectively, second) line. Since $e_{1,0}(x_1, \diamond)$ and $e_{0,1}(\diamond, x_2)$ are linear, and using [5.20], the shifted term can be rewritten as $r_{n_1}^{(1)}(x_1)$ with:

$$r_{n_1}^{(1)}(x_1) = \begin{vmatrix} \alpha_{1,1}(x_1 - S_{n_1}^{(1,1)}) & -\alpha_{2,1} S_{n_1}^{(1,2)} \\ c_1 & c_2 \end{vmatrix} c_3 \, \frac{e_{n_1,0}(x_1, \diamond)}{e_{1,0}(x_1, \diamond)}, \qquad [6.52]$$

where c_1, c_2, c_3 denote constants with respect to Δ_1. As with [5.38], $r_{n_1}^{(1)}(x_1)$ reduces to:

$$r_{n_1}^{(1)}(x_1) = \big[c_2\alpha_{1,1}(x_1 - S_{n_1}^{(1,1)}) + c_1\alpha_{2,1}S_{n_1}^{(1,2)}\big] \, c_3 \, \frac{e_{n_1,0}(x_1, \diamond)}{e_{1,0}(x_1, \diamond)}$$

$$= c_4\Big[1 - \frac{S_{n_1}^{(1,1)} - c_5 S_{n_1}^{(1,2)}}{x_1}\Big] \, e_{n_1,0}(x_1, \diamond).$$

In a condensed way, this is equivalent to:

$$r_{n_1}^{(1)}(x_1) = c_4\big(1 - \frac{W_{n_1}}{x_1}\big) \, e_{n_1,0}(x_1, \diamond), \qquad [6.53]$$

where W_{n_1} is the sum of n_1 i.i.d. random variables, denoted Y_j say. Clearly, [6.53] has exactly the structure of the term $\{\ldots\}$ in [6.32]. Therefore, applying

Δ_1, we obtain as in [6.33]:

$$\Delta_1 r_{n_1}^{(1)}(x_1) = c_4\big(1 - \frac{W_{n_1}}{x_1}\frac{n_1-1}{n_1}\big)\, e_{n_1-1,0}(x_1,\diamond). \qquad [6.54]$$

Let us decompose the term $(\dots)$ in [6.54] into the sum of two terms as before. The second will again play no role, and for the first, we obtain:

$$\big(1 - \frac{W_{n_1}-Y_1}{x_1}\big)\, e_{n_1-1,0}(x_1,\diamond),$$

which gives, after doing the previous calculation in reverse:

$$\begin{vmatrix} e_{1,0}(x_1 - S_{n_1}^{(1,1)} + X_1^{(1,1)}, x_2 - S_{n_1}^{(1,2)} + X_1^{(1,2)}) & e_{0,1}(-S_{n_1}^{(1,1)} + X_1^{(1,1)}, -S_{n_1}^{(1,2)} + X_1^{(1,2)}) \\ e_{1,0}(-S_{n_2}^{(2,1)}, -S_{n_2}^{(2,2)}) & e_{0,1}(x_1 - S_{n_2}^{(2,1)}, x_2 - S_{n_2}^{(2,2)}) \end{vmatrix}$$

$$\frac{e_{n_1-1,n_2}(x_1,x_2)}{e_{1,1}(x_1,x_2)}. \qquad [6.55]$$

We are now ready to consider $\Delta_1 L_{n_1,n_2}$. For the exponential [6.47], we use the decomposition [6.51]. For the shift operator, we write it as:

$$\begin{aligned} \mathcal{S}^{\mathbf{S}_{n_1}^{(1)}+\mathbf{S}_{n_2}^{(2)}} &= \mathcal{S}^{\mathbf{S}_{n_1}^{(1)}-\mathbf{X}_1^{(1)}+\mathbf{S}_{n_2}^{(2)}}\,\mathcal{S}^{\mathbf{X}_1^{(1)}} \\ &= \mathcal{S}^{\tilde{\mathbf{S}}_{n_1-1}+\mathbf{S}_{n_2}^{(2)}}\,\mathcal{S}^{\mathbf{X}_1^{(1)}}. \end{aligned} \qquad [6.56]$$

The application of $\mathcal{S}^{\mathbf{X}_1^{(1)}}$ to [6.55] yields:

$$\begin{vmatrix} e_{1,0}(x_1 + X_1^{(1,1)} - \tilde{S}_{n_1-1}^{(1,1)}, x_2 + X_1^{(1,2)} - \tilde{S}_{n_1-1}^{(1,2)}) & e_{0,1}(-\tilde{S}_{n_1-1}^{(1,1)}, -\tilde{S}_{n_1-1}^{(1,2)}) \\ e_{1,0}(-S_{n_2}^{(2,1)}, -S_{n_2}^{(2,2)}) & e_{0,1}(x_1 + X_1^{(1,1)} - S_{n_2}^{(2,1)}, x_2 + X_1^{(1,2)} - S_{n_2}^{(2,2)}) \end{vmatrix}$$

$$\frac{e_{n_1-1,n_2}(x_1 + X_1^{(1,1)}, x_2 + X_1^{(1,2)})}{e_{1,1}(x_1 + X_1^{(1,1)}, x_2 + X_1^{(1,2)})}, \qquad [6.57]$$

a quantity that we denote $l(x_1 + X_1^{(1,1)}, x_2 + X_1^{(1,2)})$ to simplify. Thus, using [6.56] and [6.57], we obtain:

$$\begin{aligned} \Delta_1 L_{n_1,n_2}(x_1,x_2) &= \mathbb{E}\big[e^{<\mathbf{a}^{(1)},\mathbf{X}_1^{(1)}>}\,\mathbb{E}_{\mathbf{X}_1^{(1)}}\{e^{<\mathbf{a}^{(1)},\tilde{\mathbf{S}}_{n_1-1}^{(1)}>+<\mathbf{a}^{(2)},\mathbf{S}_{n_2}^{(2)}>} \\ &\quad l(x_1 + X_1^{(1,1)}, x_2 + X_1^{(1,2)})\}\big] \\ &= \mathbb{E}\big[e^{a^{(1,1)}X_1^{(1,1)}+a^{(1,2)}X_1^{(1,2)}}\,L_{n_1-1,n_2}(x_1 + X_1^{(1,1)}, x_2 + X_1^{(1,2)})\big], \end{aligned}$$

which is equivalent to [6.48] as announced. □

Thanks to this result, we are able to extend proposition 6.4 as follows.

PROPOSITION 6.5.– *Under the hypotheses of property 6.2:*

$$\mathbb{E}\big[e^{<\mathbf{a}^{(1)},\mathbf{S}^{(1)}_{n_1}>+<\mathbf{a}^{(2)},\mathbf{S}^{(2)}_{n_2}>}\,\bar{G}_{n_1,n_2}(x_1,x_2|U^{(1)},U^{(2)})\big]$$

$$= \mathbb{E}\big[e^{<\mathbf{a}^{(1)},\mathbf{S}^{(1)}_{n_1}>+<\mathbf{a}^{(2)},\mathbf{S}^{(2)}_{n_2}>}\,\bar{A}_{n_1,n_2}(x_1,x_2|\mathbf{S}^{(1)}_{n_1},\mathbf{S}^{(2)}_{n_2})\big],\quad (n_1,n_2)\in\mathbb{N}_0^2. \qquad [6.58]$$

Proof. From property 6.2, we see that if $H_{k_1,k_2} - L_{k_1,k_2} = 0$ for $0 \leq k_1 \leq n_1$, $0 \leq k_2 \leq n_2$ with $k_1 + k_2 < n_1 + n_2$:

$$\Delta_1(H_{n_1,n_2} - L_{n_1,n_2})(x_1,x_2) = 0,\ \text{ and }\ \Delta_2(H_{n_1,n_2} - L_{n_1,n_2})(x_1,x_2) = 0. \qquad [6.59]$$

Since $H_{n_1,n_2}(0,0) = L_{n_1,n_2}(0,0) = 0$, [6.59] then implies that $H_{n_1,n_2}(x_1,x_2) = L_{n_1,n_2}(x_1,x_2)$ for any reals x_1, x_2. Thus, starting from $H_{0,0} = L_{0,0}$, we obtain by recursion the identity [6.58] for all $n_1, n_2 \in \mathbb{N}_0$. □

6.4. With u_i partial products of i.i.d. variables

This time, we consider the case where the family U corresponds to the sequence of partial products of i.i.d. random variables. More precisely:

$$u_i \equiv \Pi_i = X_1 \ldots X_i,\quad i \in \mathbb{N},\quad \text{with } u_0 = 1 \equiv \Pi_0, \qquad [6.60]$$

where the X_i, $i \geq 1$, are i.i.d. random variables. This modeling proves relevant in certain applications. It will also lead to an unexpected mathematical result.

More precisely, let us introduce a new associated family:

$$V = \{v_i \equiv \mathbb{E}(X_1^i),\ i \in \mathbb{N}_0\},$$

and, given any real β, let E^β be a shift operator such that $E^\beta V = \{v_{\beta+i}, i \in \mathbb{N}_0\}$. The required moments are assumed to exist.

PROPOSITION 6.6.– *For the multiplicative case [6.60], when S.I. is valid, $\mathcal{E}$ is standard at 0 and satisfies the multiplicative property, then:*

$$\mathbb{E}\big[(u_r)^\beta\,\bar{G}_r(0|U)\big] = \bar{G}_r(0|E^\beta V),\quad r \geq 1. \qquad [6.61]$$

Proof. The reasoning relies on two expansions: the first concerns $\bar{G}_r(1|U)$ wih respect to the family $\mathcal{E}$, and the second $e_r(0)$ with respect to the family $\{\bar{G}_n(0|E^\beta V),\ n \in \mathbb{N}_0\}$.

On the one hand, the Taylor-type formula [2.79] gives:

$$\bar{G}_r(1|U) = \sum_{i=0}^{r} \bar{G}_{r-i}(0|E^i U)\, e_i(1), \qquad [6.62]$$

which reduces to $\delta_{r,0}$ since u_0=1. Let us introduce the factor $(\Pi_r)^\beta$ in [6.62], and then use the multiplicative property to get:

$$(\Pi_r)^\beta \sum_{i=0}^{r} (\Pi_i)^{r-i}\, \bar{G}_{r-i}(0|\{\Pi_{i+j}/\Pi_i,\ j \in \mathbb{N}_0\})\, e_i(1), \qquad [6.63]$$

which remains equal to $\delta_{r,0}$ since it is $(\Pi_r)^\beta\, \delta_{r,0}$. We will now take the expectation of [6.63]. Note that for $0 \le i \le r$:

$$(\Pi_r)^\beta\, (\Pi_i)^{r-i} = (\Pi_i)^{\beta+r-i}\, (\Pi_r/\Pi_i)^\beta = (\Pi_i)^{\beta+r-i}\, (\tilde{\Pi}_{i,r-i})^\beta,$$

where $\tilde{\Pi}_{i,r-i} = \Pi_r/\Pi_i$ is independent of Π_i and distributed as Π_{r-i}. The same observation applies to the parameters in $\bar{G}_{r-i}$, that is, for the $\Pi_{i+j}/\Pi_i = \tilde{\Pi}_{i,j}$. So, denoting $\tilde{U}_i = \{\tilde{\Pi}_{i,j},\ j \in \mathbb{N}_0\}$, this A-G pseudopolynomial is $\bar{G}_{r-i}(0|\tilde{U}_i)$. From [6.63], we obtain by the independence between Π_i and $(\tilde{\Pi}_{i,r-i}, \tilde{U}_i)$:

$$\sum_{i=0}^{r} \mathbb{E}\big[(\Pi_i)^{\beta+r-i}\big]\, \mathbb{E}\big[(\tilde{\Pi}_{i,r-i})^\beta\, \bar{G}_{r-i}(0|\tilde{U}_i)\big]\, e_i(1) = \delta_{r,0}. \qquad [6.64]$$

In addition, we can then substitute (Π_{r-i}, U) for $(\tilde{\Pi}_{i,r-i}, \tilde{U}_i)$ in [6.64], while the other term inside the sum is:

$$\mathbb{E}\big[(\Pi_i)^{\beta+r-i}\big]\, e_i(1) = \big(v_{\beta+r-i}\big)^i\, e_i(1) = e_i(v_{\beta+r-i}).$$

Therefore, [6.64] becomes, after replacing i by $r-i$:

$$\sum_{i=0}^{r} e_{r-i}(v_{\beta+i})\, \mathbb{E}\big[(\Pi_i)^\beta\, \bar{G}_i(0|U)\big] = \delta_{r,0}. \qquad [6.65]$$

On the other hand, by the A–G expansion [2.32]:

$$e_r(0) = \sum_{i=0}^{r} e_{r-i}(v_{\beta+i})\, \bar{G}_i(0|E^\beta V), \tag{6.66}$$

which reduces to $\delta_{r,0}$ since $\mathcal{E}$ is standard at 0. Substracting [6.66] from [6.65] then yields:

$$\sum_{i=0}^{r} e_{r-i}(v_{\beta+i})\, \{\mathbb{E}\big[(\Pi_i)^\beta\, \bar{G}_i(0|U)\big] - \bar{G}_i(0|E^\beta V)\} = 0, \tag{6.67}$$

which makes it possible to recursively determine each of the terms in braces. For $i = 0$, we have $\{\dots\} = 0$, so that from [6.67], this is true for all $i \geq 0$, hence the announced result. □

In particular, [6.61] with $\beta = 0$ gives:

$$\mathbb{E}\big[\bar{G}_r(0|U)\big] = \bar{G}_r(0|E^\beta V).$$

Thus, in this case, the expectation of a *randomized* A–G pseudopolynomial simply becomes a *non-randomized* A–G pseudopolynomial: this is really very surprising. Now, instead of 0, consider any real x inside $\bar{G}_r$.

COROLLARY 6.3.– *Under the hypotheses of proposition 6.6:*

$$\mathbb{E}\big[(u_r)^\beta\, \bar{G}_r(x|U)\big] = \sum_{i=0}^{r} \bar{G}_{r-i}(0|E^\beta V)\, e_i(xv_{\beta+r-i}), \quad r \geq 1. \tag{6.68}$$

Proof. From [2.79], we write:

$$\bar{G}_r(x|U) = \sum_{i=0}^{r} \bar{G}_{r-i}(0|E^i U)\, e_i(x). \tag{6.69}$$

By the trick used in the previous proof, $\bar{G}_{r-i}(0|E^i U)$ is distributed as $(u_i)^{r-i}\, \bar{G}_{r-i}(0|\tilde{U}_i)$. Thus, after multiplication by $(u_r)^\beta$, the expectation of [6.69] is given by:

$$\mathbb{E}\big[(u_r)^{\beta}\,\bar{G}_r(x|U)\big] = \sum_{i=0}^{r} \mathbb{E}\big[(u_i)^{\beta+r-i}\,(\tilde{u}_{i,r-i})^{\beta}\,\bar{G}_{r-i}(0|\tilde{U}_i)\big]\,e_i(x)$$

$$= \sum_{i=0}^{r} \left(v_{\beta+r-i}\right)^{i}\,\mathbb{E}\big[(u_i)^{\beta}\,\bar{G}_{r-i}(0|U)\big]\,e_i(x)$$

$$= \sum_{i=0}^{r} \left(v_{\beta+r-i}\right)^{i}\,\bar{G}_{r-i}(0|E^{\beta}V)\,e_i(x),$$

by virtue of [6.61]. The formula [6.68] then follows from [2.55]. □

6.5. Multivariate multiplicative extension

Consider the bivariate A–G pseudopolynomials $\bar{G}_{n_1,n_2}(x_1,x_2|U^{(1)},U^{(2)})$ in the multiplicative case. As in section 6.4, there are initially two independent sequences of i.i.d. bidimensional random vectors:

$$\big\{(X_i^{(1,1)},X_i^{(1,2)}),\, i\in\mathbb{N}\big\},\quad \text{each being distributed as } (\mathrm{X}_1^{(1,1)},\mathrm{X}_1^{(1,2)}),$$

$$\big\{(X_j^{(2,1)},X_j^{(2,2)}),\, j\in\mathbb{N}\big\},\quad \text{each being distributed as } (\mathrm{X}_1^{(2,1)},\mathrm{X}_1^{(2,2)}).$$

Their associated partial products are:

$$\big\{(X_1^{(1,1)}\dots X_{i_1}^{(1,1)},X_1^{(1,2)}\dots X_{i_1}^{(1,2)}) \equiv (\Pi_{i_1}^{(1,1)},\Pi_{i_1}^{(1,2)}),\, i_1\in\mathbb{N}_0\big\},$$

with $\Pi_0^{(1,1)} = \Pi_0^{(1,2)} = 1$,

$$\big\{(X_1^{(2,1)}\dots X_{i_2}^{(2,1)},X_1^{(2,2)}\dots X_{i_2}^{(2,2)}) \equiv (\Pi_{i_2}^{(2,1)},\Pi_{i_2}^{(2,2)}),\, i_2\in\mathbb{N}_0\big\},$$

with $\Pi_0^{(2,1)} = \Pi_0^{(2,2)} = 1$.

The family of parameters $U = (U^{(1)},U^{(2)})$ is then defined as:

$$\begin{cases} U^{(1)} = \big\{u_{i_1,i_2}^{(1)} \equiv \Pi_{i_1}^{(1,1)}\Pi_{i_2}^{(2,1)},\ (i_1,i_2)\in\mathbb{N}_0^2\big\},\\ U^{(2)} = \big\{u_{i_1,i_2}^{(2)} \equiv \Pi_{i_1}^{(1,2)}\Pi_{i_2}^{(2,2)},\ (i_1,i_2)\in\mathbb{N}_0^2\big\}.\end{cases} \qquad [6.70]$$

Let us introduce a new family of parameters $V = (V^{(1)},V^{(2)})$ where:

$$\begin{cases} V^{(1)} = \big\{v_{i_1,i_2}^{(1)} \equiv \mathbb{E}\big[(X_1^{(1,1)})^{i_1}\,(X_1^{(1,2)})^{i_2}\big],\ (i_1,i_2)\in\mathbb{N}_0^2\big\},\\ V^{(2)} = \big\{v_{i_1,i_2}^{(2)} \equiv \mathbb{E}\big[(X_1^{(2,1)})^{i_1}\,(X_1^{(2,2)})^{i_2}\big],\ (i_1,i_2)\in\mathbb{N}_0^2\big\}.\end{cases} \qquad [6.71]$$

Given any reals β_1, β_2, denote E^{β_1,β_2} the shift operator such that:

$$E^{\beta_1,\beta_2}V = \{(v^{(1)}_{\beta_1+i_1,\beta_2+i_2}, v^{(2)}_{\beta_1+i_1,\beta_2+i_2}), (i_1, i_2) \in \mathbb{N}_0^2\}.$$

The remarkable relation [6.61] is generalized as follows using a similar reasoning.

PROPOSITION 6.7.– *Under [6.71], when S.I. is valid, $\mathcal{E}$ is standard at $(0,0)$ and satisfies the multiplicative property, then:*

$$\mathbb{E}\big[(u^{(1)}_{n_1,n_2})^{\beta_1} (u^{(2)}_{n_1,n_2})^{\beta_2} \bar{G}_{n_1,n_2}(0,0|U)\big] = \bar{G}_{n_1,n_2}(0,0|E^{\beta_1,\beta_2}V),$$
$$(n_1, n_2) \in \mathbb{N}_0^2. \qquad [6.72]$$

Proof. From the bivariate Taylor-type formula, we have:

$$\bar{G}_{n_1,n_2}(1,1|U) = \sum_{(k_1,k_2)=(0,0)}^{(n_1,n_2)} \bar{G}_{n_1-k_1,n_2-k_2}(0,0|E^{k_1,k_2}U)\, e_{k_1,k_2}(1,1), \qquad [6.73]$$

which reduces to $\delta_{n_1+n_2,0}$ by definition. Let us use the same trick as before:

$$E^{k_1,k_2}U = (u^{(1)}_{k_1,k_2}\tilde{U}^{(1)}_{k_1,k_2}, u^{(2)}_{k_1,k_2}\tilde{U}^{(2)}_{k_1,k_2}), \qquad [6.74]$$

where U and $\tilde{U}_{k_1,k_2} = (\tilde{U}^{(1)}_{k_1,k_2}, \tilde{U}^{(2)}_{k_1,k_2})$ are independent and have identical distributions. By inserting [6.74], the identity [6.73] becomes:

$$\mathbb{E}\big[(u^{(1)}_{n_1,n_2})^{\beta_1} (u^{(2)}_{n_1,n_2})^{\beta_2} \sum_{(k_1,k_2)=(0,0)}^{(n_1,n_2)} (u^{(1)}_{k_1,k_2})^{n_1-k_1} (u^{(2)}_{k_1,k_2})^{n_2-k_2}$$
$$\times \bar{G}_{n_1-k_1,n_2-k_2}(0,0|\tilde{U}_{k_1,k_2})\big] e_{k_1,k_2}(1,1) = \delta_{n_1+n_2,0}. \qquad [6.75]$$

Observe that in [6.75], $u^{(1)}_{n_1,n_2}$ can be written $u^{(1)}_{k_1,k_2}\,(u^{(1)}_{n_1,n_2}/u^{(1)}_{k_1,k_2})$, and this last parenthesis is distributed as $\tilde{u}^{(1)}_{n_1-k_1,n_2-k_2}$ and is independent of $u^{(1)}_{k_1,k_2}$. The same decomposition is valid for $u^{(2)}_{n_1,n_2}$. We can therefore transform [6.75] into:

$$\sum_{(k_1,k_2)=(0,0)}^{(n_1,n_2)} \Big\{\mathbb{E}\big[(u^{(1)}_{k_1,k_2})^{\beta_1+n_1-k_1}\,(u^{(2)}_{k_1,k_2})^{\beta_2+n_2-k_2}\big]$$

$$\mathbb{E}\big[(u^{(1)}_{n_1-k_1,n_2-k_2})^{\beta_1}\,(u^{(2)}_{n_1-k_1,n_2-k_2})^{\beta_2}$$

$$\times\,\bar{G}_{n_1-k_1,n_2-k_2}(0,0|\tilde{U}_{k_1,k_2})\big]\,e_{k_1,k_2}(1,1)\Big\} = \delta_{n_1+n_2,0}. \qquad [6.76]$$

Let us denote the second expectation in [6.76] by $M_{n_1-k_1,n_2-k_2}$. Temporarily leaving this factor aside, each term of the remaining sum is given by:

$$\begin{aligned}
&\mathbb{E}\big[(u^{(1)}_{k_1,k_2})^{\beta_1+n_1-k_1}\,(u^{(2)}_{k_1,k_2})^{\beta_2+n_2-k_2}\big]\,e_{k_1,k_2}(1,1)\\
&\quad= \mathbb{E}\big[(\Pi^{(1,1)}_{k_1}\Pi^{(2,1)}_{k_2})^{\beta_1+n_1-k_1}\,(\Pi^{(1,2)}_{k_1}\Pi^{(2,2)}_{k_2})^{\beta_2+n_2-k_2}\big]\,e_{k_1,k_2}(1,1)\\
&\quad= \big\{\mathbb{E}\big[(X^{(1,1)}_1)^{\beta_1+n_1-k_1}(X^{(1,2)}_1)^{\beta_2+n_2-k_2}\big]\big\}^{k_1}\\
&\quad\times\big\{\mathbb{E}\big[(X^{(2,1)}_1)^{\beta_1+n_1-k_1}(X^{(2,2)}_1)^{\beta_2+n_2-k_2}\big]\big\}^{k_2}e_{k_1,k_2}(1,1)\\
&\quad= \big\{v^{(1)}_{\beta_1+n_1-k_1,\beta_2+n_2-k_2}\big\}^{k_1}\big\{v^{(2)}_{\beta_1+n_1-k_1,\beta_2+n_2-k_2}\big\}^{k_2}\,e_{k_1,k_2}(1,1)\\
&\quad= e_{k_1,k_2}\big(v^{(1)}_{\beta_1+n_1-k_1,\beta_2+n_2-k_2}, v^{(2)}_{\beta_1+n_1-k_1,\beta_2+n_2-k_2}\big).
\end{aligned}$$

Thanks to this, and applying $i_1 = n_1 - k_1$, $i_2 = n_2 - k_2$, formula [6.76] thus simplifies to:

$$\sum_{(i_1,i_2)=(0,0)}^{(n_1,n_2)} e_{n_1-i_1,n_2-i_2}(v^{(1)}_{\beta_1+i_1,\beta_2+i_2}, v^{(2)}_{\beta_1+i_1,\beta_2+i_2})\,M_{i_1,i_2} = \delta_{n_1+n_2,0}. \quad [6.77]$$

We now expand $e_{n_1,n_2}(\;)$ with respect to the family $\{\bar{G}_{n_1,n_2}(\;|E^{\beta_1,\beta_2}V)$, $(n_1,n_2)\in\mathbb{N}_0^2\}$. At point $(0,0)$, we get from [5.24]:

$$\sum_{(i_1,i_2)=(0,0)}^{(n_1,n_2)} e_{n_1-i_1,n_2-i_2}(v^{(1)}_{\beta_1+i_1,\beta_2+i_2}, v^{(2)}_{\beta_1+i_1,\beta_2+i_2})$$

$$\times\,\bar{G}_{i_1,i_2}(0,0|E^{\beta_1,\beta_2}V) = \delta_{n_1+n_2,0}. \qquad [6.78]$$

Note that the desired result [6.72] is equivalent to:

$$M_{i_1,i_2} - \bar{G}_{i_1,i_2}(0, 0|E^{\beta_1,\beta_2}V) = 0, \quad (i_1, i_2) \in \mathbb{N}_0^2. \qquad [6.79]$$

To prove [6.79], it is then sufficient to take the difference between [6.77] and [6.78] and to reason recursively from $M_{0,0} - \bar{G}_{0,0}(0, 0|E^{\beta_1,\beta_2}V) = 0$. □

6.6. Additive case for exponential functions

Consider any exponential on $\mathbb{R}_+$ which has the simple form $\exp(bx)$, $b > 0$. We will continue the analysis of the additive case [6.4] when the i.i.d. random variables X_i are valued in $\mathbb{R}_+$, with finite mean μ and such that $\mathbb{E}[\exp(\tau X_1)] < \infty$ for some given $\tau > 0$.

We will also suppose that $\mathcal{E}$ is generated by $\exp[xg(s)]$, the coefficients α_i in $g(s)$ being ≥ 0. So, there exists s_b such that $b = g(s_b)$, and by [3.19]:

$$\Delta(e^{bx}) = s_b\,(e^{bx}).$$

Then, the formal A–G expansion of $\exp(bx)$ is:

$$\sum_{n=0}^{\infty} e^{-bS_n}\,(s_b)^n\,\bar{G}_n(x|\{-S_i,\, i \in \mathbb{N}_0\}). \qquad [6.80]$$

By mimicry with [6.36], one can imagine that the expectation of [6.80]:

$$\mathbb{E}\Big[\sum_{n=0}^{\infty} e^{-bS_n}\,(s_b)^n\,\bar{G}_n(x|\{-S_i,\, i \in \mathbb{N}_0\})\Big]$$

is equal to

$$\mathbb{E}\Big[\sum_{n=0}^{\infty} e^{-bS_n}\,(s_b)^n\,\bar{A}_n(x|S_n)\Big]. \qquad [6.81]$$

Two issues discussed are as follows: Does the series [6.81] converge? If yes, is its sum equal to $\exp(bx)$?

PROPERTY 6.3.– *Under [6.4], when the previous hypotheses are satisfied, then [6.81] is convergent (except possibly when $g'(s_b)\, s_b\, \mu = 1$).*

Proof. Note that since all terms in [6.81] are positive, the expectation of the sum is the sum of the expectations; we write L this expression. For $\sigma > 0$ sufficiently small, we have:

$$\begin{aligned}\bar{A}_n(x|S_n) &= \frac{x}{x+S_n}\, e_n(x+S_n) \le e_n(x+S_n) = \frac{1}{\sigma^n}\, e_n(x+S_n)\,\sigma^n \\ &< \frac{1}{\sigma^n}\sum_{r=0}^{\infty} e_r(x+S_n)\,\sigma^r = \frac{1}{\sigma^n}\, e^{(x+S_n)\,g(\sigma)},\end{aligned} \qquad [6.82]$$

thus, L is bounded by:

$$\begin{aligned}L &< \sum_{n=0}^{\infty} \mathbb{E}\big[e^{-bS_n}\,(s_b/\sigma)^n\, e^{(x+S_n)\,g(\sigma)}\big] \\ &= e^{x\,g(\sigma)} \sum_{n=0}^{\infty} (s_b/\sigma)^n \left\{\mathbb{E}\big[e^{(-b+g(\sigma))X_1}\big]\right\}^n,\end{aligned}$$

and a sufficient condition for the convergence of the r.h.s. is:

$$(s_b/\sigma)\,\mathbb{E}\big[e^{(-b+g(\sigma))X_1}\big] < 1. \qquad [6.83]$$

Let us denote $\phi(\tau) = \mathbb{E}[\exp(\tau X_1)]$, finite for $\tau < t_0\,(> 0)$ by hypothesis. Then, on $(-\infty, t_0)$, the function ϕ exists and is positive, increasing, convex with $\phi(0) = 1$, $\phi'(0) = \mu$. Obviously, [6.83] means:

$$\phi\big(-b + g(\sigma)\big) < \sigma/s_b. \qquad [6.84]$$

We choose $\sigma > 0$ to also ensure that $\sigma/s_b < \phi(t_0 - 0)$, so that we can apply ϕ^{-1} to both sides of [6.84]. This gives:

$$-b + g(\sigma) < \phi^{-1}(\sigma/s_b) \equiv \zeta \text{ say},$$

ζ being some real number (positive or not), which is equivalent to:

$$-g(s_b\phi(0)) + g(s_b\phi(\zeta)) < \zeta, \qquad [6.85]$$

after putting in the l.h.s. $b = g(s_b)$, $\phi(0) = 1$ and $\sigma = s_b\phi(\zeta)$.

Let us rewrite [6.85] as $h(\zeta) < \zeta$. We observe that the function h is increasing, convex with $h(0) = 0$, $h'(0) = g'(s_b)s_b\mu > 0$, so we have to compare $h'(0)$ with 1. If $h'(0) < 1$, [6.85] can be checked by taking a value

$\zeta > 0$. If $h'(0) > 1$, we must take a value $\zeta < 0$. Note that in all cases, $\phi(\zeta) > 0$ of course, implying an appropriate corresponding choice for σ. The case $h'(0) = 1$ however remains unresolved. These conclusions provide the desired result. □

Clearly, the convergence of [6.81] remains true when s_b is replaced by $\lambda \in (0, s_b)$. Let us move on to the question of evaluating the sum of [6.81]. For this, we define a set J by:

$$J = \{s_b \in \mathbb{R}_+ : g'(s_b)\, s_b\, \mu < 1\}. \qquad [6.86]$$

PROPOSITION 6.8.– *Under the hypotheses of property 6.3,* $\exp(bx)$ *is the sum of the series [6.81] on the set J.*

Proof. We first show that the result is true for s_b small enough. Indeed, we have:

$$e^{bx} = e^{g(s_b)x} = \sum_{n=0}^{\infty} e_n(x)\,(s_b)^n = \lim_{m\to\infty} p_m(x),$$

where $p_m(x)$ denotes the pseudopolynomial:

$$p_m(x) = \sum_{n=0}^{m} e_n(x)\,(s_b)^n, \qquad [6.87]$$

which is such that, for $0 \le i \le m$:

$$\Delta^i p_m(x) = \sum_{n=i}^{m} e_{n-i}(x)\,(s_b)^n = (s_b)^i \sum_{r=0}^{m-i} e_r(x)\,(s_b)^r = (s_b)^i\, p_{m-i}(x). \qquad [6.88]$$

From corollary 6.2 and using [6.87], we see that p_m admits the expansion:

$$p_m(x) = \mathbb{E}\Big[\sum_{n=0}^{m} p_{m-n}(-S_n)\,(s_b)^n\,\bar{A}_n(x|S_n)\Big]. \qquad [6.89]$$

When $m \to \infty$, provided that the operators $\lim_{m\to\infty}$ and $\mathbb{E}$ commute with each other, [6.89] will imply the desired result:

$$e^{bx} = \mathbb{E}\Big[\sum_{n=0}^{\infty} e^{-bS_n}\,(s_b)^n\,\bar{A}_n(x|S_n)\Big]. \qquad [6.90]$$

It now remains to justify the permutation of operators. From [6.87], we have:

$$|p_m(-S_n)| \le \sum_{r=0}^{m} |e_r(-S_n)|\,(s_b)^r \le \sum_{r=0}^{m} e_r(S_n)\,(s_b)^r \le e^{bS_n}.$$

Thus, arguing as with [6.82] for any $\sigma > 0$, we bound the r.h.s. of [6.89] by:

$$\mathbb{E}\Big[\sum_{n=0}^{m} |p_{m-n}(-S_n)|\,(s_b)^n\,\bar{A}_n(x|S_n)\Big] \le \mathbb{E}\Big[\sum_{n=0}^{m} e^{bS_n}\,(s_b/\sigma)^n\,e^{(x+S_n)\,g(\sigma)}\Big]$$

$$\le e^{x\,g(\sigma)} \sum_{n-0}^{\infty} (s_b/\sigma)^n \left\{\mathbb{E}\big[e^{(b+g(\sigma))X_1}\big]\right\}^n,$$

a series that will certainly converge if:

$$(s_b/\sigma)\,\mathbb{E}\big[e^{(b+g(\sigma))X_1}\big] < 1 \qquad [6.91]$$

this condition is verified when s_b is taken sufficiently small. To extend this case, we will use analytical continuation.

Considering the complex plane, let $s_b \in \mathbb{C}$ satisfying a condition stronger than [6.91], namely:

$$(|s_b|/\sigma)\,\mathbb{E}\big[e^{(g(|s_b|)+g(\sigma))X_1}\big] < 1 \qquad [6.92]$$

the condition is satisfied on a convenient open disk $D(0,\epsilon) \subset \mathbb{C}$. Thus, on this disk, $\exp[xg(s_b)]$ is the sum of the expansion [6.81]. Let us now take any s_b fixed in J defined by [6.86], with $J \subset \mathbb{R} \subset \mathbb{C}$. There is in $\mathbb{C}$ an open disk $D(s_b, \epsilon_{s_b})$ on which the series [6.81] is still convergent (i.e. it converges in $s_b + \eta$, $0 < |\eta| < \epsilon_{s_b}$). To show this, we replace [6.84], that is, $(s_b/\sigma)\,\phi[-g(s_b) + g(\sigma)] < 1$, with the stronger condition:

$$(|s_b + \eta|/\sigma)\,\phi\big[\Re\big(-g(s_b+\eta) + g(\sigma)\big)\big] < 1, \qquad [6.93]$$

where $\Re$ indicates the real part of a complex number. Due to the continuity, [6.84] implies [6.93] when $|\eta|$ is small enough, therefore $D(s_b, \epsilon_{s_b})$ does exist. In addition, [6.81] is also convergent on the domain:

$$W = \cup_{s_b \in J} D(s_b, \epsilon_{s_b}),$$

and since $W \cap D(0, \epsilon) \neq \emptyset$, the analytic continuation works from $D(0, \epsilon)$ to $W \cup D(0, \epsilon)$, and in particular to $J \subset W \cup D(0, \epsilon)$. Since the sum of [6.81] on $D(0, \epsilon)$ is $\exp[xg(s_b)]$, we deduce that it is the same on the set J. □

To close, we derive a simplified expression for the series [6.81] when the random variables X_i have an arithmetic distribution $\{q_j, j \geq 1\}$, with a generating function denoted $h(s)$.

PROPERTY 6.4.– *In this discrete case:*

$$\mathbb{E}[\sum_{n=0}^{\infty} e^{-bS_n} (s_b)^n \bar{A}_n(x|S_n)] = \sum_{k=0}^{\infty} e^{-bk} \bar{A}_k(x|\tilde{\mathcal{E}}, k), \qquad [6.94]$$

where $\tilde{\mathcal{E}} = \{\tilde{e}_n, n \in \mathbb{N}_0\}$ *is the family generated by* $\exp[xg(s_b h(s))]$, *instead of* $\exp[xg(s)]$.

Proof. We start by expressing $\tilde{e}_n$, $n \geq 0$, in terms of the e_k. According to the definition of $\tilde{\mathcal{E}}$ and since X_i takes values in $\mathbb{N}$, we have:

$$\sum_{n=0}^{\infty} \tilde{e}_n(x)\, s^n = e^{xg(s_b h(s))} = \sum_{k=0}^{\infty} e_k(x)\, (s_b)^k\, [h(s)]^k$$

$$= \sum_{k=0}^{\infty} e_k(x)\, (s_b)^k\, [\sum_{j=k}^{\infty} q_j^{*k}\, s^j] = \sum_{j=0}^{\infty} s^j [\sum_{k=0}^{j} e_k(x)\, (s_b)^k\, q_j^{*k}],$$

so that

$$\tilde{e}_n(x) = \sum_{k=0}^{n} e_k(x)\, (s_b)^k\, q_n^{*k}. \qquad [6.95]$$

Now, the l.h.s. of [6.94], L say, is given in explicit form by:

$$L = \sum_{n=0}^{\infty} [\sum_{k=0}^{n} e^{-bk}\, (s_b)^n\, \frac{x}{x+k}\, e_n(x+k)\, q_k^{*n}]$$

$$= \sum_{k=0}^{\infty} e^{-bk}\, \frac{x}{x+k} [\sum_{n=0}^{k} e_n(x+k)\, (s_b)^n\, q_k^{*n}],$$

and using [6.95]:

$$L = \sum_{k=0}^{\infty} e^{-bk} \frac{x}{x+k} \tilde{e}_k(x+k),$$

which corresponds precisely to the r.h.s. of [6.94]. □

7

First Meeting Level with a Lower Boundary

To illustrate the general theory by stochastic applications, we will first examine several problems of first meeting the trajectory of a Poisson-type process with a given continuous lower boundary (increasing without real loss of generality). This topic, as well as the parallel problem of first crossing an upper boundary, have long been the subject of much research in the field of applied probability. Most often, the methods used until now have relied on fairly standard probabilistic methods, notably Laplace transforms (see the monograph by Zacks (2017)[1]). On the contrary, the approach that we will develop here will be mainly algebraic in nature and will be based on the application of the A–G pseudopolynomials.

7.1. Return to a classical Poisson process

As mentioned in section 1.5, consider the trajectory of a Poisson process with rate λ that starts at level 0 at time x (possibly negative); let $\{N_{t-x},\ t \geq x\}$ be this process. We examine the problem of first crossing of its trajectory with a continuous increasing lower bound $f(t)$, $t \geq x$, that starts below 0 at $t = x$. Due to this hypothesis, the crossing is necessarily a meeting, and the useful part of the boundary is reduced to the set F given by:

$$F \equiv \{(v_i, i),\ i \in \mathbb{N}_0\},$$

1 Zacks, S. (2017). *Sample Path Analysis and Distributions of Boundary Crossing Times*. Springer, Cham.

where v_i is the first moment when the boundary reaches level i. The level of the first meeting of the Poisson trajectory with the lower boundary is a random variable denoted as K.

1) We established in [1.17] that the distribution of K can be expressed as:

$$\mathbb{P}(K = n|x, V) = e^{-\lambda(v_n - x)}(-\lambda)^n \, G_n(x|V), \quad n \geq 0,$$

where V is the family $\{v_i, i \in \mathbb{N}_0\}$ and G_n are the associated *A–G polynomials*. Recall that the proof method used the total probability law with respect to the first jump time T_0 of the trajectory.

2) Let us adopt a different reasoning based again on the total probability law but this time with respect to the trajectory level at time v_0, that is, with respect to $N_{v_0 - x}$. We will show that the distribution of K can be alternatively expressed in terms of A–G pseudopolynomials. We start by writing:

$$\mathbb{P}(K = n|x, V) = \sum_{r=0}^{n} \mathbb{P}(N_{v_0 - x} = r)$$

$$\mathbb{P}(\text{a trajectory initiated at level } r \text{ at time } v_0 \text{ meets the boundary for the first time at level } n), \quad n \geq 0. \qquad [7.1]$$

The unknown probability in the r.h.s. of [7.1] is identical to the probability in the l.h.s. of the first meeting level in $n - r$, after substituting v_0 for x and the shifted family $E^r V$ for V. In other words, [7.1] is simply:

$$\mathbb{P}(K = n|x, V) = \sum_{r=0}^{n} \mathbb{P}(N_{v_0 - x} = r) \, \mathbb{P}(K = n - r|v_0, E^r V). \qquad [7.2]$$

By writing as in [1.20]:

$$\mathbb{P}(K = n|x, V) = e^{-\lambda(v_n - x)} \, R_n(x|V), \quad n \geq 0, \qquad [7.3]$$

[7.2] gives $R_0(x|V) = 1$, $R_n(v_0|V) = \delta_{n,0}$ and from [7.3]:

$$R_n(x|V) = \sum_{r=0}^{n} \frac{\mathbb{P}(N_{v_0 - x} = r)}{e^{-\lambda(v_n - x)}} \, e^{-\lambda(v_n - v_0)} \, R_{n-r}(v_0|E^r V)$$

$$= \sum_{r=0}^{n} \frac{[\lambda(v_0 - x)]^r}{r!} \, R_{n-r}(v_0|E^r V), \quad n \geq 1, \qquad [7.4]$$

thanks to the Poisson process hypothesis.

Formula [7.4] shows a strong similarity with a Taylor-type formula. In fact, by virtue of [2.79], any A–G pseudopolynomial $\bar{G}_n(x|\mathcal{E}, V)$ can be expanded around a as:

$$\bar{G}_n(x|\mathcal{E}, V) = \sum_{r=0}^{n} \bar{G}_{n-r}(a|\mathcal{E}, E^r V)\, e_r(x-a). \qquad [7.5]$$

Since [7.4] recursively gives all R_n, one could hope to express them as $\bar{G}_n$ provided that [7.4] and [7.5] can be identified. Two options are available to us.

2.a) Let us work with the family $\hat{\mathcal{E}}$ of functions:

$$\hat{e}_n(t) = (-\lambda t)^n/n!, \quad n \in \mathbb{N}_0. \qquad [7.6]$$

In that case, $\hat{e}_r(x - v_0) = [\lambda(v_0 - x)]^r/r!$, and we obtain:

$$R_n(x|V) = \bar{G}_n(x|\hat{\mathcal{E}}, V).$$

Therefore, [7.3] gives the distribution of K in terms of *A–G pseudopolynomials*.

PROPOSITION 7.1.– *With $\hat{\mathcal{E}}$ defined by [7.6]:*

$$\mathbb{P}(K = n|x, V) = e^{-\lambda(v_n - x)}\, \bar{G}_n(x|\hat{\mathcal{E}}, V), \quad n \geq 0. \qquad [7.7]$$

2.b) Let us change the notation as in [1.22] of section 1.5, that is:

$$v_i = -u_i \text{ for all } i \text{ (so, } V = -U\text{)}, \; x = -y \text{ and } R_n(x|V) = \lambda^n \tilde{R}_n(y|U). \qquad [7.8]$$

In that case, [7.4] leads to:

$$\tilde{R}_n(y|U) = \sum_{r=0}^{n} \frac{[\lambda(y - u_0)]^r}{r!} \tilde{R}_{n-r}(u_0|E^r U)$$

$$= \sum_{r=0}^{n} e_r(y - u_0)\, \tilde{R}_{n-r}(u_0|E^r U), \qquad [7.9]$$

after choosing the family $\mathcal{E}$ of functions:

$$e_n(t) = t^n/n!, \quad n \in \mathbb{N}_0. \tag{7.10}$$

Clearly, [7.9] corresponds to a Taylor expansion, hence the identification:

$$\tilde{R}_n(y|U) = G_n(y|\mathcal{E}, U), \quad n \geq 0.$$

Thus, with $\mathcal{E}$ defined by [7.10], we see that [7.3] with [7.8] give again [1.17] for the distribution of K in terms of standard A–G polynomials.

7.2. For a compound Poisson process

This time we deal with a process $\{N_{t-x}, t \geq x\}$ initiated at time x which is a compound Poisson process with rate λ and arithmetic distribution $\{q_j, j \in \mathbb{N}\}$ with $q_1 > 0$ for the successive i.i.d. jump sizes. Thus:

$$\mathbb{P}(N_{t-x} = n) = e^{-\lambda(t-x)} \sum_{k=0}^{n} \frac{[\lambda(t-x)]^k}{k!} q_n^{*k}, \quad t \geq x, \; n \geq 0.$$

We focus on the first meeting of this process with an increasing lower boundary $f(t)$, $t \geq x$, which, in practice, reduces again to the set $F = \{(v_i, i), i \in \mathbb{N}_0\}$. Let K be the first meeting level of the compound Poisson trajectory with the lower boundary.

1) First, we will adapt the reasoning of section 7.1 applied to the Poisson process case. Of course, [7.2] is still valid, and writing here:

$$\mathbb{P}(K = n|x, V) = \mathbb{P}(N_{v_0-x} = 0)\, R_n(x|V), \quad n \geq 0, \tag{7.11}$$

we get, instead of [7.4]:

$$R_n(x|V) = \sum_{r=0}^{n} \frac{\mathbb{P}(N_{v_0-x} = r)}{\mathbb{P}(N_{v_0-x} = 0)} R_{n-r}(v_0|E^r V), \quad n \geq 1. \tag{7.12}$$

Here again, two options are possible.

1.a) Let us define the family $\bar{\mathcal{E}}$ of functions as:

$$\bar{e}_n(-t) = \frac{\mathbb{P}(N_t = n)}{\mathbb{P}(N_t = 0)} = \sum_{k=0}^{n} \frac{(\lambda t)^k}{k!} q_n^{*k}, \quad n \in \mathbb{N}_0. \tag{7.13}$$

Then, [7.12] takes the form:

$$R_n(x|V) = \sum_{r=0}^{n} R_{n-r}(v_0|E^r V)\, \bar{e}_r(x - v_0),$$

and by the Taylor formula:

$$R_n(x|V) = \bar{G}_n(x|\bar{\mathcal{E}}, V).$$

Therefore, [7.11] gives a result similar to [7.7], namely, the distribution of K is expressed in terms of A–G pseudopolynomials by the following:

PROPOSITION 7.2.– *With $\bar{\mathcal{E}}$ defined by [7.13]:*

$$\mathbb{P}(K = n|x, V) = e^{-\lambda(v_n - x)}\, \bar{G}_n(x|\bar{\mathcal{E}}, V), \quad n \geq 0. \qquad [7.14]$$

1.b) We make almost the change of notation made in [7.8] by writing:

$v_i = -u_i$ for all i (so, $V = -U$), $x = -y$ and $R_n(x|V) = \bar{R}_n(y|U)$, [7.15]

so that [7.12] becomes:

$$\bar{R}_n(y|V) = \sum_{r=0}^{n} \frac{\mathbb{P}(N_{y\ \ u_0} = r)}{\mathbb{P}(N_{y-u_0} = 0)}\, \bar{R}_{n-r}(u_0|E^r U), \quad n \geq 1. \qquad [7.16]$$

This leads us to choose the family $\mathcal{E}$ of functions defined by:

$$e_n(t) = \bar{e}_n(-t), \quad n \in \mathbb{N}_0, \qquad [7.17]$$

where $\bar{e}_n(-t)$ is defined by [7.13], and to reexpress [7.16] as:

$$\bar{R}_n(y|V) = \sum_{r=0}^{n} e_r(y - u_0)\, \bar{R}_{n-r}(u_0|E^r U),$$

hence the identification

$$\bar{R}_n(y|U) = \bar{G}_n(y|\mathcal{E}, U), \quad n \geq 0.$$

Thus, [7.11] with [7.15] now give the distribution of K in terms of A–G pseudopolynomials in the following form.

PROPOSITION 7.3.– *With $\mathcal{E}$ defined by [7.17]:*

$$\mathbb{P}(K = n|x, V) = e^{-\lambda(v_n - x)}\, \bar{G}_n(-x|\mathcal{E}, -V), \quad n \geq 0. \qquad [7.18]$$

REMARK.– The two methods above differ regarding the choice of the two families $\bar{\mathcal{E}}$ or $\mathcal{E}$. In a certain sense, $\mathcal{E}$ seems to be more canonical: indeed, by [7.13]:

$$\bar{e}_n(t) = \mathbb{P}(N_{-t} = n)/\mathbb{P}(N_{-t} = 0),$$

an expression without direct probabilistic interpretation, while by [7.17] and [7.13], $e_n(t)$ has the simpler form:

$$e_n(t) = \mathbb{P}(N_t = n)/\mathbb{P}(N_t = 0).$$

The presence of the $-$ sign in the final results [7.14] and [7.18] appears somewhat surprising but it cannot be avoided in any case.

2) Alternatively, we will now follow the reasoning applied in the Introduction for the Poisson process. Of course, [1.18] is still true, and [1.17] is replaced this time by:

$$\mathbb{P}(K = n|x, V) = \int_x^{v_0} \sum_{j=1}^{n} q_j \, \mathbb{P}(K = n - j|t, E^j V) \, \lambda e^{-\lambda(t-x)} dt, \quad n \geq 1. \tag{7.19}$$

Similarly to [1.20], we write:

$$\mathbb{P}(K = n|x, V) = e^{-\lambda(v_n - x)} \, H_n(x|V), \quad n \geq 0, \tag{7.20}$$

and using the identity

$$e^{-\lambda(v_n - x)} = e^{-\lambda(v_{j+n-j} - t)} \, e^{-\lambda(t-x)},$$

we get from [7.19] that:

$$H_n(x|V) = \int_x^{v_0} \sum_{j=1}^{n} \lambda q_j \, H_{n-j}(t|E^j V) dt, \quad n \geq 1. \tag{7.21}$$

Equivalently, [7.20] gives $H_0 = 1$, $H_n(v_0|V) = \delta_{n,0}$ and after differentiation of [7.21]:

$$H_n'(x|V) = \sum_{j=1}^{n} (-\lambda q_j) \, H_{n-j}(x|E^j V). \tag{7.22}$$

Let us make the notation change:

$$\tilde{\alpha}_j = -\lambda q_j \text{ and } \bar{g}(s) = \sum_{j=1}^{\infty} \tilde{\alpha}_j \, s^j,$$

then let us construct a family of functions $\tilde{\mathcal{E}}$ through the generating function:

$$\sum_{n=0}^{\infty} \tilde{e}_n(x) \, s^n = e^{x\bar{g}(s)}. \qquad [7.23]$$

In that case, we know from [3.22] that:

$$D = \sum_{j=1}^{\infty} \tilde{\alpha}_j \, \Delta^j,$$

and therefore:

$$\bar{G}_n'(x|\tilde{\mathcal{E}}, V) = \sum_{j=1}^{n} \tilde{\alpha}_j \, \bar{G}_{n-j}(x|\tilde{\mathcal{E}}, E^j V). \qquad [7.24]$$

By comparing [7.22] and [7.24], we deduce that:

$$H_n(x|V) = \bar{G}_n(x|\tilde{\mathcal{E}}, V),$$

hence by [7.20]:

$$\mathbb{P}(K = n|x, V) = e^{-\lambda(v_n - x)} \, \bar{G}_n(x|\tilde{\mathcal{E}}, V), \quad n \geq 0. \qquad [7.25]$$

This formula [7.25] is identical to [7.14] since the generating function on $\bar{\mathcal{E}}$ is $\exp(-\lambda x \sum_{j=1}^{\infty} q_j \, s^j)$, which is the same as the generating function [7.23] of $\tilde{\mathcal{E}}$.

An illustration in queue modeling: We are interested in the duration of the first busy period in an M/D/1 type queue. Specifically, customers arrive according to a Poisson process, with parameter λ, initiated at time x, and these arrivals are made in batches of i.i.d. sizes distributed according to the law $\{q_j, \, j \in \mathbb{N}\}$. There is a single server that starts serving at time $v_0 \geq x$, and each i-th serving time is of constant duration v_i. Note that traditionally, all services have the same duration, v say, but the current methodology does not require this constraint. For example, the service could request periodic maintenance operations planned in advance and of fixed durations.

A queue forms due to customers arriving before time v_0 and of course after v_0. The server is busy until the queue is exhausted, and the number of customers served provides a natural measure of its first busy period. It is then clear that this first busy measure is given by the random variable K whose distribution has been derived above.

7.3. Related first passage problems

We continue with a compound Poisson initiated at time x with level j (denoted by $A = (x, j)$), and expected to reach at time τ the level n (denoted by $B = (\tau, n)$), where of course, $\tau > x$ and $n \geq j$. So, using the family $\mathcal{E}$ defined by [7.17], that is,

$$e_n(t) = \frac{\mathbb{P}(N_t = n)}{\mathbb{P}(N_t = 0)} = \sum_{k=0}^{n} \frac{(\lambda t)^k}{k!} q_n^{*k}, \quad n \in \mathbb{N}_0,$$

we have (in obvious notation):

$$\mathbb{P}(A \to B) = \mathbb{P}(N_{\tau-x} = n - j) = e^{-\lambda(\tau-x)} e_{n-j}(\tau - x). \qquad [7.26]$$

The boundary is always an increasing function $f(t), t \geq x$, defined via the set $F = \{(v_i, i), i \in \mathbb{N}_0\}$. We will address here several problems related to the first passage of the Poisson trajectory with F. Three cases are examined according to the position of points A and B relative to the boundary.

7.3.1. *When A and B are above F*

In this case, any first crossing is necessarily a meeting.

PROPOSITION 7.4.–

$$\mathbb{P}(A \to B \textit{ without meeting } F) = e^{-\lambda(\tau-x)} \bar{G}_{n-j}(-x|\mathcal{E}, -E^j V^{(\tau)}), \qquad [7.27]$$

where $V^{(\tau)}$ is the family of parameters:

$$V^{(\tau)} = \{v_0, v_1, \ldots, v_{k_0}, \tau, \tau, \ldots\}, \textit{ with } v_{k_0} < \tau \leq v_{k_0+1}. \qquad [7.28]$$

Therefore:

$$\mathbb{P}(A \to B \textit{ with meetings of } F) = e^{-\lambda(\tau - x)}$$

$$\times \left[e_{n-j}(\tau - x) - \bar{G}_{n-j}(-x|\mathcal{E}, -E^j V^{(\tau)})\right]. \qquad [7.29]$$

Proof. Consider the time τ, the abscissis of point B. We modify the lower boundary by changing it into a vertical line from the point $(\tau, f(\tau))$. Let $V^{(\tau)}$ be this modified boundary, which is thus given by [7.28]. We observe that the event ($A \to B$ without meeting F) is equivalent to the event (the trajectory has its first meeting with $V^{(\tau)}$ at point B). Thus, the probability of this event corresponds to $\mathbb{P}(K = n - j|x, E^j V^{(\tau)})$, hence [7.27] by virtue of [7.18]. The formula [7.29] is obtained by substracting [7.27] from [7.26]. □

7.3.2. *When A and B are below F*

Here, a crossing is no longer necessarily a meeting.

PROPOSITION 7.5.–

$$\mathbb{P}(A \to B \textit{ without crossing } F) = e^{-\lambda(\tau - x)}\, \bar{G}_{n-j}(0|\mathcal{E}, -W^{(\tau - x)}), \qquad [7.30]$$

where $W^{(\tau-x)}$ is the family of parameters:

$$W^{(\tau-x)} = \{w_0, w_1, \ldots, w_{r_0}, \tau - x, \tau - x, \ldots\}, \textit{ with } w_{r_0} < \tau - x \leq w_{r_0+1}, \qquad [7.31]$$

after denoting:

$$w_i = \tau - v_{n-i}, \quad 0 \leq i \leq n. \qquad [7.32]$$

Therefore:

$$\mathbb{P}(A \to B \textit{ with crossings of } F) = e^{-\lambda(\tau - x)}$$

$$\times \left[e_{n-j}(\tau - x) - \bar{G}_{n-j}(0|\mathcal{E}, -W^{(\tau - x)})\right]. \qquad [7.33]$$

Proof. This time, we put the origin at point B and we reverse the orientation of the two axes (i.e. for time and level). On the one hand, a trajectory $A = (x, j) \to B = (\tau, n)$ for the original compound Poisson process $\{N_t,\ t \geq x\}$

becomes a trajectory $C = (0,0) \to D = (\tau - x, n - j)$ for the reversed process $\{N_s^*, 0 \le s \le \tau - x\}$, which is a similar compound Poisson process. Any time interval $(t, \tilde{t})$, $t < \tilde{t}$, is thus transformed into the time interval $(\tau - \tilde{t}, \tau - t)$, $\tau - \tilde{t} < \tau - t$, and the event $(N_{\tilde{t}-t} = r)$ becomes $(N^*_{\tau - t - (\tau - \tilde{t})} = N^*_{\tilde{t}-t} = r)$ with the same probability.

On the other hand, thanks to this operation, the boundary F which is upper for the trajectory of $\{N_t\}$ is transformed into a lower boundary F^* for the trajectory of $\{N_s^*\}$. The coordinates of the point (v_i, i) become $(\tau - v_i, n - i)$, which we note $(w_{n-i}, n - i)$ for convenience. Let $W = \{w_i, 0 \le i \le n\}$ where thus $w_i = \tau - v_{n-i}$ as indicated in [7.32].

We finally returned to the case (1) discussed above. Thus, we have to modify the boundary F^* at the old point A, which now has abscissa $\tau - x$. This leads us to construct a new boundary $F^{*(\tau - x)}$, which corresponds to the family of parameters $W^{(\tau - x)}$ defined by [7.31]. Furthermore, we see from [7.17] that as the processes $\{N_t\}$ and $\{N_s^*\}$ are indistinguishable, they provide the same family $\mathcal{E}$. We therefore obtain [7.30] and [7.33] simply by substituting in [7.27] and [7.29] the elements $(0,0)$ for (x, j) and $W^{(\tau - x)}$ for $E^j V^{(\tau)}$, the duration $\tau - x$ and the level difference $n - j$ remaining unchanged. □

7.3.3. *When A is above F while B is below*

No meeting is impossible and there are many possible passages through F: for example, only one meeting, or two meetings separated by one crossing, and so on.

PROPOSITION 7.6.–

$$\mathbb{P}(A \to B \textit{ with a single meeting with } F \textit{ at level } k)$$
$$= e^{-\lambda(\tau - x)}\, \bar{G}_{k-j}(-x|\mathcal{E}, -E^j V)\, \bar{G}_{n-k}(0|\mathcal{E}, -W^{(\tau - v_k)}), \quad j \le k \le n. \qquad [7.34]$$

Proof. From [7.18], the probability of a first meeting at level k (and at time v_k) is:

$$e^{-\lambda(v_k - x)}\, \bar{G}_{k-j}(-x|\mathcal{E}, -E^j V).$$

After this meeting, the trajectory remains horizontal for a certain time, say ϵ, and it thus reaches the point $M = (v_k + \epsilon, k)$. We then require that $M \to B$ without again meeting (or crossing) F, whose probability is given by [7.30] when (x, j) is replaced by $(v_k + \epsilon, k)$. It is then sufficient to let ϵ tend toward 0, which leads to [7.34]. □

7.4. With the number of Poisson jumps

The previous compound Poisson process $\{N_{t-x},\, t \geq x\}$ can be represented as:

$$N_{t-x} = \sum_{i=1}^{M_{t-x}} X_i,$$

where $\{M_{t-x},\, t \geq x\}$ is a Poisson process of parameter λ, and the X_i are i.i.d. random variables distributed as $\{q_j,\, j \in \mathbb{N}\}$ and independent of this Poisson process. In section 7.2, we looked for the distribution of $K = N_T$, the level of the compound Poisson process at the first meeting time T with a lower increasing boundary. Here, we want to derive the joint distribution of the vector (L, K) where $L = M_T$ is the random variable counting the jumps made by the Poisson process until the same time T.

We first obtain the probability generating function of L jointly with any value for K.

PROPOSITION 7.7.– *For $s \in [0, 1]$:*

$$\sum_{l=0}^{n} \mathbb{P}(L = l, K = n | x, V)\, s^l = e^{-\lambda(v_n - x)}\, \bar{G}_n(-sx | \mathcal{E}, -sV), \quad n \geq 0. \qquad [7.35]$$

Proof. Since $q_0 = 0$ and $q_1 > 0$, we know that $N_{t-x} \geq M_{t-x}$ for all t, thus $K \geq L$, and $N_{t-x} = 0$ iff $M_{t-x} = 0$. Clearly, when $l = 0$:

$$\mathbb{P}(L = 0, K = n | x, V) = e^{-\lambda(v_0 - x)}\, \delta_{0,n}, \qquad [7.36]$$

and when $l \geq 1$:

$$\mathbb{P}(L = l, K = n|x, V) = \int_x^{v_0} \sum_{j=1}^{n} q_j$$

$$\times\, \mathbb{P}(L = l - 1, K = n - j|t, E^j V)\, \lambda e^{-\lambda(t-x)}\, dt. \qquad [7.37]$$

Let us introduce the function $R_{l,n}(x|V)$ defined by:

$$\mathbb{P}(L = l, K = n|x, V) = e^{-\lambda(v_n - x)}\, R_{l,n}(x|V), \quad 0 \leq l \leq n,\ n \geq 0. \qquad [7.38]$$

From [7.36], $R_{0,n}(x|V) = \delta_{0,n}$, and from [7.37], when $l \geq 1$:

$$R_{l,n}(x|V) = \int_x^{v_0} \sum_{j=1}^{n} \lambda q_j\, R_{l-1,n-j}(t|E^j V)\, dt.$$

We now consider the generating function of $R_{l,n}$ with respect to l and of parameter s, which is denoted as $Q_n(x|s, V)$:

$$Q_n(x|s, V) = \sum_{l=0}^{\infty} R_{l,n}(x|V)\, s^l = \delta_{0,n} + \sum_{l=1}^{\infty} R_{l,n}(x|V)\, s^l. \qquad [7.39]$$

Then, we find that $Q_0(x|s, V) = 1$, and for $n \geq 1$:

$$Q_n(x|s, V) = \int_x^{v_0} \sum_{j=1}^{n} s\lambda q_j \left[\sum_{l=1}^{\infty} R_{l-1,n-j}(t|E^j V)\, s^{l-1}\right] dt$$

$$= \int_x^{v_0} \sum_{j=1}^{n} s\lambda q_j\, Q_{n-j}(t|s, E^j V)\, dt. \qquad [7.40]$$

Comparing [7.40] with [7.21], we can use [7.18] (among others) to assert that:

$$Q_n(x|s, V) = \bar{G}_n(-x|\breve{\mathcal{E}}, -V), \qquad [7.41]$$

where $\breve{\mathcal{E}}$ is the family of functions:

$$\breve{e}_n(t) = \sum_{k=0}^{n} \frac{(s\lambda t)^k}{k!}\, q_n^{*k} = e_n(st), \quad n \in \mathbb{N}_0. \qquad [7.42]$$

From [7.38], [7.39], [7.41] and [7.42], we deduce [7.35] by virtue of the (explicit) formula [3.64] of $\bar{G}_n$. □

As expected, [7.35] when $s = 1$ reduces to [7.18]. Therefore, conditional on $K = n$, the generating function of $\mathbb{P}(L = l|x, V)$, $0 \leq l \leq n$, is equal to:

$$\bar{G}_n(-sx|\mathcal{E}, -sV)/\bar{G}_n(-x|\mathcal{E}, -V).$$

Let us examine two simpler particular cases: on the one hand, when the boundary is affine, and, on the other hand, when the trajectory begins at time $x = 0$.

COROLLARY 7.1.– *If $v_i = a + bi$ for all i:*

$$\mathbb{P}(L = l, K = n|x, V) = e^{-\lambda(v_n - x)}\, \lambda^l\, \frac{(v_0 - x)(v_n - x)^{l-1}}{l!}\, q_n^{*l}, \quad 0 \leq l \leq n. \qquad [7.43]$$

Proof. When the u_i are affine, $\bar{G}_n(x|U)$ is given explicitly by [3.83]. Thus, here we have:

$$\bar{G}_n(-sx|\mathcal{E}, -sV) = \frac{-sx + sv_0}{-sx + sv_n}\, e_n(-sx + sv_n),$$

so that from [7.35]:

$$\sum_{l=0}^{n} s^l\, \mathbb{P}(L = l, K = n|x, V) = e^{-\lambda(v_n - x)}\, \frac{v_0 - x}{v_n - x}\, e_n(-sx + sv_n)$$

$$= e^{-\lambda(v_n - x)}\, \frac{v_0 - x}{v_n - x} \sum_{k=0}^{n} \frac{[s\lambda(v_n - x)]^k}{k!}\, q_n^{*k}, \qquad [7.44]$$

after inserting [7.42]. It remains to replace k by l in the r.h.s. of [7.44], which then becomes:

$$\sum_{l=0}^{n} s^l\, e^{-\lambda(v_n - x)}\, \lambda^l\, \frac{v_0 - x}{v_n - x}\, \frac{(v_n - x)^l}{l!}\, q_n^{*l},$$

hence [7.43] by identification of the two generating functions. □

COROLLARY 7.2.– *If $x = 0$:*

$$\mathbb{P}(L = l, K = n|0, V) = e^{-\lambda v_n}\, \frac{1}{n} \sum_{k=1}^{n} (\sigma_k)^{n-l}\, \bar{G}_n(0|\mathcal{E}, -\sigma_k V), \quad 1 \leq l \leq n, \qquad [7.45]$$

where σ_k is a nth root of unity, that is:

$$\sigma_k = \exp(2k\pi i/n) \in \mathbb{C}, \quad 0 \leq k \leq n. \qquad [7.46]$$

Proof. First, we have from [7.46]:

$$\sum_{k=1}^{n} (\sigma_k)^j = \sum_{k=1}^{n} (\sigma_j)^k = n\,\delta_{j,n}. \qquad [7.47]$$

Now, consider [7.35] for $x = 0$ and write:

$$\bar{G}_n(0|\mathcal{E}, -sV) = \sum_{j=1}^{n} b_{j,n}\, s^j, \qquad [7.48]$$

so that we are looking for $b_{l,n}$. Choose any $\alpha \in \mathbb{N}_0$. From [7.48], we then see that:

$$\sum_{k=1}^{n} (\sigma_k)^\alpha\, \bar{G}_n(0|\mathcal{E}, -\sigma_k V) = \sum_{j=1}^{n} b_{j,n} \big[\sum_{k=1}^{n} (\sigma_k)^{\alpha+j}\big]$$

$$= \sum_{j=1}^{n} b_{j,n} \big[\sum_{k=1}^{n} (\sigma_{\alpha+j})^k\big] = \sum_{j=1}^{n} b_{j,n}\, n\, \delta_{\alpha+j,n} = n\, b_{n-\alpha,n}, \qquad [7.49]$$

the second line resulting from [7.47]. Thus, to obtain $b_{l,n}$, we just need to take $\alpha = n - l$ in [7.49], which leads to [7.45]. □

7.5. For a linear birth process with immigration

A natural counting process that generalizes the Poisson process is the classical linear birth process with immigration. This is a Markovian process in which the rate of occurrence is no longer constant but depends linearly on the current state. Thus, in state $n \geq 0$, this rate λ_n is given by $\lambda_n = \lambda_0 + bn$ where λ_0 and b represent, respectively, the immigration and birth rates. A priori, $\lambda > 0$ and $b > 0$, but $b < 0$ is admitted provided that there is $n_0 \in \mathbb{N}$ such that $\lambda_0 + bn_0 = 0$. The set of possible rates is denoted as:

$$\Lambda = \{\lambda_n = \lambda_0 + bn,\ n \in \mathbb{N}_0\},$$

and this set will be incorporated in the notation. Of course, the shift operator E will act on Λ in the same way as on V.

The boundary is represented as previously by the set $F = \{(v_i, i), i \in \mathbb{N}_0\}$. We wish to again determine the distribution of K, the first meeting level of such a birth trajectory with this boundary.

1) To this end, we first choose to reason as in section 1.5 for the Poisson process. Thus, instead of [1.18] and [1.19], we have:

$$\mathbb{P}(K = 0|x, V, \Lambda) = e^{-\lambda_0(v_0 - x)}, \qquad [7.50]$$

and for $n \geq 1$:

$$\mathbb{P}(K = n|x, V, \Lambda) = \int_x^{v_0} \mathbb{P}(K = n-1|t, EV, E\Lambda)\, \lambda_0 e^{-\lambda_0(t-x)} dt. \qquad [7.51]$$

Let us write these probabilities in the form:

$$\mathbb{P}(K = n|x, V, \Lambda) = (\lambda_0 \dots \lambda_{n-1})\, e^{\lambda_0 x - \lambda_n v_n}\, H_n(x|V, \Lambda), \quad n \geq 0, \qquad [7.52]$$

where the factor $(\dots)$ is equal to 1 for $n = 0$. Inserting [7.52] into [7.50] and [7.51], we find that $H_0(x|V, \Lambda) = 1$, $H_n(v_0|V, \Lambda) = \delta_{n,0}$, and:

$$H_n(x|V, \Lambda) = \int_x^{v_0} e^{bt}\, H_{n-1}(t|EV, E\Lambda) dt, \quad n \geq 1, \qquad [7.53]$$

since $(\lambda_0 \dots \lambda_{n-1}) = \lambda_0(\lambda_1 \dots \lambda_{n-1})$ and $\lambda_1 = \lambda_0 + b$ by assumption, so that:

$$e^{\lambda_0 x - \lambda_n v_n} = e^{\lambda_1 t - \lambda_{1+(n-1)} v_{1+(n-1)}}\, e^{-\lambda_0(t-x)}\, e^{-bt}.$$

Therefore, the $H_n(x|V, \Lambda)$ can be determined by recursion from [7.53] and boundary conditions. Note that they are obviously independent of Λ, but the reference to Λ was retained for later situations where this is no longer true.

Now, by differentiation of [7.53], we get:

$$H_n'(x|V) = -e^{bx}\, H_{n-1}(x|EV), \quad n \geq 1. \qquad [7.54]$$

Let us try to construct a family $\mathcal{E}$ of functions $e_n(x)$ that satisfy [7.54], that is:

$$e_n'(x) = -e^{bx}\, e_{n-1}(x), \quad n \geq 1. \qquad [7.55]$$

Two simple solutions of [7.55] are:

$$\tilde{e}_n(x) = (1/n!)\big(-e^{bx}/b\big)^n, \qquad [7.56]$$

so $\tilde{\mathcal{E}}$ is not standard at $x = 0$:

$$\hat{e}_n(x) = (1/n!)\big[(1 - e^{bx})/b\big]^n, \qquad [7.57]$$

so $\hat{\mathcal{E}}$ is standard at $x = 0$ but more complicated. There is no canonical choice, but it does not matter since $\tilde{\mathcal{E}}$ and $\hat{\mathcal{E}}$ are copies of each other. Indeed:

$$(e^{bx})^n = (e^{bx} - 1 + 1)^n = \sum_{i=0}^{n} \binom{n}{i} 1^{n-i} (e^{bx} - 1)^i,$$

implying that:

$$(1/n!)\big(-e^{bx}/b\big)^n = \sum_{i=0}^{n}(1/(n-i)!)\big(-e^{b0}/b\big)^{n-i}(1/i!)\big[(1-e^{bx})/b\big]^i,$$

or equivalently:

$$\tilde{e}_n(x) = \sum_{i=0}^{n} \tilde{e}_{n-i}(0)\, \hat{e}_i(x).$$

By definition 2.2, this means that $\tilde{\mathcal{E}}$ and $\hat{\mathcal{E}}$ are copies of each other.

From property 2.1, we know that both families generate the same operator Δ, and here $\Delta = -\exp(-bx)\, D$. Therefore, $D = -\exp(bx)\, \Delta$, and [7.54] is satisfied by $H_n(x|v) \equiv \bar{G}_n(x|V)$ since Δ acts on $\bar{G}_n$ as it does on $\tilde{e}_n$. After substitution in [7.52], we obtain the following distribution of K in terms of A–G pseudopolynomials.

PROPOSITION 7.8.– *With $\mathcal{E}$ defined by [7.56] or [7.57]:*

$$\mathbb{P}(K = n|x, V, \Lambda) = (\lambda_0 \dots \lambda_{n-1})\, e^{\lambda_0 x - \lambda_n v_n}\, \bar{G}_n(x|\mathcal{E}, V), \quad n \geq 0. \qquad [7.58]$$

2) Actually, the distribution of K can be reexpressed in terms of standard *A–G polynomials* by arguing like this. Let us write the A–G expansion of $\tilde{e}_n(x)$ defined by [7.56] with respect to the family of A–G pseudopolynomials $\{\bar{G}_i(x|\tilde{\mathcal{E}}, V),\, i \in \mathbb{N}_0\}$. By [2.33], we have:

$$\bar{G}_n(x|\tilde{\mathcal{E}}, V) = \tilde{e}_n(x) - \sum_{i=0}^{n-1} \tilde{e}_{n-i}(v_i)\, \bar{G}_i(x|\tilde{\mathcal{E}}, V), \quad n \geq 0. \qquad [7.59]$$

Similarly, consider the A–G expansion of $y^n/n!$ with respect to the family of A–G polynomials $\{G_i(y|W),\, i \in \mathbb{N}_0\}$ where W is some family of parameters $\{w_0, w_1, \ldots\}$. This gives:

$$G_n(y|W) = \frac{y^n}{n!} - \sum_{i=0}^{n-1} \frac{(w_i)^{n-i}}{(n-i)!} \bar{G}_i(y|W), \quad n \geq 0. \qquad [7.60]$$

When $y = \exp(bx)$ and $w_i = \exp(bv_i)$ so that:

$$W = \{e^{bv_i},\, i \in \mathbb{N}_0\} = e^{bV}, \text{ say,}$$

we get from [7.60], after division by $(-b)^n$, that:

$$(-1/b)^n\, G_n(e^{bx}|e^{bV}) = (1/n!)\big(-e^{bx}/b\big)^n - \sum_{i=0}^{n-1}(1/(n-i)!)$$

$$\times \big(-e^{bv_i}/b\big)^{n-i}(-1/b)^i\, G_i(e^{bx}|e^{bV})$$

$$= \tilde{e}_n(x) - \sum_{i=0}^{n-1} \tilde{e}_{n-i}(v_i)\,(-1/b)^i\, G_i(e^{bx}|e^{bV}). \qquad [7.61]$$

Now take the difference between [7.59] and [7.61]. Denoting:

$$L_i(x) = \bar{G}_i(x|\tilde{\mathcal{E}}, V) - (-1/b)^i\, G_i(e^{bx}|e^{bV}), \quad 0 \leq i \leq n,$$

we obtain:

$$L_n(x) = -\sum_{i=0}^{n-1} \tilde{e}_{n-i}(v_i)\, L_i(x), \quad n \geq 0. \qquad [7.62]$$

Since $L_0(x) = 1 - 1 = 0$, [7.62] shows recursively that $L_i(x) = 0$ for all i, that is:

$$\bar{G}_i(x|\tilde{\mathcal{E}}, V) = (-1/b)^i\, G_i(e^{bx}|e^{bV}), \quad 0 \leq i \leq n.$$

From [7.58], this directly yields the result below.

PROPOSITION 7.9.–

$$\mathbb{P}(K = n|x, V, \Lambda) = (-1/b)^n\,(\lambda_0 \ldots \lambda_{n-1})\, e^{\lambda_0 x - \lambda_n v_n}\, G_n(e^{bx}|e^{bV}), \quad n \geq 0. \qquad [7.63]$$

REMARK.– Suppose that the boundary is a vertical line of abscissis v_0. Then, the first meeting with F is a meeting with this straight line and K corresponds to N_{v_0-x}, the level of the trajectory at time v_0. The latter distribution can be found in books but it is not often memorized. From [7.63] and putting $V^* = \{v_0, v_0, \dots\}$, we find:

$$\mathbb{P}(N_{v_0-x} = n) = \mathbb{P}(K = n|x, V^*, \Lambda) = (\lambda_0 \dots \lambda_{n-1})$$
$$\times\, e^{\lambda_0 x - \lambda_n v_0} \, (-1/b)^n \, G_n(e^{bx}|e^{bV^*})$$
$$= (\lambda_0 \dots \lambda_{n-1}) \, e^{\lambda_0 x - \lambda_n v_0} \left[-(e^{bx} - e^{bv_0})/b \right]^n / n!, \quad n \geq 0,$$

since by [2.83]:

$$G_n(e^{bx}|e^{bV^*}) = e_n(e^{bx} - e^{bv_0}) = (e^{bx} - e^{bv_0})/n!.$$

3) We now apply the alternative strategy based on the decomposition [7.1]. Thus, instead of [7.2], we obtain:

$$\mathbb{P}(K = n|x, V, \Lambda) = \sum_{r=0}^{n} \mathbb{P}(N_{v_0-x} = r) \, \mathbb{P}(K = n - r|v_0, E^r V, E^r \Lambda), \quad [7.64]$$

and after denoting:

$$\mathbb{P}(K = n|x, V, \Lambda) = e^{\lambda_0 x - \lambda_n v_n} \, R_n(x|V, \Lambda), \quad n \geq 0, \quad [7.65]$$

[7.64] gives $R_0(x|V, \Lambda) = 1$, $R_n(v_0|V, \Lambda) = \delta_{n,0}$ and from [7.65]:

$$R_n(x|V, \Lambda) = \sum_{r=0}^{n} \frac{\mathbb{P}(N_{v_0-x} = r)}{\mathbb{P}(N_{v_0-x} = 0)} \, e^{brv_0} \, R_{n-r}(v_0|E^r V, E^r \Lambda), \quad n \geq 1. \quad [7.66]$$

Since the distribution of N_{v_0-x} is known and recalled in the remark above, [7.66] becomes:

$$R_n(x|V, \Lambda) = \sum_{r=0}^{n} (\lambda_0 \dots \lambda_{r-1}) \, \frac{[-(e^{bx} - e^{bv_0})/b]^r}{r!}$$
$$\times\, R_{n-r}(v_0|E^r V, E^r \Lambda), \quad n \geq 1. \quad [7.67]$$

Let us change the notation again and write:

$$R_n(x|V,\Lambda) = (\lambda_0 \ldots \lambda_{n-1})\,(-1/b)^n\, S_n(x|V,\Lambda), \quad n \geq 0. \qquad [7.68]$$

Obviously, we have:

$$\lambda_0 \ldots \lambda_{n-1} = (\lambda_0 \ldots \lambda_{r-1})(\lambda_r \ldots \lambda_{n-1}) = (\lambda_0 \ldots \lambda_{r-1})E^r(\lambda_0 \ldots \lambda_{n-r-1}),$$

so that [7.67] can be rewritten as:

$$S_n(x|V,\Lambda) = \sum_{r=0}^{n} \frac{(e^{bx} - e^{bv_0})^r}{r!}\, S_{n-r}(v_0|E^rV, E^r\Lambda). \qquad [7.69]$$

By the classic Taylor formula, we have, for any family $W = \{w_i,\ i \in \mathbb{N}_0\}$:

$$G_n(y|W) = \sum_{r=0}^{n} G_{n-r}(w_0|E^rW)\, \frac{(y - w_0)^r}{r!}.$$

So, by choosing $y = \exp(bx)$ and $w_i = \exp(bv_i)$ for all i, we deduce from [7.69] that:

$$S_n(x|V,\Lambda) = G_n(x|e^{bV}), \quad n \geq 0.$$

Combined with [7.65] and [7.68], this provides the formula [7.63] obtained previously.

7.6. Extension allowing multiple births

We again consider a linear birth process with immigration, but as in section 7.2 for the compound Poisson process, births can now occur in batches of varying sizes j with probabilities q_j, $j \geq 1$, where $q_1 > 0$. The problem is always to determine the distribution of the level K of first meeting with the boundary F.

Following the previous reasoning, we have:

$$\mathbb{P}(K = 0|x, V, \Lambda) = e^{-\lambda_0(v_0 - x)}, \qquad [7.70]$$

and for $n \geq 1$:

$$\mathbb{P}(K = n|x, V, \Lambda) = \int_x^{v_0} \sum_{j=1}^{\infty} q_j \, \mathbb{P}(K = n - j|t, E^j V, E^j \Lambda) \, \lambda_0 e^{-\lambda_0(t-x)} dt. \quad [7.71]$$

Keeping the notation [7.52], we see from [7.70] and [7.71] that $H_0(x|V, \Lambda) = 1$, $H_n(v_0|V, \Lambda) = \delta_{n,0)}$, and:

$$H_n(x|V, \Lambda) = \int_x^{v_0} \sum_{j=1}^{\infty} q_j \, e^{bjt} \, H_{n-j}(t|E^j V, E^j \Lambda) dt, \quad n \geq 1,$$

which gives by differentiation, in place of [7.54]:

$$H'_n(x|V) = -\sum_{j=1}^{\infty} q_j \, e^{bjx} \, H_{n-j}(x|E^j V), \quad n \geq 1. \quad [7.72]$$

Note that here too, Λ is superfluous in the definition of H_n.

To construct a family $\mathcal{E}$, we take inspiration from [7.72] to write:

$$e'_n(x) = -\sum_{j=1}^{\infty} q_j \, e^{bjx} \, e_{n-j}(x), \quad n \geq 1. \quad [7.73]$$

This will lead to the operator on $\mathcal{E}$ given by:

$$D = -\sum_{j=1}^{\infty} q_j \, e^{bjx} \, \Delta^j,$$

and therefore, [7.72] will be satisfied by the family $\{\bar{G}_n(x|\mathcal{E}, V\}$. It remains to find a solution to [7.73], which can be achieved through the generating function:

$$\sum_{n=0}^{\infty} e_n(x) \, s^n = \exp\big[-\sum_{j=1}^{\infty}(q_j/bj) \, e^{bjx} \, s^j\big]. \quad [7.74]$$

since [7.74] implies:

$$\sum_{n=0}^{\infty} e'_n(x) \, s^n = \big[-\sum_{j=1}^{\infty} q_j \, e^{bjx} \, s^j\big] \sum_{n=0}^{\infty} e_n(x) \, s^n,$$

hence [7.73]. Furthermore, these e_n can be expressed very simply from [7.74]. Indeed, taking $x = 0$ and substituting $s \exp(bx)$ for s, we get:

$$\sum_{n=0}^{\infty} e_n(0)\, e^{bnx}\, s^n = \exp\big[-\sum_{j=1}^{\infty}(q_j/bj)\,(se^{bx})^j\big] = \sum_{n=0}^{\infty} e_n(x)\, s^n,$$

so that:

$$e_n(x) = e_n(0)\, e^{bnx}, \quad n \geq 0, \qquad [7.75]$$

where the $e_n(0)$ are provided by [7.74] for $x = 0$. Using [7.52], we then deduce the following result.

PROPOSITION 7.10.– *With $\mathcal{E}$ defined by [7.75] with [7.74] at $x = 0$:*

$$\mathbb{P}(K = n|x, V, \Lambda) = (\lambda_0 \ldots \lambda_{n-1})\, e^{\lambda_0 x - \lambda_n v_n}\, \bar{G}_n(x|\mathcal{E}, V), \quad n \geq 0. \qquad [7.76]$$

REMARK.– Another possibility would be to work with the family $\hat{\mathcal{E}}$ of functions generated by:

$$\sum_{n=0}^{\infty} \hat{e}_n(x)\, s^n = \exp\big[-\sum_{j=1}^{\infty}(q_j/bj)\,(e^{bjx} - 1)\, s^j\big] \qquad [7.77]$$

instead of [7.74]. Indeed, [7.73] would then be satisfied but $\hat{\mathcal{E}}$, although standard at 0, is no longer as simple as $\mathcal{E}$ defined via [7.74]. Writing:

$$\exp\big[\sum_{j=1}^{\infty}(q_j/bj)\, s^j\big] = \sum_{r=0}^{\infty} \beta_r\, s^r,$$

[7.77] becomes:

$$\sum_{n=0}^{\infty} \hat{e}_n(x)\, s^n = \big(\sum_{n=0}^{\infty} e_n(x)\, s^n\big)\,\big(\sum_{r=0}^{\infty} \beta_r\, s^r\big),$$

which implies that:

$$\hat{e}_n(x) = \sum_{r=0}^{n} \beta_{n-r}\, e_r(x) = \sum_{r=0}^{n} \beta_{n-r}\, e_r(0)\, e^{brx}, \quad n \geq 0.$$

thanks to [7.75]. In any case, we have $\bar{G}_n(x|\hat{\mathcal{E}}, V) = \bar{G}_n(x|\mathcal{E}, V)$, so considering $\mathcal{E}$ is more direct.

7.7. For a nonlinear birth process

We turn to a birth process defined by arbitrary rates. It is thus a Markovian counting process in which the rate λ_n in the current state $n \geq 0$ can depend on the state n in a general way, rather than linearly as in section 7.6. For clarity, all rates are taken distinct, that is, $\lambda_i \neq \lambda_j$ when $i \neq j$; they are included in the set $\Lambda = \{\lambda_n, n \in \mathbb{N}_0\}$. Our goal is to derive the distribution of K, the first meeting level with an identical boundary F.

We start again with [7.50] and [7.51], but instead of [7.52] which was:

$$\mathbb{P}(K = n|x, V, \Lambda) = (\lambda_0 \dots \lambda_{n-1})\, e^{\lambda_0 x - \lambda_n v_n}\, H_n(x|V, \Lambda),$$

we write:

$$\mathbb{P}(K = n|x, V, \Lambda) = n!\,\Big(\prod_{i=0}^{n-1} \frac{\lambda_i}{\lambda_i - \lambda_n}\Big)\, e^{\lambda_0 x - \lambda_n v_n}\, \bar{H}_n(x|V, \Lambda), \quad n \geq 0, \tag{7.78}$$

in order to be closer to the final result. Note that in the linear case:

$$n!\,\Big(\prod_{i=0}^{n-1} \frac{\lambda_i}{\lambda_i - \lambda_n}\Big) = \prod_{i=0}^{n-1} \frac{\lambda_i}{\lambda_i - \lambda_n}(n-i) = \frac{\lambda_0 \dots \lambda_{n-1}}{(-b)^n}.$$

Using [7.78], we find that $\bar{H}_0(x|V, \Lambda) = 1$, $\bar{H}_n(v_0|V, \Lambda) = \delta_{n,0}$ and:

$$\bar{H}_n(x|V, \Lambda) = \frac{\lambda_0 - \lambda_n}{n} \int_x^{v_0} e^{(\lambda_1 - \lambda_0)t}\, \bar{H}_{n-1}(t|EV, E\Lambda)dt, \quad n \geq 1, \tag{7.79}$$

since

$$n!\,\Big(\prod_{i=0}^{n-1} \frac{\lambda_i}{\lambda_i - \lambda_n}\Big) = n \frac{\lambda_0}{\lambda_0 - \lambda_n}\,(n-1)!\,\Big(\prod_{i=0}^{n-2} \frac{\lambda_{i+1}}{\lambda_{i+1} - \lambda_{n-1+1}}\Big),$$

and

$$\left(e^{\lambda_0 x - \lambda_n v_n}\right)^{-1} e^{\lambda_1 t - \lambda_{1+(n-1)} v_{1+(n-1)}}\, e^{-\lambda_0 (t-x)} = e^{(\lambda_1 - \lambda_0)t}.$$

By differentiation, [7.79] gives:

$$\bar{H}'_n(x|V,\Lambda) = \frac{\lambda_n - \lambda_0}{n}\, e^{(\lambda_1-\lambda_0)x}\, \bar{H}_{n-1}(x|EV, E\Lambda), \quad n \geq 1, \quad [7.80]$$

which has a unique solution given the boundary conditions.

To obtain this solution in pseudopolynomial form $\bar{G}_n(\ |\mathcal{E}, W(V,\Lambda))$, we first choose $\mathcal{D} = \mathrm{I\!R}^{\mathrm{I\!N}}$ and the family of functions $\mathcal{E}$ by:

$$e_n(\mathbf{z}) = (1/n!)\, z_1 z_2 \dots z_n, \quad \mathbf{z} \in \mathrm{I\!R}^{\mathrm{I\!N}}. \quad [7.81]$$

We will then determine the family of parameters $W(V,\Lambda) = \{\mathbf{w}_i(V,\Lambda),\ i \in \mathrm{I\!N}_0\}$, where $\mathbf{w}_i(V,\Lambda) = \{w_{i,k}(V,\Lambda),\ k \geq 1\}$. So, the strategy followed here is the opposite of that used previously when we were looking for $\mathcal{E}$ while V was known.

Let us take for $\mathbf{z}$ the sequence defined by:

$$\mathbf{z}_{x,\Lambda} = \big\{ \exp\big[(\lambda_i - \lambda_{i-1})x\big],\ i \geq 1 \big\}. \quad [7.82]$$

From [7.81] with [7.82], we get:

$$e_n(\mathbf{z}_{x,\Lambda}) = (1/n!)\, \exp[(\lambda_n - \lambda_0)x], \quad [7.83]$$

so that by differentiation with respect to x:

$$\begin{aligned} e'_n(\mathbf{z}_{x,\Lambda}) &= (1/n!)\,(\lambda_n - \lambda_0)\, \exp[(\lambda_n - \lambda_0)x] \\ &= (1/n)\,(\lambda_n - \lambda_0)\, \exp[(\lambda_1 - \lambda_0)x]\,(1/(n-1)!) \\ &\quad \times \exp[(\lambda_{1+n-1} - \lambda_1)x] \\ &= (1/n)\,(\lambda_n - \lambda_0)\, \exp[(\lambda_1 - \lambda_0)x]\, e_{n-1}(\mathbf{z}_{x,E\Lambda}), \end{aligned} \quad [7.84]$$

where E is the shift operator that operates on Λ as it does on V. Therefore, [7.84] shows that $e_n(\mathbf{z}_{x,\Lambda})$ (given by [7.83]) is a solution of [7.80] (which is independent of V).

We are going to prove that $\bar{G}_n(\mathbf{z}_{x,\Lambda}|\mathcal{E}, W(V,\Lambda))$ is also the solution of [7.80] provided that the $\mathbf{w}_i(V,\Lambda)$ satisfy the conditions [7.86].

LEMMA 7.1.–

$$\bar{G}'_n\big(\mathbf{z}_{x,\Lambda}|\mathcal{E}, W(V,\Lambda)\big) = \frac{\lambda_n - \lambda_0}{n} e^{(\lambda_1-\lambda_0)x}$$
$$\times \bar{G}_{n-1}\big(\mathbf{z}_{x,E\Lambda}|\mathcal{E}, W(EV, E\Lambda)\big), \quad n \geq 1, \qquad [7.85]$$

if and only if the $\mathbf{w}_i(V,\Lambda)$ *are such that:*

$$e_{n-i}\big(\mathbf{w}_i(V,\Lambda)\big)\,\frac{n}{i}\frac{\lambda_i-\lambda_0}{\lambda_n-\lambda_0} = e_{n-i}\big(\mathbf{w}_{i-1}(EV,E\Lambda)\big),$$
$$\leq i \leq n-1, n \geq 1. \qquad [7.86]$$

Proof. From [7.84], we have the desired formula [7.85] for $n = 1$ (whatever V). Proceeding by recurrence, we suppose that [7.85] holds for the indices $1, \ldots, n-1$, and we will check when it is then true for the index n.

From [2.33], we know that:

$$\bar{G}_n\big(\mathbf{z}_{x,\Lambda}|\mathcal{E}, W(V,\Lambda)\big) = e_n(\mathbf{z}_{x,\Lambda}) - \sum_{i=0}^{n-1} e_{n-i}\big(\mathbf{w}_i(V,\Lambda)\big)\,\bar{G}_i\big(\mathbf{z}_{x,\Lambda}|\mathcal{E}, W(V,\Lambda)\big),$$

so that using [7.84] and by induction:

$$\bar{G}'_n\big(\mathbf{z}_{x,\Lambda}|\mathcal{E}, W(V,\Lambda)\big) = \frac{\lambda_n-\lambda_0}{n} e^{(\lambda_1-\lambda_0)x}\, e_{n-1}(\mathbf{z}_{x,E\Lambda})$$
$$-\sum_{i=1}^{n-1} e_{n-i}\big(\mathbf{w}_i(V,\Lambda)\big)\,\frac{\lambda_i-\lambda_0}{i} e^{(\lambda_1-\lambda_0)x}\,\bar{G}_{i-1}\big(\mathbf{z}_{x,E\Lambda}|\mathcal{E}, W(EV,E\Lambda)\big),$$

which can be rewritten as:

$$\bar{G}'_n\big(\mathbf{z}_{x,\Lambda}|\mathcal{E}, W(V,\Lambda)\big) = \frac{\lambda_n-\lambda_0}{n} e^{(\lambda_1-\lambda_0)x}$$
$$\Bigg[e_{n-1}(\mathbf{z}_{x,E\Lambda}) - \sum_{i=1}^{n-1} e_{n-i}\big(\mathbf{w}_i(V,\Lambda)\big)\,\frac{n}{i}\frac{\lambda_i-\lambda_0}{\lambda_n-\lambda_0} e^{(\lambda_1-\lambda_0)x}$$
$$\times \bar{G}_{i-1}\big(\mathbf{z}_{x,E\Lambda}|\mathcal{E}, W(EV,E\Lambda)\big)\Bigg]. \qquad [7.87]$$

Therefore, [7.87] becomes the desired formula [7.85] provided that the term $[\ldots]$ is equivalent to $\bar{G}_{n-1}(\mathbf{z}_{x,E\Lambda}|\mathcal{E}, W(EV,E\Lambda))$. Again from [2.33], we see that this condition is verified iff the $\mathbf{w}_i(V,\Lambda)$ satisfy the relations [7.86). □

Note that [7.86] gives the $w_{i,k}(V,\Lambda)$, $i \geq 1$, as a function of the $w_{0,k}(V,\Lambda)$ which can be chosen arbitrarily. Let us now determine their expressions for appropriate values of $w_{0,k}(V,\Lambda)$.

LEMMA 7.2.– *Under [7.86] and for* $w_{0,k}(V,\Lambda) = e^{(\lambda_k - \lambda_{k-1})v_0}$*:*

$$w_{i,k}(V,\Lambda) = \frac{k}{i+k}\Big(\prod_{r=0}^{i-1} \frac{\lambda_{i+k}-\lambda_r}{\lambda_{i+k-1}-\lambda_r}\Big)\, e^{(\lambda_{i+k}-\lambda_{i+k-1})v_i}, \quad i \geq 0,\, k \geq 1. \tag{7.88}$$

Proof. On the one hand, the definition [7.81] of $\mathcal{E}$ gives $e_n(\mathbf{z})/e_{n-1}(\mathbf{z}) = z_n/n$, $n \geq 1$, thus:

$$e_{n-i}(\mathbf{w}_i)/e_{n-1-i}(\mathbf{w}_i) = w_{i,n-i}/(n-i), \quad 1 \leq i \leq n-1. \tag{7.89}$$

On the other hand, [7.86] implies:

$$\frac{e_{n-i}\big(\mathbf{w}_i(V,\Lambda)\big)}{e_{n-1-i}\big(\mathbf{w}_i(V,\Lambda)\big)} = \frac{n-1}{n}\,\frac{\lambda_n-\lambda_0}{\lambda_{n-1}-\lambda_0}\,\frac{e_{n-i}\big(\mathbf{w}_{i-1}(EV,E\Lambda)\big)}{e_{n-1-i}\big(\mathbf{w}_{i-1}(EV,E\Lambda)\big)}. \tag{7.90}$$

Together, [7.89] and [7.90] lead to:

$$\frac{w_{i,n-i}(V,\Lambda)}{n-i} = \frac{n-1}{n}\,\frac{\lambda_n-\lambda_0}{\lambda_{n-1}-\lambda_0}\,\frac{w_{i-1,n-i}(EV,E\Lambda)}{n-i}.$$

By setting $n = i+k$ with $k \geq 1$, this yields:

$$w_{i,k}(V,\Lambda) = \frac{i+k-1}{i+k}\,\frac{\lambda_{i+k}-\lambda_0}{\lambda_{i+k-1}-\lambda_0}\,w_{i-1,k}(EV,E\Lambda), \tag{7.91}$$

which is symbolically denoted:

$$w_{i,k} = b_{i+k}\,Ew_{i-1,k}. \tag{7.92}$$

From [7.92], we obtain by recursion on i:

$$\begin{aligned} w_{i,k} &= b_{i+k}\,E(b_{i-1+k}Ew_{i-2,k}) = b_{i+k}\,(Eb_{i-1+k})\,E^2 w_{i-2,k} \\ &= b_{i+k}\,(Eb_{i-1+k})\,(E^2 b_{i-2+k})\,E^3 w_{i-3,k} \\ &= \Big(\prod_{r=0}^{i-1} E^r b_{i-r+k}\Big)\,E^i w_{i-i,k}. \end{aligned} \tag{7.93}$$

Using [7.91] and [7.92], we see that:

$$E^r b_{i-r+k} = E^r\left(\frac{i-r+k-1}{i-r+k}\,\frac{\lambda_{i-r+k}-\lambda_0}{\lambda_{i-r+k-1}-\lambda_0}\right)$$
$$= \frac{i+k-r-1}{i+k-r}\,\frac{\lambda_{i+k}-\lambda_r}{\lambda_{i+k-1}-\lambda_r},$$

so that in [7.93]:

$$\prod_{r=0}^{i-1} E^r b_{i-r+k} = \frac{k}{i+k}\prod_{r=0}^{i-1}\frac{\lambda_{i+k}-\lambda_r}{\lambda_{i+k-1}-\lambda_r}. \qquad [7.94]$$

Note that in [7.93], $w_{i-i,k} = w_{0,k}$ is arbitrary. As announced, we choose $w_{0,k} = \exp[(\lambda_k - \lambda_{k-1})v_0]$, so that:

$$E^i w_{0,k} = e^{(\lambda_{i+k}-\lambda_{i+k-1})v_i}. \qquad [7.95]$$

Therefore, inserting [7.94] and [7.95] in [7.93], we deduce the formula [7.88] for $w_{i,k}$. □

The family W having been found, we can assert that:

$$\bar{H}_n(x|V,\Lambda) = \bar{G}_n(\mathbf{z}_{x,\Lambda}|\mathcal{E}, W(V,\Lambda)),$$

so that [7.78] provides the following result.

PROPOSITION 7.11.– *With $\mathcal{E}$ defined by [7.81] and $\mathbf{z}_{x,\Lambda}$ by [7.82]:*

$$\mathbb{P}(K=n|x,V,\Lambda) = n!\left(\prod_{i=0}^{n-1}\frac{\lambda_i}{\lambda_i-\lambda_n}\right)e^{\lambda_0 x-\lambda_n v_n}$$
$$\times\, \bar{G}_n\big(\mathbf{z}_{x,\Lambda}|\mathcal{E}, W(V,\Lambda)\big), \quad n\geq 0, \qquad [7.96]$$

where $W(V,\Lambda) = \{w_{i,k},\ i\geq 0,\ k\geq 1\}$ is given by [7.88].

To evaluate by recursion $\bar{G}_n$ in [7.96], an easy way is to use the A–G expansion of $e_n(\mathbf{z}_{x,\Lambda})$ (given by [7.83]) with respect to the $\bar{G}_i$, $0\leq i\leq n$, that is:

$$(1/n!)\exp[(\lambda_n-\lambda_0)x] = \sum_{i=0}^{n} e_{n-i}\big(\mathbf{w}_i(V,\Lambda)\big)\,\bar{G}_i\big(\mathbf{z}_{x,\Lambda}|\mathcal{E}, W(V,\Lambda)\big),$$

in which thanks to [7.88]:

$$\begin{aligned}
e_{n-i}\big(\mathbf{w}_i(V,\Lambda)\big) &= w_{i,1}w_{i,2}\dots w_{i,n-i}/(n-i)! \\
&= \frac{i!}{n!}\prod_{k=1}^{n-i}\Big[\Big(\prod_{r=0}^{i-1}\frac{\lambda_{i+k}-\lambda_r}{\lambda_{i+k-1}-\lambda_r}\Big)\, e^{(\lambda_{i+k}-\lambda_{i+k-1})v_i}\Big] \\
&= \frac{i!}{n!}\Big(\prod_{r=0}^{i-1}\frac{\lambda_n-\lambda_r}{\lambda_i-\lambda_r}\Big)\, e^{(\lambda_n-\lambda_i)v_i}.
\end{aligned}$$

8

Less Standard First Meeting Models

In the first part of this section, we examine several problems of first meeting level between two trajectories of counting processes when they are independent. More precisely, we will discuss the cases of a compound Poisson process with a lower renewal process, then a linear birth process with immigration with a lower renewal process, finally a nonlinear death process with a lower nonlinear birth process.

In the second part, we return to the problem of the first meeting with a lower boundary, this time working on a discrete time scale. We thus consider a binomial process, which can be seen as a discrete-time version of the Poisson process, then its generalization given by a compound binomial process. We also briefly address the related issue of two independent trajectories.

8.1. Compound Poisson process with a renewal process

We start with a compound Poisson process as in section 7.2 but which is here initiated at point $(0, 0)$. Instead of a fixed lower increasing boundary, we consider this time an independent renewal process, which is initiated later at point $(x, 0)$ $(x > 0)$. This renewal process is defined by a sequence of i.i.d. random variables X_j, $j \geq 1$. So, the ith event, $i \geq 1$, will occur at time $x + S_i$ where $S_i = X_1 + \ldots + X_i$. We denote by S the set $\{S_i,\ i \in \mathbb{N}_0\}$ with $S_0 \equiv 0$.

Let us introduce the family $\mathcal{E}$ of functions given by [7.17], that is:

$$e_n(t) = \sum_{k=0}^{n} \frac{(\lambda t)^k}{k!} q_n^{*k}, \quad n \in \mathbb{N}_0$$

The distribution of the first meeting level K is obtained directly by conditioning on the renewal process (R, say). Recall the previous notation $\mathbb{P}(K = n|x, V)$ (so, the sign | here does not mean a condition). A similar notation will be adopted in section 8.2.

PROPOSITION 8.1.– *With $\mathcal{E}$ defined by [7.17]:*

$$\mathbb{P}(K = n|x, R) = \mathbb{E}\big[e^{-\lambda(x+S_n)}\, \bar{G}_n(x|\mathcal{E}, -S)\big] \quad [8.1]$$

$$= \mathbb{E}\big[e^{-\lambda(x+S_n)}\, \bar{A}_n(x|\mathcal{E}, S_n)\big], \quad n \geq 0. \quad [8.2]$$

Proof. When we place a condition on the renewal trajectory, this becomes an increasing lower bound, as in section 7.2. Therefore, we can apply the result [7.18] to the corresponding conditional probability, denoted by $\mathbb{P}_R$. After adjustment to the current situation, we thus obtain:

$$\mathbb{P}_R(K = n|0, x + S) = e^{-\lambda(x+S_n)}\, \bar{G}_n(0|\mathcal{E}, -x - S), \quad n \geq 0. \quad [8.3]$$

Taking the expectation of [8.3] directly gives the formula [8.1].

Now, we meet by [8.1] randomized A–G pseudopolynomials where S is the sequence of partial sums of i.i.d. random variables. This case was studied in detail in Chapter 6. As $\mathcal{E}$ satisfies S.I. and is standard at 0, it is then sufficient to use the identity [6.36] to deduce the equivalent formula [8.2]. □

REMARK.– Obviously, the first meeting time T between the two trajectories satisfies $\mathbb{P}(T < \infty) \leq 1$. From [8.2], this implies the following inequality:

$$\sum_{n=0}^{\infty} \mathbb{E}\big[e^{-\lambda S_n}\, \bar{A}_n(x|\mathcal{E}, S_n)\big] \leq e^{\lambda x}. \quad [8.4]$$

Note that by definition of $\mathcal{E}$:

$$\sum_{n=0}^{\infty} e_n(t)\, s^n = e^{tg(s)}, \text{ where } g(s) = \lambda \sum_{j=1}^{\infty} q_j s^j,$$

so that $g(1) = \lambda$. Referring to section 6.6, the solution s_b of the equation $g(s_b) = b$ then gives 1 when $b = \lambda$. Therefore, the series in [8.4] is indeed convergent by virtue of proposition 6.3.

COROLLARY 8.1.–

$$\mathbb{P}(K = n|x, R) = \sum_{j=1}^{n} (j/n)\, \mathbb{E}\big[\mathbb{P}\big(N_x = j,\ N_{x+S_n} = n\big)\big], \quad n \geq 0. \qquad [8.5]$$

Proof. Thanks to proposition 6.3, the r.h.s. of [8.1] can be expressed as:

$$\mathbb{E}\big[e^{-\lambda(x+S_n)}\, \bar{G}_n(x|\mathcal{E}, -S)\big] = \sum_{j=1}^{n} (j/n)\, e^{-\lambda x}\, e_j(x)\, \mathbb{E}[e^{-\lambda S_n}\, e_{n-j}(S_n)]$$

$$= \sum_{j=1}^{n} (j/n)\, \mathbb{P}(N_x = j)\, \mathbb{E}\big[\mathbb{P}_{S_n}(N_{S_n} = n - j)\big], \qquad [8.6]$$

where the second equality comes from the distribution of the Poisson process $\{N_t,\, t \geq 0\}$, and $\mathbb{P}_{S_n}$ is the conditional probability given S_n. Moreover, [8.6] is equivalent to:

$$\mathbb{E}\big[e^{-\lambda(x+S_n)}\, \bar{G}_n(x|\mathcal{E}, -S)\big] = \sum_{j=1}^{n} (j/n)\, \mathbb{E}\big[\mathbb{P}(N_x = j)$$

$$\times\, \mathbb{P}_{S_n, N_x = j}(N_{S_n} = n - N_x)\big]$$

$$= \sum_{j=1}^{n} (j/n)\, \mathbb{E}\big[\mathbb{P}_{S_n}(N_x = j,\ N_{x+S_n} = n)\big],$$

hence the announced formula [8.5]. □

COROLLARY 8.2.–

$$\mathbb{P}(K = n, T > t|x, R) = \mathbb{E}\big[e^{-\lambda(x+S_n)}\, \bar{A}_n(x|\mathcal{E}, S_n)\, \mathbf{1}_{x+S_n > t}\big],$$

$$n \geq 0,\ t \in \mathbb{R}_+. \qquad [8.7]$$

Proof. From [8.1], we obviously get:

$$\mathbb{P}(K = n, T > t|x, R) = \mathbb{P}(K = n|x, R)\, \mathbb{P}(T > t|K = n, x, R)$$

$$= e^{-\lambda(x+S_n)}\, \bar{G}_n(x|\mathcal{E}, -S)\, \mathbb{P}(T > t|K = n, x, R). \qquad [8.8]$$

By definition, $T = v_K = v_n = x + S_n$, so that conditionally on S, the event $(x + S_n > t)$ is true or not, that is, its probability is 1 or 0. Now, take the conditional expectation $\mathbb{E}_{S_n}$ on (8.8). Using [6.38], we obtain:

$$\mathbb{E}_{S_n}\big[\mathbb{P}(K = n, T > t|x, R)\big] = e^{-\lambda(x+S_n)}\, \bar{A}_n(x|\mathcal{E}, S_n)\, \mathbf{1}_{x+S_n>t},$$

hence the expected result [8.7]. □

8.2. Linear birth process with a renewal process

We adopt the same notation as previously, for the first meeting of a linear birth process of rates $\lambda_n = \lambda_0 + bn$, $n \in \mathbb{N}_0$, starting at $(0, 0)$ with a lower renewal process starting at $(x, 0)$, $x > 0$.

PROPOSITION 8.2.– *With $\mathcal{E}$ defined by [7.17]:*

$$\mathbb{P}(K = n|x, R, \Lambda) = (-1/b)^n\, (\lambda_0 \ldots \lambda_{n-1}) \\ \times e^{-\lambda_0 x}\, \mathbb{E}\big[e^{-\lambda_n S_n}\, G_n(e^{-bx}|e^{bS})\big], \quad n \geq 0, \quad [8.9]$$

where $\exp(bS)$ *is the family of variables* $\{\exp(bS_i),\ i \in \mathbb{N}_0\}$.

Proof. Here too, we condition on the renewal trajectory. From [7.62], we then get:

$$\mathbb{P}_R(K = n|x, R, \Lambda) = (-1/b)^n\, (\lambda_0 \ldots \lambda_{n-1}) \\ \times e^{-\lambda_n(x+S_n)}\, G_n(1|e^{b(x+S)}), \quad n \geq 0,$$

which gives [8.9] after elimination of the conditioning. □

We meet by [8.9] randomized A–G pseudopolynomials where $\exp(bS)$ is a sequence of partial products of i.i.d. random variables. Recall that this case was also studied in Chapter 6. A key property is that if S.I. is valid, $\mathcal{E}$ is standard at 0 and satisfies the multiplicative property, then the expectation of such pseudopolynomials reduces to A–G pseudopolynomials. Note that we are working here with standard A–G polynomials.

COROLLARY 8.3.– *For $n \geq 0$:*

$$\mathbb{P}(K = n|x, R, \Lambda) = (-1/b)^n\, (\lambda_0 \ldots \lambda_{n-1}) \\ \times \sum_{j=0}^{n} \frac{e^{-\lambda_j x}}{j!} \left[\mathbb{E}(e^{-\lambda_j X_1})\right]^j G_{n-j}(0|E^{-\lambda_n/b}V), \quad [8.10]$$

where V is the family of parameters:

$$\{v_i = \mathbb{E}(e^{bX_1\,i}),\ i \in \mathbb{N}_0\}.$$

Proof. We first expand the A–G pseudopolynomial in [8.9] around 0 using the simple Taylor formula:

$$G_n(e^{-bx}|e^{bS}) = \sum_{j=0}^{n} \frac{(e^{-bx})^j}{j!}\, G_{n-j}(0|E^j e^{bS}). \qquad [8.11]$$

Consider the expectation in the r.h.s. of [8.9] and let us insert [8.11] to write:

$$\mathbb{E}\big[e^{-\lambda_n S_n}\, G_n(e^{-bx}|e^{bS})\big] = \mathbb{E}\big[e^{-\lambda_n S_n} \sum_{j=0}^{n} \frac{e^{-bxj}}{j!}\, G_{n-j}(0|E^j e^{bS})\big]$$

$$= \sum_{j=0}^{n} \frac{e^{-bxj}}{j!} \mathbb{E}\big[e^{-\lambda_n S_n}\, (e^{bS_j})^{n-j}\, G_{n-j}(0|e^{b\tilde{S}})\big], \qquad [8.12]$$

where $\exp(b\tilde{S})$ is the family of variables:

$$\{e^{b\tilde{S}_i} = e^{b(S_{i+j}-S_j)},\ i \in \mathbb{N}_0\},$$

which are distributed as e^{bS_i} and independent of e^{bS_j}. Since:

$$e^{-\lambda_n S_n} = (e^{bS_n})^{-\lambda_n/b} = (e^{bS_j}\, e^{b\tilde{S}_{n-j}})^{-\lambda_n/b},$$

[8.12] is equivalent to:

$$\mathbb{E}\big[e^{-\lambda_n S_n}\, G_n(e^{-bx}|e^{bS})\big] = \sum_{j=0}^{n} \frac{e^{-bxj}}{j!} \mathbb{E}\big[(e^{bS_j})^{n-j-\lambda_n/b}$$

$$\times\, (e^{b\tilde{S}_{n-j}})^{-\lambda_n/b} G_{n-j}(0|e^{b\tilde{S}})\big]$$

$$= \sum_{j=0}^{n} \frac{e^{-bxj}}{j!}\, \mathbb{E}\big[(e^{bS_j})^{n-j-\lambda_n/b}\big]$$

$$\times\, \mathbb{E}\big[(e^{b\tilde{S}_{n-j}})^{-\lambda_n/b}\, G_{n-j}(0|e^{b\tilde{S}})\big]. \qquad [8.13]$$

By proposition 6.6, we know that in [8.13]:

$$\mathbb{E}\big[(e^{b\tilde{S}_{n-j}})^{-\lambda_n/b}\, G_{n-j}(0|e^{b\tilde{S}})\big] = G_{n-j}(0|E^{-\lambda_n/b}V), \qquad [8.14]$$

where V is the family of parameters v_i defined above. Furthermore, we see that:

$$\begin{aligned}\mathbb{E}\big[(e^{bS_j})^{n-j-\lambda_n/b}\big] &= \{\mathbb{E}\big[(e^{bX_1})^{n-j-\lambda_n/b}\big]\}^j \\ &= \{\mathbb{E}\big[e^{bX_1(-\lambda_0/b-j)}\big]\}^j = \big[\mathbb{E}(e^{-\lambda_j X_1})\big]^j \qquad [8.15]\end{aligned}$$

Incorporating [8.14] and [8.15] into [8.13] and then [8.9] leads us to [8.10]. □

Note that inside the A–G polynomial of formula [8.10]:

$$E^{-\lambda_n/b}V = \{\mathbb{E}\big(e^{-X_1(\lambda_n-bi)}\big),\ i \in \mathbb{N}_0\}.$$

Now, the case where the renewal process starts from a different position is easily deduced. Suppose the starting point is $(y, -m)$ where $y \in \mathbb{R}_+$ and $m \in \mathbb{N}_0$.

COROLLARY 8.4.–

$$\begin{aligned}\mathbb{P}(K = n|x, R, \Lambda) = (-1/b)^n\,(\lambda_0 \ldots \lambda_{n-1}) \sum_{j=0}^{n} \frac{e^{-\lambda_j y}}{j!} \\ \times \big[\mathbb{E}(e^{-\lambda_j X_1})\big]^{m+j}\, G_{n-j}(0|E^{-\lambda_n/b}V)\big], \quad n \geq 0.\end{aligned}$$

In particular, when the renewals occur according to a Poisson process of parameter μ:

$$\begin{aligned}\mathbb{P}(K = n|x, R, \Lambda) = (-1/b)^n\,(\lambda_0 \ldots \lambda_{n-1}) \sum_{j=0}^{n} \frac{e^{-\lambda_j y}}{j!} \Big(\frac{\mu}{\mu+\lambda_j}\Big)^{m+j} \\ \times G_{n-j}\Big(0|\{\frac{\mu}{\mu+\lambda_{n-i}},\ i \in \mathbb{N}_0\}\Big).\end{aligned}$$

Proof. The renewal trajectory will meet the time axis at time τ satisfying $\tau - y = S_m$ (in distribution). As this time, τ plays the role of x, and we have to replace $\exp(-\lambda_j x)$ in [8.10] by $\exp(-\lambda_j y)\,\mathbb{E}[\exp(-\lambda_j S_m)]$, hence the desired result. The particular case follows directly. □

8.3. Nonlinear death process with a birth process

This time, we consider the meeting between a nonlinear death process starting at $(0, m)$ and a lower nonlinear birth process starting at $(0, r)$ where $r \leq m$. The death rates are denoted by μ_i, $0 \leq i \leq m$ (with $\mu_0 = 0$), the birth rates by λ_i, $i \geq r$.

The level of the first meeting between the two trajectories is an integer random variable N. Our goal is to obtain an expression for the probability generating function of N. For clarity, it is written as:

$$\mathbb{E}(x^N) = \psi(x|r, m), \quad 0 \leq x \leq 1.$$

Since $r \leq N \leq m$ by definition, $\mathbb{E}(x^N)$ is a polynomial of degree m in x.

The reasoning that will be followed is quite different from the previous ones. Le us look at the first event that will occur over time. It is either a birth with the probability $\lambda_r/(\lambda_r + \mu_m)$ or a death with the probability $\mu_m/(\lambda_r + \mu_m)$, whatever the moment of this event. Thus, a renewal argument gives when $r < m$:

$$\psi(x|r, m) = \frac{\lambda_r}{\lambda_r + \mu_m}\,\psi(x|r+1, m) + \frac{\mu_m}{\lambda_r + \mu_m}\,\psi(x|r, m-1), \quad [8.16]$$

and when $r = m$:

$$\psi(x|m, m) = x^m. \quad [8.17]$$

Note that x is simply a parameter in these equations. We first want to find elementary solutions, say $\theta_k(r, m)$, which depend on an arbitrary integer $k \in \mathbb{N}$. More precisely, let us try product solutions of the form:

$$\theta_k(r, m) = m_{[k]}\, a_m(k)\, b_r(k), \quad [8.18]$$

where $a_m(k) \neq 0$, $a_k(k) = b_k(k) = 1$ and $m_{[k]} \equiv m(m-1)\ldots(m-k+1)$, so $m_{[k]} = 0$ for $k \geq m+1$. Then, we get from [8.16]:

$$(\lambda_r + \mu_m)\,\theta_k(r, m) = \lambda_r\,\theta_k(r+1, m) + \mu_m\,\theta_k(x|r, m-1), \quad r < m, \quad [8.19]$$

a relation which will now be discussed by taking k fixed. It will be shown that the case $m = k$ provides the $b_r(k)$, while the case $m > k$ gives the $a_m(k)$.

Indeed, when $m = k$, [8.19] becomes, using [8.18]:

$$(\lambda_r + \mu_k)\, k_{[k]}\, a_k(k)\, b_r(k) = \lambda_r\, k_{[k]}\, a_k(k)\, b_{r+1}(k) + \mu_k 0,$$

since $(k-1)_{[k]} = 0$, that is:

$$(\lambda_r + \mu_k)\, b_r(k) = \lambda_r\, b_{r+1}(k),$$

which gives recursively starting from $b_k(k) = 1$:

$$b_r(k) = \prod_{i=1}^{r-k} \left(1 + \frac{\mu_k}{\lambda_{k+i-1}}\right). \qquad [8.20]$$

A convention is however necessary when $r - k < 0$ in [8.20]. When this happens, we choose any integer $c > 0$ large enough to have $-c < r - k$ and rewrite the product involved as:

$$\prod_{i=1}^{r-k} = \prod_{i=-c}^{r-k} \Big/ \prod_{i=-c}^{0},$$

in which all the terms have a meaning.

When $m > k$, [8.19] with [8.18] becomes:

$$\begin{aligned} \lambda_r + \mu_m &= \lambda_r \frac{b_{r+1}(k)}{b_r(k)} + \frac{\mu_m}{m} \frac{a_{m-1}(k)}{a_m(k)} \\ &= \lambda_r + \mu_k + \frac{\mu_m}{m} \frac{a_{m-1}(k)}{a_m(k)}. \end{aligned} \qquad [8.21]$$

Under the hypothesis that the μ_i are all distinct, this gives:

$$\frac{a_m(k)}{a_{m-1}(k)} = \frac{\mu_m}{m(\mu_m - \mu_k)},$$

hence recursively starting from $a_k(k) = 1$:

$$a_m(k) = \prod_{i=1}^{m-k} \frac{\mu_{k+i}}{(k+i)(\mu_{k+i} - \mu_k)}. \qquad [8.22]$$

Now, it is clear that [8.16] and [8.17] admit a unique solution. As this solution is a polynomial of degree m in x, we seek for $\psi(x|r,m)$ an expression of the form:

$$\psi(x|r,m) = \sum_{k=0}^{m} \theta_k(r,m)\, H_k(x) = \sum_{k=0}^{m} m_{[k]}\, a_m(k)\, b_r(k)\, H_k(x), \qquad [8.23]$$

where $H_k(x)$ is a polynomial of degree k. For [8.23] to satisfy [8.17], it is necessary that:

$$x^m = \sum_{k=0}^{m} m_{[k]}\, a_m(k)\, b_m(k)\, H_k(x), \qquad [8.24]$$

that is, after inserting [8.20] and [8.22]:

$$\frac{x^m}{m!} = \sum_{k=0}^{m} \frac{1}{(m-k)!} \left\{ \prod_{i=1}^{m-k} \left[\frac{\mu_{k+i}}{(k+i)(\mu_{k+i}-\mu_k)} \left(1 + \frac{\mu_k}{\lambda_{k+i-1}}\right) \right] \right\} H_k(x). \qquad [8.25]$$

Let us define:

$$u_{k,i} = \frac{\mu_{k+i}}{(k+i)(\mu_{k+i}-\mu_k)} \left(1 + \frac{\mu_k}{\lambda_{k+i-1}}\right), \qquad [8.26]$$

so that [8.25] is written as:

$$\frac{x^m}{m!} = \sum_{k=0}^{m} \frac{u_{k,1} \ldots u_{k,m-k}}{(m-k)!}\, H_k(x). \qquad [8.27]$$

The structure of $\psi(x|r,m)$ is easily recognized by choosing, as in section 7.7, $\mathcal{D} = \mathbb{R}^{\mathbb{N}}$ and the family of functions $\mathcal{E}$ by:

$$e_n(\mathbf{z}) = z_1 z_2 \ldots z_n / n!, \quad \mathbf{z} \in \mathbb{R}^{\mathbb{N}}. \qquad [8.28]$$

In particular, for $\mathbf{z}_x \equiv \{x, x, \ldots\}$, then $e_n(\mathbf{z}_x) = x^n/n!$. Moreover, let U be the family of parameters $\{\mathbf{u}_k,\ k \in \mathbb{N}_0\}$, where $\mathbf{u}_k = \{u_{k,i},\ i \geq 1\}$ and $u_{k,i}$ are defined in [8.26]. The A-G expansion of $e_m(\mathbf{z})$ gives:

$$e_m(\mathbf{z}) = \sum_{k=0}^{m} e_{m-k}(\mathbf{u}_k)\, \bar{G}_k(\mathbf{z}|\mathcal{E}, U),$$

and thus for $\mathbf{z} = \mathbf{z}_x$:

$$\frac{x^m}{m!} = \sum_{k=0}^{m} \frac{u_{k,1} \ldots u_{k,m-k}}{(m-k)!} \bar{G}_k(\mathbf{z}_x|\mathcal{E}, U). \qquad [8.29]$$

By comparing [8.27] and [8.29], we deduce that:

$$H_k(x) = \bar{G}_k(\mathbf{z}_x|\mathcal{E}, U),$$

so that [8.23] leads to the result [8.30].

PROPOSITION 8.3.– *When all the μ_i are distinct:*

$$\mathbb{E}(x^N) = \sum_{k=0}^{m} m_{[k]}\, a_m(k)\, b_r(k)\, \bar{G}_k(\mathbf{z}_x|\mathcal{E}, U), \qquad [8.30]$$

where $b_r(k)$ is given by [8.20], $a_m(k)$ by [8.22], $\mathbf{z}_x$ by $\{x, x, \ldots\}$ and $U = \{\mathbf{u}_k,\ k \in \mathbb{N}_0\}$ with $\mathbf{u}_k = \{u_{k,i},\ i \geq 1\}$ given by [8.26].

Note that on $\{\mathbf{z}_x,\ x \in \mathbb{R}\}$, the operator Δ is the standard derivative. In fact, writing:

$$\bar{G}_k(\ |\mathcal{E}, U) = \sum_{i=0}^{k} b_{k,i}\, e_i(\ ),$$

we immediately have, for $0 \leq j \leq k$:

$$\begin{aligned}
\frac{d^j}{dx^j}\bar{G}_k(\mathbf{z}_x|\mathcal{E}, U) &= \sum_{i=0}^{k} b_{k,i} \frac{d^j}{dx^j} e_i(\mathbf{z}_x) = \sum_{i=0}^{k} b_{k,i} \frac{x^{i-j}}{(i-j)!} \\
&= \sum_{i=0}^{k} b_{k,i}\, e_{i-j}(\mathbf{z}_x) \\
&= \Delta^j\Big(\sum_{i=0}^{k} b_{k,i}\, e_i(\mathbf{z}_x)\Big) = \Delta^j \bar{G}_k(\mathbf{z}_x|\mathcal{E}, U) \\
&= \bar{G}_{k-j}(\mathbf{z}_x|\mathcal{E}, E^j U).
\end{aligned}$$

Thanks to this property, the distribution of N and its moments are easily obtained from the generating function [8.30].

REMARK.– The assumption of distinct μ_i is restrictive only to a certain extent. Suppose, for example, that $\mu_{m_0} = \mu_k$ for some $m_0 \geq k+1$. In this case, [8.21] implies $a_{m_0-1}(k) = 0$ and $a_{m_0}(k)$ is indeterminate.

Let us examine the difficulty encountered in more detail. Our strategy is to identify [8.24] with [8.27], that is:

$$a_m(k)\, b_m(k) \text{ with } u_{k,1}u_{k,2}\dots u_{k,m-k}.$$

Equivalently, we search for $u_{k,m-k}$ such that:

$$a_m(k)\, b_m(k) = a_{m-1}(k)\, b_{m-1}(k)\, u_{k,m-k}. \qquad [8.31]$$

If $m_0 - 1 \geq k+1$, there is no problem with the first values of $m = k+1, \dots, m_0 - 1$. When $m = m_0$, we saw above that $a_{m_0-1}(k) = 0$, so [8.31] implies u_{k,m_0-k} arbitrary and $a_{m_0}(k) = 0$. Of course, the same situation occurs for $m > m_0$, but we can give the value 0 to the indeterminate $u_{k,i}$ in order to have $\mathbf{u}_k$ well defined. Therefore, the only difficulty is when $m_0 = k+1$, that is, if $\mu_k = \mu_{k+1}$. Then, $a_k(k) = 0$ which is in contradiction with the hypothesis made for [8.18]: the identification [8.31] becomes impossible.

8.4. Binomial process with a lower boundary

Let us move on to the original meeting problem but formulated here in a discrete time scale $t \in \mathbb{Z}$. More precisely, at each unit of time, a fixed event can occur either with probability β or not with probability $1-\beta$, independently at different times. Such an event could, for example, be the arrival of a customer waiting for service in a queue. Now, from an initial time $x \in \mathbb{Z}$, the process that counts the events occurring up to time t, $\{N_{t-x}, t \geq x\}$, is traditionally called a binomial process. Of course:

$$\mathbb{P}(N_{t-x} = n) = \binom{t-x}{n} \beta^n (1-\beta)^{t-x-n}, \quad n \geq 0, \qquad [8.32]$$

hence the name given to this process. After extension on $\mathbb{R}_+$, its trajectories can be chosen right continuous. Intuitively, it is clear that $\{N_{t-x}, t \geq x\}$ can be viewed as a discrete-time version of a Poisson process initiated at time x. On the other hand, the lower boundary F starts at time x below 0 and it is still defined through a family $V = \{v_i \in \mathbb{R}, i \in \mathbb{N}_0\}$ where $v_i \leq v_{i+1}$ for all i. Let K be the first meeting level of a trajectory with the boundary.

To begin, we assume for F that all the $v_i \in \mathbb{Z}$, which will then allow us to derive the general case. Looking at the trajectory during the time interval (x, v_0), we get from [8.32]:

$$\mathbb{P}(K = 0|x, V) = \mathbb{P}(N_{v_0 - x} = 0) = (1 - \beta)^{v_0 - x}, \qquad [8.33]$$

$$\mathbb{P}(K = n|x, V) = \sum_{k=0}^{n} \binom{v_0 - x}{k} \beta^k (1 - \beta)^{v_0 - x - k}$$

$$\times \mathbb{P}(K = n - k|v_0, E^k V), \quad n \geq 1. \qquad [8.34]$$

Let us write:

$$\mathbb{P}(K = n|x, V) = (1-\beta)^{v_n - x} [\beta/(1-\beta)]^n Q_n(x|V), \quad n \geq 0, \qquad [8.35]$$

where $Q_n(x|V)$ is not yet determined. Then, [8.33] gives $Q_0(x|V) = 1$ and [8.34] becomes:

$$Q_n(x|V) = \sum_{k=0}^{n} \binom{v_0 - x}{k} Q_{n-k}(v_0|E^k V), \quad n \geq 1, \qquad [8.36]$$

with $Q_n(v_0|V) = \delta_{n,0}$ for all n.

To obtain a Taylor-type formula, we choose the family $\mathcal{E}$ of functions (see section 2.1):

$$e_n(t) = \binom{t}{n}, \quad n \geq 0, \qquad [8.37]$$

which satisfy S.I. and are standard at 0. In addition, we change the notation by:

$$\begin{cases} \tilde{v}_i = -v_i \text{ (thus, } \tilde{V} = -V), \quad \tilde{x} = -x, \\ \text{and } \tilde{Q}_n(\tilde{x}|\tilde{V}) = Q_n(x|V), \end{cases} \qquad [8.38]$$

similarly to section 1.5. Using [8.38], we obtain for [8.36]:

$$\tilde{Q}_n(\tilde{x}|\tilde{V}) = \sum_{k=0}^{n} \binom{\tilde{x} - \tilde{v}_0}{k} \tilde{Q}_{n-k}(\tilde{v}_0|E^k \tilde{V}). \qquad [8.39]$$

On the other hand, Taylor's formula [2.79] yields:

$$\bar{G}_n(\tilde{x}|\mathcal{E},\tilde{V}) = \sum_{k=0}^{n} \Delta^k \bar{G}_n(\tilde{v}_0|\mathcal{E},\tilde{V})\, e_k(\tilde{x}-\tilde{v}_0). \qquad [8.40]$$

From [8.39] and [8.40] with [8.37], we deduce that:

$$\tilde{Q}_n(\tilde{x}|\tilde{V}) = \bar{G}_n(\tilde{x}|\mathcal{E},\tilde{V}), \quad n \geq 0,$$

so that [8.35] gives [8.41], after substitution of [8.38].

LEMMA 8.1.– *If $v_i \in \mathbb{Z}$ for all i:*

$$\mathbb{P}(K = n|x, V) = (1-\beta)^{v_n - x}\,(\beta/(1-\beta))^n\, \bar{G}_n(-x|\mathcal{E}, -V), \quad n \geq 0, \qquad [8.41]$$

where the family $\mathcal{E}$ is defined by [8.37].

To get rid of the assumption $v_i \in \mathbb{Z}$, we first observe that since the trajectories are right continuous, changing some v_i to v_i' where $v_i < v_i' < v_{i+1}$ and $v_i' - v_i < 1$, has no effect on what happens at level i. Indeed, if a meeting took place at point (v_i, i), this meeting will persist at point (v_i', i). If there was no meeting at point (v_i, i), the trajectory at instant v_i was either at a level $j \neq i$ and nothing could happen, or at the level i with an instantaneous jump $(v_i, i) \to (v_i, i+1)$ and a meeting at the level i became impossible.

Note that we have implicitly assumed that $v_i < v_{i+1}$. The case $v_i = v_{i+1}$ can however be treated in an analogous manner by grouping all the v_j that have the same value v_i and proceeding on this block of v_j as was done with a single v_i.

Now, let us make the changes indicated above at all levels. The original boundary F is then replaced by a new boundary F' associated with the family $V' = \{v_i', i \in \mathbb{N}_0\}$ where $v_i' \leq v_{i+1}'$ for all i. The status of meetings is the same with F or F': no meetings added, no meetings deleted. The level of first meeting K is not modified, only its instant of occurrence T becomes T' with $T' > T$. Therefore, the formula [8.41] remains applicable, it suffices to change the notation v_i to $\lceil v_i' \rceil \in \mathbb{Z}$ for all i. Denoting $V'_{\lceil\rceil} = \{\lceil v_i' \rceil, i \in \mathbb{N}_0\}$, this gives:

$$\mathbb{P}(K = n|x, V'_{[\,]}) = (1-\beta)^{[v'_n]-x}\left(\beta/(1-\beta)\right)^n \bar{G}_n(-x|\mathcal{E}, -V'_{[\,]}), \quad n \geq 0. \tag{8.42}$$

Finally, consider any family $W = \{w_i \in \mathbb{R}, \text{i} \in \mathbb{N}_0\}$ with $w_i \leq w_{i+1}$ for all i, and let $W_{[\,]}$ be defined by:

$$W_{[\,]} = \{[w_i], \, i \in \mathbb{N}_0\}, \tag{8.43}$$

with, as before, $w_i - [w_i] < 1$ and $[w_i] \leq [w_{i+1}]$. We can then use [8.41] and [8.42] where w_i plays the role of v'_i and $[w_i]$ the role of v_i, hence [8.44].

PROPOSITION 8.4.– *If $w_i \in \mathbb{R}$ for all i:*

$$\mathbb{P}(K = n|x, W) = (1-\beta)^{[w_n]-x}\left(\beta/(1-\beta)\right)^n \bar{G}_n(-x|\mathcal{E}, -W_{[\,]}), \quad n \geq 0, \tag{8.44}$$

where $W_{[\,]}$ and $\mathcal{E}$ are defined by [8.43] and [8.37], respectively.

Suppose that x is a real number instead of an integer.

COROLLARY 8.5.– *If in addition $x \in \mathbb{R}$:*

$$\mathbb{P}(K = n|x, W) = (1-\beta)^{[w_n-x]}\left(\beta/(1-\beta)\right)^n \bar{G}_n(0|\mathcal{E}, -\hat{W}_{[\,]}), \quad n \geq 0, \tag{8.45}$$

where $\hat{W}_{[\,]} = \{[w_i - x], \, i \in \mathbb{N}_0\}$.

Proof. Given any real x, let $\alpha = x - [x]$ and translate the origin of the time axis to the right of α. Then, x becomes $[x]$ (denoted $x \to [x]$), $w_i \to w_i - \alpha$, $[w_n] - x \to [w_n - (x - [x])] - [x] = [w_n - x]$ and $x - W_{[\,]} \to [x] - \{[w_i - x + [x]]\} = -\{[w_i - x]\}$, so that [8.44] gives [8.45]. □

8.5. For a compound binomial process

Referring to queues, for example, this means that each event here is the possible arrival, with probability β, of a batch of customers with arithmetic size distribution $\{q_j, \, j \in \mathbb{N}\}$.

The reasoning is very similar to that of the previous section. Let us initially assume that the starting time is $x \in \mathbb{Z}$ and that the boundary is provided by $V = \{v_i \in \mathbb{Z},\ i \in \mathbb{N}_0\}$ $(v_i \le v_{i+1})$. Considering the time interval (x, v_0), we see that:

$$\mathbb{P}(K = n|x, V) = \sum_{k=0}^{n} \mathbb{P}(N_{v_0 - x} = k)\, \mathbb{P}(K = n - k|v_0, E^k V). \quad [8.46]$$

Write this time:

$$\mathbb{P}(K = n|x, V) = (1 - \beta)^{v_n - x}\, R_n(x|V), \quad n \ge 0, \quad [8.47]$$

with $R_n(x|V)$ to be determined. Then, [8.46] gives $R_0(x|V) = 1$ and:

$$R_n(x|V) = \sum_{k=0}^{n} \frac{\mathbb{P}(N_{v_0 - x} = k)}{\mathbb{P}(N_{v_0 - x} = 0)}\, R_{n-k}(v_0|E^k V), \quad n \ge 1, \quad [8.48]$$

with $R_n(v_0|V) = \delta_{n,0}$ for all n.

Now, we introduce the family of functions $\mathcal{E}$ by:

$$e_n(t) = \frac{\mathbb{P}(N_t = n)}{\mathbb{P}(N_t = 0)}, \quad n \ge 0. \quad [8.49]$$

So, denoting $g(s) = \sum_{j=1}^{\infty} q_j\, s^j$, their generating function is:

$$\sum_{n=0}^{\infty} e_n(t)\, s^n = \frac{\mathbb{E}(s^{N_t})}{\mathbb{E}(0^{N_t})} = \frac{(1 - \beta + \beta g(s))^t}{(1 - \beta)^t} = \big(1 + \frac{\beta}{1-\beta} g(s)\big)^t,$$

which gives by expansion:

$$\sum_{n=0}^{\infty} e_n(t)\, s^n = \sum_{l=0}^{\infty} \binom{t}{l} \big(\frac{\beta}{1-\beta}\big)^l (g(s))^l = \sum_{l=0}^{\infty} \binom{t}{l} \big(\frac{\beta}{1-\beta}\big)^l \sum_{n=l}^{\infty} q_n^{*l}\, s^j,$$

and therefore:

$$e_n(t) = \sum_{l=0}^{\infty} \binom{t}{l} \big(\frac{\beta}{1-\beta}\big)^l q_n^{*l}, \quad n \ge 0. \quad [8.50]$$

Then, [8.48] becomes:

$$R_n(x|V) = \sum_{k=0}^{n} e_k(v_0 - x)\, R_{n-k}(v_0|E^k V), \quad n \ge 1,$$

and arguing as before, we find that:

$$R_n(x|V) = \bar{G}_n(-x|\mathcal{E}, -V), \quad n \geq 0.$$

Finally, we can again get rid of the simplifying hypotheses on x and v_i, hence the result [8.51] in terms of $x \in \mathbb{R}$ and any family $W = \{w_i \in \mathbb{R}, i \in \mathbb{N}_0\}$ $(w_i \leq w_{i+1})$.

PROPOSITION 8.5.– *If x and $w_i \in \mathbb{R}$ for all i:*

$$\mathbb{P}(K = n|x, W) = (1-\beta)^{[w_n - x]}\, \bar{G}_n(0|\mathcal{E}, -\hat{W}_{[\,]}), \quad n \geq 0, \qquad [8.51]$$

where $\hat{W}_{[\,]} = \{[w_i - x], i \in \mathbb{N}_0\}$ and $\mathcal{E}$ is defined by [8.49], that is, [8.50].

8.6. Compound binomial process with a renewal process

We continue with a compound binomial process which is initiated at point $(0, 0)$, and an independent renewal process which is initiated later at point $(x, 0)$, $x \in \mathbb{N}$. The renewal process (R) is defined by a sequence of i.i.d. random variables X_j, $j \geq 1$. We denote by S the set of partial sums $\{S_i = X_1 + \ldots + X_i, i \in \mathbb{N}_0\}$ where $S_0 \equiv 0$.

Using proposition 8.5, we obtain:

$$\mathbb{P}(K = n|x, R) = \mathbb{E}\Big[(1-\beta)^{x+S_n}\, \bar{G}_n(x|\mathcal{E}, -S)\Big], \quad n \geq 0, \qquad [8.52]$$

where the family $\mathcal{E}$ is given by [8.50]. Then, from [6.36], formula [8.52] is equivalent to:

$$\mathbb{P}(K = n|x, R) = \mathbb{E}\Big[(1-\beta)^{x+S_n}\, \bar{A}_n(x|\mathcal{E}, S_n)\Big],$$

and by virtue of [6.6]:

$$\mathbb{P}(K = n|x, R) = \mathbb{E}\Big[(1-\beta)^{x+S_n}\, \frac{x}{x+S_n}\, e_n(x+S_n)\Big]. \qquad [8.53]$$

By inserting [8.50] into [8.53], we deduce that:

$$\mathbb{P}(K = n|x, R) = \mathbb{E}\Big[(1-\beta)^{x+S_n}\, x \sum_{l=0}^{n} \frac{(S_n + x - 1)_{[l-1]}}{l!} \big(\frac{\beta}{1-\beta}\big)^l q_n^{*l}\Big]. \qquad [8.54]$$

Finally, let ϕ_n be the real function of u:

$$\phi_n(u) = x\,\mathbb{E}\big(u^{S_n+x-1}\big) = xu^{x-1}\,\big[\mathbb{E}(u^{X_1})\big]^n, \qquad [8.55]$$

and let $\phi_n^{(k)}$ denote its kth derivative. The formula [8.54] can thus be rewritten in the more compact form [8.56].

PROPOSITION 8.6.–

$$\mathbb{P}(K = n|x, R) = \sum_{l=0}^{n} \frac{\beta^l}{l!}\,\big[\phi_n^{(l-1)}(1-\beta)\big]\, q_n^{*l}, \quad n \geq 0, \qquad [8.56]$$

where ϕ_n is defined by [8.55].

9

Martingales and A–G Pseudopolynomials

Very often, the exploitable information that can be obtained in the study of a stochastic model comes from a family of martingales that depend on a certain number of parameters of the model. An application of the optional stopping theorem then leads to a particular system of algebraic relations that can have a complicated structure. The analytical question we now face is whether it is possible to extract from these relationships the information we seek about the model, especially the probability distributions of interest and their moments or characteristic functions.

9.1. Motivation via damage-type models

We are initially interested in epidemic and mortality processes occurring in a closed homogeneous population. Such processes describe patterns of damage that affect a limited set of similar individuals. A formulation and analysis in the context of damage processes would be possible but this is beyond the scope of the present study. The extension to a heterogeneous population will be discussed later.

9.1.1. *Reed–Frost epidemic process*

Section 1.6 considers the (discrete-time) epidemic Reed–Frost model $\{(S_t, I_t),\ t \in \mathbb{N}_0\}$ where S_t and I_t denote, respectively, the numbers of susceptibles and infectives at time t. The dynamics of this epidemic process is defined by [1.24] and [1.25], starting from $(S_0, I_0) = (n, m)$. Recall that we derived the family of martingales:

$$M_{t,k} \equiv S_{t,[k]}\, q^{k(S_t+I_t)}, \quad k = 1, \ldots, n, \qquad [9.1]$$

which gave at the end of the epidemic (i.e. when $I_T = 0$) the relations [1.30].

$$\mathbb{E}\big[S_{T,[k]}\, q^{kS_T}\big] = n_{[k]}\, q^{k(n+m)}, \quad k = 1, \ldots, n. \qquad [9.2]$$

Looking for the generating function $\phi(x)$ of S_T, we transformed [9.2] into:

$$\phi^{(k)}(q^k) = \mathbb{E}[S_{T,[k]}\, q^{k(S_T-k)}] = n_{[k]}\, q^{k(n+m-k)}, \quad k = 1, \ldots, n, \qquad [9.3]$$

so that the A–G expansion led us to:

$$\phi(x) = \sum_{k=0}^{n} \phi^{(k)}(q^k)\, G_k(x|\{q^i,\, i \in \mathbb{N}_0\})$$

$$= \sum_{k=0}^{n} n_{[k]}\, q^{k(n+m-k)}\, G_k(x|\{q^i,\, i \in \mathbb{N}_0\}), \qquad [9.4]$$

thanks to the relations [9.3]. This is precisely the result [1.31], which expresses the unknown $\phi(x)$ in terms of standard A–G polynomials.

9.1.2. *Nonlinear death process*

The upcoming section 9.4 considers a (continuous-time) death process $\{X_t,\, t \in \mathbb{R}_+\}$ with nonlinear rates μ_i, starting at $X_0 = n$. It covers, for example, the case of the so-called simple epidemic that corresponds to a scheme $S \to I$ with rates $\mu_i = (n-i)i$ with $I(0) = m$.

We will prove that a family of martingales for this process is:

$$M_{t,k} \equiv X_{t,[k]} \Big(\prod_{j=1}^{X_t-k} u_{k,j}\Big) Y_t, \quad k = 0, \ldots, n, \qquad [9.5]$$

where $Y_t = \exp[-\int_0^t g(X_\tau)d\tau]$ for some function g, and the $u_{k,j}$ are parameters that depend on the rates μ_i and the function g. Let T be some stopping time to be defined and such that Y_0 is known. By the stopping time theorem applied to [9.5], we obtain the relations:

$$\mathbb{E}\Big[X_{T,[k]} \Big(\prod_{j=1}^{X_T-k} u_{k,j}\Big) Y_T\Big] = n_{[k]} \Big(\prod_{j=1}^{n-k} u_{k,j}\Big) Y_0, \quad k = 0, \ldots, n. \qquad [9.6]$$

In the previous symbolism, choose $\mathcal{D} = \mathbb{R}^{\mathbb{N}}$, $\mathbf{U}$ the set of parameters $\{\mathbf{u}_k,\, k \in \mathbb{N}_0\}$ with $\mathbf{u}_k = \{u_{k,j},\, j \geq 1\}$, and $\mathcal{E}$ the family of functions $e_n(\mathbf{x}) = x_1 x_2 \ldots x_n / n!$ with $e_0 = 1$. Then, the martingales [9.5] can be rewritten as:

$$M_{t,k} = X_t!\, e_{X_t - k}(\mathbf{u}_k)\, Y_t, \quad k = 0, \ldots, n, \tag{9.7}$$

hence the relations at T:

$$\mathbb{E}\big[X_T!\, e_{X_T - k}(\mathbf{u}_k)\, Y_T\big] = n!\, e_{n-k}(\mathbf{u}_k)\, Y_0 \equiv c_k, \quad k = 0, \ldots, n. \tag{9.8}$$

Now, since $0 \leq X_T \leq n$, the A–G expansion gives:

$$e_{X_T}(\mathbf{x}) = \sum_{k=0}^{n} e_{X_T - k}(\mathbf{u}_k)\, \bar{G}_k(\mathbf{x}|\mathcal{E}, \mathbf{U}),$$

so that using [9.8]:

$$\begin{aligned} \mathbb{E}\big[X_T!\, e_{X_T}(\mathbf{u}_k)\, Y_T\big] &= \sum_{k=0}^{n} \mathbb{E}\big[X_T!\, e_{X_T - k}(\mathbf{u}_k)\, Y_T\big]\, \bar{G}_k(\mathbf{x}|\mathcal{E}, \mathbf{U}) \\ &= \sum_{k=0}^{n} c_k\, \bar{G}_k(\mathbf{x}|\mathcal{E}, \mathbf{U}), \end{aligned} \tag{9.9}$$

or equivalently:

$$\mathbb{E}\big(x_1 x_2 \ldots x_{X_T}\, Y_T\big) = \sum_{k=0}^{n} c_k\, \bar{G}_k(\mathbf{x}|\mathcal{E}, \mathbf{U}). \tag{9.10}$$

In particular, if all the x_j are equal to z, this yields:

$$\mathbb{E}\big(z^{X_T}\, e^{-\int_0^T g(X_s) ds}\big) = \sum_{k=0}^{n} c_k\, \bar{G}_k(z, \ldots, z|\mathcal{E}, \mathbf{U}). \tag{9.11}$$

9.1.3. *Combined general and fatal epidemics*

Many Markovian models describing the spread of SIR (susceptible-infected-removed) epidemics can be studied using a similar martingale approach. This is the case, inter alia, for the (continuous time) general and fatal epidemic processes $\{(S_t, I_t), t \in \mathbb{R}_+\}$ starting from $(S_0, I_0) = (n, m)$. For the general epidemic, the instantaneous infection rate is $\beta si/(n+m)$ when there are s susceptibles and i infectives present. For the fatal epidemic, only the $s+i$ surviving individuals can contact each other so that the infection rate is $\beta si/(s+i)$. In both cases, the instantaneous removal rate is μi.

These well-known models can be generalized by defining the infinitesimal transition probabilities during any time interval $(t, t+dt)$ as:

$$\begin{cases} \mathbb{P}(S_{t+dt} = s-1, I_{t+dt} = i | S_t = s, I_t = i) = \beta[f(s)i/g(s+i)]\, dt + o(dt), \\ \mathbb{P}(S_{t+dt} = s, I_{t+dt} = i-1 | S_t = s, I_t = i) = \mu i\, dt + o(dt). \end{cases} \quad [9.12]$$

Here, $f(s)$ is a positive function with $f(0) = 0$ (e.g. $f(s) = s^\alpha$ could translate a group effect of the susceptibles), and $g(s+i) > 0$ takes account of the population size $s+i$ still present (in particular, $g(s+i) = n+m$ or $s+i$ for the general or fatal epidemic).

In section 9.5, we will derive for this process the family of martingales:

$$M_{t,k} \equiv S_{t,[k]} \Big(\prod_{j=1}^{S_t-k} a_{k,j} \prod_{j=1}^{S_t+I_t-k} b_{k,j} \Big) Y_t, \quad k = 0, \dots, n, \quad [9.13]$$

where $Y_t = \exp[-\theta \int_0^t h(S_\tau + I_\tau) I_\tau d\tau]$ for $\theta \geq 0$ and some function h, and the parameters $a_{k,j}$ and $b_{k,j}$ depend on the coefficients β, μ and the functions $f, g, \theta h$. At the end T of the epidemic, $I_T = 0$ and from [9.13], we get:

$$\mathbb{E}\Big[S_{T,[k]} \Big(\prod_{j=1}^{S_T-k} a_{k,j}\, b_{k,j} \Big) Y_T\Big] = n_{[k]} \Big(\prod_{j=1}^{n-k} a_{k,j} \prod_{j=1}^{n+m-k} b_{k,j} \Big). \quad [9.14]$$

Denote $u_{k,j} = a_{k,j} b_{k,j}$ for all k, j. Then, [9.14] can be rewritten using the same $\mathcal{D}$, U and $\mathcal{E}$ as for [9.6], which will lead again to the formulas [9.9] and [9.10].

9.2. Unified treatment by A–G pseudopolynomials

In the above models and many others, the main variable of interest, denoted here X_T, has a bounded range of $\{0, 1, \dots, n\}$ and satisfies relations of the form:

$$\mathbb{E}\big[a(X_T, k)\, Y_T\big] = c_k, \quad 0 \leq k \leq n, \tag{9.15}$$

where Y_T is independent of k with $Y(0) = 1$, while $a(s,k)$, $0 \leq s \leq n$, is some function such that:

$$a(s,k) = 0 \text{ iff } s < k, \tag{9.16}$$

and with say $a(k,k) = k!$, an assumption which is not restrictive due to the linearity of [9.8].

No reference is made so far to the families $\mathcal{E}$ and $\mathbf{U}$, but we can transform [9.15] and [9.16] into [9.8] as before. Specifically, we choose again $e_n(\mathbf{x}) = x_1 x_2 \ldots x_n / n!$ with $e_0 = 1$, and then look for the $u_{k,j}$ that guarantee:

$$a(s,k) = s!\, e_{s-k}(\mathbf{u}_k), \quad \text{for all } s, k. \tag{9.17}$$

This is indeed feasible since by [9.17], $a(s-1,k) = (s-1)!\, e_{s-1-k}(\mathbf{u}_k)$, thus:

$$\frac{a(s,k)}{a(s-1,k)} = \frac{s}{s-k}\, u_{k,s-k}, \quad s \geq k+1,$$

which yields:

$$u_{k,j} = \frac{j}{k+j} \frac{a(k+j,k)}{a(k+j-1,k)}, \quad j \geq 1. \tag{9.18}$$

Therefore, we can express [9.15] as:

$$\mathbb{E}\big[a(X_T,k)\, Y_T\big] = \mathbb{E}\big[X_T!\, e_{X_T - k}(\mathbf{u}_k)\, Y_T\big] = c_k, \tag{9.19}$$

while by A-G expansion:

$$\mathbb{E}\big[X_T!\, e_{X_T}(\mathbf{x})\, Y_T\big] = \sum_{k=0}^{n} \mathbb{E}\big[X_T!\, e_{X_T-k}(\mathbf{u}_k)\, Y_T\big]\, \bar{G}_k(\mathbf{x}|\mathcal{E}, \mathbf{U}). \tag{9.20}$$

Combining [9.20] with [9.19], we obtain the result [9.21].

PROPOSITION 9.1.– *Under [9.15]:*

$$\mathbb{E}\big[X_T!\, e_{X_T}(\mathbf{x})\, Y_T\big] = \sum_{k=0}^{n} c_k\, \bar{G}_k(\mathbf{x}|\mathcal{E}, \mathbf{U}), \tag{9.21}$$

where $\mathcal{E} = \{e_n(\mathbf{x}) = (1/n!) x_1 \ldots x_n,\ n \in \mathbb{N}_0\}$ *and* $\mathbf{U} = \{\mathbf{u}_k,\ k \in \mathbb{N}_0\}$ *with* $u_{k,j}$ *given by [9.18].*

By [9.21], we need to determine the A–G pseudopolynomials $\bar{G}_k(\mathbf{x}|\mathcal{E},\mathbf{U})$. This can be done recursively by one of the methods described in section 3.3. Note that using the A–G expansion [3.50] requires knowing the quantities $e_{s-k}(\mathbf{u}_k) = a(s,k)/s!$ (recall [9.17]), but the exact values of $\mathbf{u}_k$ can be ignored (only their existence is important).

Of course, if all x_j are equal to the same real z, we get from [9.21] a formula for $\mathbb{E}(z^{X_T}\,Y_T)$. In fact, we can deduce more. Suppose we want to determine the expectation $\mathbb{E}[l(X_T)\,Y_T]$, where $l(j) \neq 0$ for $0 \leq j \leq n$. Then, by choosing:

$$x_j = l(j)/l(j-1), \quad j \geq 1, \qquad [9.22]$$

we have:

$$l(j) = l(0)x_1x_2\ldots x_j = l(0)\,j!\,e_j(\mathbf{x}), \quad j \geq 1.$$

From [9.21], we thus obtain:

$$\begin{aligned}\mathbb{E}\big[l(X_T)\,Y_T\big] &= l(0)\,\mathbb{E}\big(X_T!\,e_{X_T}(\mathbf{x})\,Y_T\big)\\ &= l(0)\sum_{k=0}^{n}\mathbb{E}\big(X_T!\,e_{X_T-k}(\mathbf{u}_k)\,Y_T\big)\,\bar{G}_k(\mathbf{x}|\mathcal{E},\mathbf{U})\\ &= l(0)\sum_{k=0}^{n} c_k\,\bar{G}_k(\mathbf{x}|\mathcal{E},\mathbf{U}).\end{aligned} \qquad [9.23]$$

EXAMPLE 9.1.– Of course, [9.21] for $l(j) = z^j$ gives $\mathbb{E}(z^{X_T}\,Y_T)$, in particular the generating function $\mathbb{E}(z^{X_T})$ of the final state X_T when $Y_T = 1$.

For $l(j) = z_1^j z_2^{j^2}$, we get from [9.23] $\mathbb{E}(z_1^{X_T}\,z_2^{X_T^2}\,Y_T)$, thus the generating function $\mathbb{E}(z_2^{X_T^2})$ of the variable X_T^2 when $z_1 = 1, Y_T = 1$.

For $l(j) = 1 + \delta_{j,r}$ with any r fixed in $\{0,\ldots,n\}$, [9.23] gives $\mathbb{E}(Y_T) + \mathbb{E}(\mathbf{1}_{\{X_T=r\}}\,Y_T)$. Since $\mathbb{E}(Y_T)$ follows from [9.21] when $z = 1$, this provides, that is, $\mathbb{P}(X_T = r)$ when $Y_T = 1$.

For $l(j) = \delta_{j,0} + (1-\delta_{j,0})j^\alpha$ with α fixed in $\mathbb{R}_+$, we get $\mathbb{E}(\mathbf{1}_{\{X_T=0\}}\,Y_T) + \mathbb{E}(X_T^\alpha\,Y_T)$, in particular $\mathbb{E}(X_T^\alpha)$ when $Y_T = 1$.

9.3. Reed–Frost multipopulation epidemic

Let us generalize the Reed–Frost epidemic of section 1.6 to a heterogeneous population subdivided into two homogeneous subpopulations (for simplicity). Such an extension will make it possible to take into account certain differences, of sex or age, for example, existing within susceptible and infected individuals. So, the epidemic spreads in each subgroup according to the pattern described in section 9.1, and all infectives are capable of transmitting the disease to any susceptible present in the two subgroups.

For each subpopulation $j = 1, 2$, consider the (discrete-time) process $\{(S_t^{(j)}, I_t^{(j)}),\ t \in \mathbb{N}_0\}$ where $S_t^{(j)}$ and $I_t^{(j)}$ count the susceptibles and infectives at time t. Initially, $(S_0^{(j)}, I_0^{(j)}) = (n^{(j)}, m^{(j)})$. We suppose that any infective of type 1 can contact each susceptible present of type 1 with probability $p_{1,1} = 1 - q_{1,1}$ and of type 2 with probability $p_{1,2} = 1 - q_{1,2}$. In the same way, any infective of type 2 can contact each susceptible present of type 1 with probability $p_{2,1} = 1 - q_{2,1}$ and of type 2 with probability $p_{2,2} = 1 - q_{2,2}$. The possible contacts are all assumed independent. Therefore, instead of [1.24] we now have:

$$S_{t+1}^{(j)} =_d Bin\big(S_t^{(j)}, q_{1,j}^{I_t^{(1)}} q_{2,j}^{I_t^{(2)}}\big), \quad t \in \mathbb{N}_0. \qquad [9.24]$$

Any infective is permanently removed after each unit of time, so:

$$I_{t+1}^{(j)} = S_t^{(j)} - S_{t+1}^{(j)}, \quad t \in \mathbb{N}_0. \qquad [9.25]$$

The epidemic continues as long as there are infectives in the global population. In other words, it ends at time $T = \inf\{t : I_t^{(1)} = I_t^{(2)} = 0\}$. At this time, $(S_T^{(1)}, S_T^{(2)})$ represents the number of susceptibles who ultimately escaped the disease in both subgroups. We wish to determine precisely the exact distribution of this final state vector.

To this end, we construct a family of martingales for the epidemic model with respect to the σ−fields $\mathcal{F}_t$ generated by $\{(S_\tau^{(j)}, I_\tau^{(j)}), j = 1, 2, 0 \le \tau \le t\}$. This result generalizes lemma 1.1 (and its proof) obtained in the case of a homogeneous population.

LEMMA 9.1.– *For* $0 \le k_1 \le n^{(1)}$ *and* $0 \le k_2 \le n^{(2)}$ *with* $k_1 + k_2 > 1$, *the process:*

$$M_{t,k_1,k_2} \equiv S^{(1)}_{t,[k_1]} \left(q^{k_1}_{1,1} q^{k_2}_{1,2}\right)^{S^{(1)}_t + I^{(1)}_t} S^{(2)}_{t,[k_2]} \left(q^{k_1}_{2,1} q^{k_2}_{2,2}\right)^{S^{(2)}_t + I^{(2)}_t}, \quad t \in \mathbb{N}_0, \tag{9.26}$$

constitutes a martingale with respect to $\mathcal{F}_t$.

Proof. From [9.24], the generating function of $(S^{(1)}_{t+1}, S^{(2)}_{t+1})$, with arguments $x_1, x_2 \in \mathbb{R}$, is:

$$\mathbb{E}\left(x_1^{S^{(1)}_{t+1}} x_2^{S^{(2)}_{t+1}} | \mathcal{F}_t\right) = \left[1 + (x_1 - 1) q_{1,1}^{I^{(1)}_t} q_{2,1}^{I^{(2)}_t}\right]^{S^{(1)}_t} \left[1 + (x_2 - 1) q_{1,2}^{I^{(1)}_t} q_{2,2}^{I^{(2)}_t}\right]^{S^{(2)}_t}. \tag{9.27}$$

Let us differentiate [9.27] k_1 (respectively, k_2) times with respect to x_1 (respectively, x_2) and put $x_1 = 1$ (respectively, $x_2 = 1$). We then obtain:

$$\begin{aligned}\mathbb{E}\left(S^{(1)}_{t+1,[k_1]} S^{(2)}_{t+1,[k_2]} | \mathcal{F}_t\right) &= S^{(1)}_{t,[k_1]} \left(q_{1,1}^{I^{(1)}_t} q_{2,1}^{I^{(2)}_t}\right)^{k_1} S^{(2)}_{t,[k_2]} \left(q_{1,2}^{I^{(1)}_t} q_{2,2}^{I^{(2)}_t}\right)^{k_2} \\ &= S^{(1)}_{t,[k_1]} \left(q^{k_1}_{1,1} q^{k_2}_{1,2}\right)^{I^{(1)}_t} S^{(2)}_{t,[k_2]} \left(q^{k_1}_{2,1} q^{k_2}_{2,2}\right)^{I^{(2)}_t}. \end{aligned} \tag{9.28}$$

Finally, we mutiply both sides of [9.28] by:

$$\left(q^{k_1}_{1,1} q^{k_2}_{1,2}\right)^{S^{(1)}_t} \left(q^{k_1}_{2,1} q^{k_2}_{2,2}\right)^{S^{(2)}_t},$$

and using the formula [9.25]:

$$\begin{aligned}&\mathbb{E}\left[S^{(1)}_{t+1,[k_1]} \left(q^{k_1}_{1,1} q^{k_2}_{1,2}\right)^{S^{(1)}_{t+1} + I^{(1)}_{t+1}} S^{(2)}_{t+1,[k_2]} \left(q^{k_1}_{2,1} q^{k_2}_{2,2}\right)^{S^{(2)}_{t+1} + I^{(2)}_{t+1}} | \mathcal{F}_t\right] \\ &= \mathbb{E}\left[S^{(1)}_{t,[k_1]} \left(q^{k_1}_{1,1} q^{k_2}_{1,2}\right)^{S^{(1)}_t + I^{(1)}_t} S^{(2)}_{t,[k_2]} \left(q^{k_1}_{2,1} q^{k_2}_{2,2}\right)^{S^{(2)}_t + I^{(2)}_t}\right],\end{aligned}$$

hence the martingale [9.26]. □

Applying the optional stopping theorem at the end time T, we deduce from [9.26] the following relations in number $(n^{(1)} + n^{(2)} - 1)$.

PROPOSITION 9.2.– *For $0 \leq k_1 \leq n^{(1)}$ and $0 \leq k_2 \leq n^{(2)}$ with $k_1 + k_2 \geq 1$:*

$$\begin{aligned}&\mathbb{E}\left[S^{(1)}_{T,[k_1]} \left(q^{k_1}_{1,1} q^{k_2}_{1,2}\right)^{S^{(1)}_T} S^{(2)}_{T,[k_2]} \left(q^{k_1}_{2,1} q^{k_2}_{2,2}\right)^{S^{(2)}_T}\right] \\ &= n^{(1)}_{[k_1]} \left(q^{k_1}_{1,1} q^{k_2}_{1,2}\right)^{n^{(1)} + m^{(1)}} n^{(2)}_{[k_2]} \left(q^{k_1}_{2,1} q^{k_2}_{2,2}\right)^{n^{(2)} + m^{(2)}}.\end{aligned} \tag{9.29}$$

Let us now derive the generating function $\phi(x_1, x_2)$, $x_1, x_2 \in \mathbb{R}$, of the final state $(S_T^{(1)}, S_T^{(2)})$. Using the relations [9.29], we will show that this function can be expressed in terms of multivariate A–G polynomials as presented in section 5.3 for pseudopolynomials.

COROLLARY 9.1.– *For* $x_1, x_2 \in \mathbb{R}$:

$$\phi(x_1, x_2) = \sum_{(k_1,k_2)=(0,0)}^{(n^{(1)},n^{(2)})} n_{[k_1]}^{(1)} \left(u_{k_1,k_2}^{(1)}\right)^{n^{(1)}+m^{(1)}-k_1}$$

$$\times\, n_{[k_2]}^{(2)} \left(u_{k_1,k_2}^{(2)}\right)^{n^{(2)}+m^{(2)}-k_2} G_{k_1,k_2}(x_1, x_2 | U^{(1)}, U^{(2)}), \qquad [9.30]$$

where $U^{(j)}$, $j = 1, 2$, *is the family* $\{u_{i_1,i_2}^{(j)},\ (i_1, i_2) \in \mathbb{N}_0^2\}$ *given by:*

$$u_{i_1,i_2}^{(1)} = q_{1,1}^{i_1} q_{1,2}^{i_2}, \quad \textit{and} \quad u_{i_1,i_2}^{(2)} = q_{2,1}^{i_1} q_{2,2}^{i_2}. \qquad [9.31]$$

Proof. Since $\phi(x_1, x_2)$ is a polynomial of degree $n^{(1)}$ in x_1 and $n^{(2)}$ in x_2, its A–G expansion is given by [5.28], which here becomes:

$$\phi(x_1, x_2) = \sum_{(k_1,k_2)=(0,0)}^{(n_1,n_2)} \phi^{(k_1,k_2)}(u_{k_1,k_2}^{(1)}, u_{k_1,k_2}^{(2)})\, G_{k_1,k_2}(x_1, x_2 | U^{(1)}, U^{(2)}).$$

[9.32]

Choose the $u_{i_1,i_2}^{(j)}$, $j = 1, 2$, defined by [9.31]. We then get:

$$\phi^{(k_1,k_2)}(u_{k_1,k_2}^{(1)}, u_{k_1,k_2}^{(2)}) = \mathbb{E}\left[S_{T,[k_1]}^{(1)} \left(q_{1,1}^{k_1} q_{1,2}^{k_2}\right)^{S_T^{(1)}-k_1} S_{T,[k_2]}^{(2)} \left(q_{2,1}^{k_1} q_{2,2}^{k_2}\right)^{S_T^{(2)}-k_2}\right]$$

$$= n_{[k_1]}^{(1)} \left(q_{1,1}^{k_1} q_{1,2}^{k_2}\right)^{n^{(1)}+m^{(1)}-k_1} n_{[k_2]}^{(2)} \left(q_{2,1}^{k_1} q_{2,2}^{k_2}\right)^{n^{(2)}+m^{(2)}-k_2}, \qquad [9.33]$$

after inserting [9.29]. Thus, [9.32] combined with [9.33] yields the formula [9.30]. □

9.4. Nonlinear death process

We consider a Markovian death process, $\{X_t,\ t \in \mathbb{R}_+\}$, that starts in state n and whose successive death rates μ_i, $i = n, \ldots, 1$, are positive and allowed to be nonlinear, with $\mu_0 = 0$.

Let us begin by constructing a family of martingales that depend on any integer k fixed in $\{0, \ldots, n\}$. Let $\mathcal{F}_t$ be the $\sigma-$fields generated by $\{X_\tau, 0 \leq \tau \leq t\}$. For simplicity, we will use the notation l_i or $l(i)$ indistinctly.

LEMMA 9.2.– *Consider any real function h defined on $\{k, \ldots, n\}$ such that:*

$$h_k + \mu_k = 0, \quad \textit{with} \quad h_i + \mu_i \neq 0 \textit{ for } k+1 \leq i \leq n. \qquad [9.34]$$

Next, define the family of parameters:

$$u_{k,j} = \frac{j}{k+j} \frac{\mu_{k+j}}{\mu_{k+j} + h_{k+j}}, \quad 1 \leq j \leq n-k. \qquad [9.35]$$

Then, the process:

$$M_{t,k} \equiv X_{t,[k]} \Big(\prod_{j=1}^{X_t - k} u_{k,j} \Big) \; e^{-\int_0^t h(X_\tau) d\tau}, \quad t \in \mathbb{R}_+, \qquad [9.36]$$

is a martingale with respect to $\mathcal{F}_t$.

Proof. We are looking for a martingale of the form:

$$M_{t,k} = a(X_t) \; e^{-\int_0^t h(X_\tau) d\tau}, \qquad [9.37]$$

where a and h are two real functions on $\{0, \ldots, n\}$, so far unknown, which may depend on k. To do this, simply choose these functions to be able to satisfy the martingale conditions:

$$\frac{d}{dt} \mathbb{E}(M_{t,k} | \mathcal{F}_\tau) = 0, \quad t \geq \tau. \qquad [9.38]$$

After expansion, [9.38] becomes:

$$(\mu_i + h_i) \, a_i = \mu_i \, a_{i-1}, \quad 0 \leq i \leq n. \qquad [9.39]$$

Because of [9.34], a particular solution to [9.39] is:

$$a_0 = \ldots = a_{k-1} = 0, \quad \text{and} \quad a_i = k! \prod_{j=k+1}^{i} \frac{\mu_j}{\mu_j + h_j}, \quad k \leq i \leq n,$$

which, using the definition [9.35] of $u_{k,j}$, can be expressed as:

$$a_i = i_{[k]} \prod_{j=1}^{i-k} u_{k,j}, \quad 0 \leq i \leq n. \qquad [9.40]$$

It remains to insert [9.40] into [9.37] to obtain [9.36]. Note that in [9.36], the values of h on $\{0, \ldots, k-1\}$ are not used. □

We now introduce a level $v \in \mathbb{R}_+$, and a positive real function f on $\{k, \ldots, n\}$ to define a stopping time T by:

$$T = \inf\{s \geq 0 : \int_0^s f(X_\tau)d\tau \geq v\}. \quad [9.41]$$

Thus, for an instantaneous cost of f_i in state i, the variable T will represent the first instant where the total cost incurred reaches a given amount v.

PROPOSITION 9.3.– *Let γ_i be the parameters:*

$$\gamma_i = \mu_i / f_i, \quad k \leq i \leq n. \quad [9.42]$$

Next, for any non-negative real function g defined on $\{k, \ldots, n\}$, let β_i be the parameters:

$$\beta_i = \frac{\mu_i + \theta g_i}{f_i}, \quad k \leq i \leq n, \quad [9.43]$$

where $\theta \geq 0$ and $\beta_k \neq \beta_i$ for $i \geq k+1$. Finally, we denote:

$$u_{k,j} = \frac{j}{k+j} \frac{\gamma_{k+j}}{\beta_{k+j} - \beta_k}, \quad 1 \leq j \leq n-k. \quad [9.44]$$

Then, the following $n+1$ relations apply at the stopping time T:

$$\mathbb{E}\left[X_{T,[k]} \left(\prod_{j=1}^{X_T - k} u_{k,j} \right) e^{\theta \int_0^T g(X_\tau)d\tau}\right]$$

$$= n_{[k]} \left(\prod_{j=1}^{n-k} u_{k,j} \right) e^{-\beta_k v}, \quad k = 0, \ldots, n. \quad [9.45]$$

Proof. Define a new function h by:

$$h_i = \theta g_i - \beta_k f_i, \quad k \leq i \leq n, \quad [9.46]$$

so that from [9.43]:

$$h_i + \mu_i = f_i(\beta_i - \beta_k), \quad k \leq i \leq n. \quad [9.47]$$

Given the assumptions made above, this function h satisfies the required conditions [9.34]. Therefore, by lemma 9.2, $M_{t,k}$ given by [9.36] is a martingale, with the $u_{k,j}$ defined by [9.35] reducing to [9.44].

By applying the optional theorem to this martingale, we obtain for each time $s > 0$:

$$\mathbb{E}\big(M_{T\wedge s,k}\big) = M_{0,k}. \qquad [9.48]$$

Observe that $f > 0$ implies $T < \infty$ and $\int_0^T f(X_\tau)d\tau = v$. Since $g \geq 0$, [9.46] then implies:

$$\begin{aligned}\exp\Big[-\int_0^{T\wedge s} h(X_\tau)d\tau\Big] &= \exp\Big[-\theta\int_0^{T\wedge s} g(X_\tau)d\tau + \beta_k\int_0^{T\wedge s} f(X_\tau)d\tau\Big]\\ &\leq \exp\Big[\beta_k\int_0^{T\wedge s} f(X_\tau)d\tau\Big] = \exp\big(\beta_k v\big),\end{aligned}$$

so that $M_{T\wedge s,k}$ remains bounded as $s \to \infty$. Therefore, we deduce from [9.48] that:

$$\mathbb{E}\big(M_{T,k}\big) = M_{0,k}, \qquad [9.49]$$

which is equivalent to the desired result [9.45]. We also note that f_i, g_i, $i = 0, \ldots, k-1$, play no role in [9.45]. □

REMARK.– In the special case where $f_i = 1$ for all i, [9.41] gives $T = v$ and [9.45] then follows directly from [9.49], making here the hypothesis $g \geq 0$ superfluous.

The case $f_i = i$ with $v \in \{0, \ldots, n\}$ is of limited interest because $\int_0^T g(X_\tau)d\tau = g_n T_n + \ldots + g_{v+1}T_{v+1}$ where the T_i are independent exponential variables of parameters μ_i.

Now, denoting c_k the r.h.s. of [9.45], we have from [9.10]:

$$\mathbb{E}\big(x_1 \ldots x_{X_T}\, e^{-\theta\int_0^T g(X_\tau)d\tau}\big) = \sum_{k=0}^{n} c_k\, \bar{G}_k(\mathbf{x}|\mathcal{E}, \mathbf{U}), \qquad [9.50]$$

where $\mathcal{D} = \mathbb{R}^{\mathbb{N}}$ and $e_n(\mathbf{x}) = x_1 x_2 \ldots x_n/n!$, $n \geq 1$. In particular, [9.50] provides the joint generating function and Laplace transform:

$$\mathbb{E}\big(z^{X_T}\, e^{-\theta\int_0^T g(X_\tau)d\tau}\big) = \sum_{k=0}^{n} c_k\, \bar{G}_k(z, \ldots, z|\mathcal{E}, \mathbf{U}).$$

We will show that this latter result can be expanded as follows.

COROLLARY 9.2.– *For* $r = 0, \ldots, n$*:*

$$\mathbb{E}\Big(\mathbf{1}_{\{X_T = r\}}\, e^{-\theta \int_0^T g(X_\tau)d\tau}\Big) = \frac{1}{\gamma_r} \sum_{k=r}^{n} \gamma_k\, e^{-\beta_k v} \Big(\prod_{i=r,\, i \neq k}^{n} \frac{\gamma_i}{\beta_i - \beta_k} \Big). \qquad [9.51]$$

Proof. For the special form [9.44] of $u_{k,j}$, we have seen in section 3.5 (part 3) that the A–G pseudopolynomials are explicitly given by [3.100], that is:

$$\bar{G}_k(\mathbf{x}|\mathcal{E}, \mathbf{U}) = \frac{\gamma_k}{k!} \sum_{r=0}^{k} \frac{1}{\gamma_r} \Big(\prod_{i=r}^{k-1} \frac{\gamma_i}{\beta_i - \beta_k} \Big)\, x_1 \ldots x_r. \qquad [9.52]$$

Moreover, each c_k can be rewritten as:

$$c_k = k! \Big(\prod_{i=k+1}^{n} \frac{\gamma_i}{\beta_i - \beta_k} \Big)\, e^{-\beta_k v}. \qquad [9.53]$$

Inserting [9.52] and [9.53] into [9.50], we obtain, after some simplifications:

$$\mathbb{E}\big(x_1 \ldots x_{X_T}\, e^{-\theta \int_0^T g(X_\tau)d\tau}\big)$$

$$= \sum_{k=0}^{n} \gamma_k\, e^{-\beta_k v} \Big[\sum_{r=0}^{k} \frac{1}{\gamma_r} \Big(\prod_{i=r,\, i \neq k}^{n} \frac{\gamma_i}{\beta_i - \beta_k} \Big)\, x_1 \ldots x_r \Big]. \qquad [9.54]$$

Finally, we permute the two sums inside [9.54], which leads to:

$$\mathbb{E}\big(x_1 \ldots x_{X_T}\, e^{-\theta \int_0^T g(X_\tau)d\tau}\big)$$

$$= \sum_{r=0}^{n} x_1 \ldots x_r\, \frac{1}{\gamma_r} \Big[\sum_{k=r}^{n} \gamma_k\, e^{-\beta_k v} \Big(\prod_{i=r,\, i \neq k}^{n} \frac{\gamma_i}{\beta_i - \beta_k} \Big) \Big].$$

The coefficients of each term $x_1 \ldots x_r$ can now be identified, hence the formula [9.51]. □

9.5. Combined general and fatal epidemics

We return to the Markovian epidemic process $\{(S_t, I_t),\, t \in \mathbb{R}_+\}$ of section 9.1, which allows us to combine the classical so-called general and fatal epidemic models. Its infinitesimal infection and removal rates in state (s, i) are defined in [9.12] as $\beta f(s)i/g(s+i)$ and μi, respectively.

Let us define the variable:

$$Y_t = e^{-\theta \int_0^t h(S_\tau + I_\tau) I_\tau d\tau}, \quad t \geq 0,$$

where $\theta \geq 0$ and h is any positive real function. Note that if this function h reduces to the constant 1, then $Y_t = \exp(-\theta \int_0^t I_\tau d\tau)$ is the Laplace transform of the area under the infectious curve up to time t. This is an important epidemic statistic that is generally called the severity of the infection.

We will construct a family of martingales indexed by an integer $k \in \{0, \ldots, n\}$. The σ-fields $\mathcal{F}_t$ are here generated by $\{(S_\tau, I_\tau), 0 \leq \tau \leq t\}$, and the space of the epidemic states $\mathcal{S}$ is the set $\{(s, i) : s = 0, \ldots, n, \, i = 0, \ldots, n + m - s\}$. For convenience, the values of s, i will often be placed as indices below.

LEMMA 9.3.– *Define the two families of parameters:*

$$a_{k,j} = \frac{j}{k+j} \frac{f_{k+j}}{f_{k+j} - f_k}, \quad \textit{and}$$

$$b_{k,j} = \frac{\mu g_{k+j}}{(\mu + \theta h_{k+j}) g_{k+j} + \beta f_k}, \quad 1 \leq j \leq n-k, \, 0 \leq k \leq n. \quad [9.55]$$

Then, the process [9.13], that is,

$$M_{t,k} \equiv S_{t,[k]} \Big(\prod_{j=1}^{S_t - k} a_{k,j} \prod_{j=1}^{S_t + I_t - k} b_{k,j} \Big) Y_t, \quad k = 0, \ldots, n, \quad [9.56]$$

is a martingale with respect to $\mathcal{F}_t$.

Proof. As with lemma 9.2, we look for a martingale of the form:

$$M_t = l(S_t, I_t) \, Y_t, \quad [9.57]$$

where l is a real bivariate function on $\mathcal{S}$ to be determined. The condition:

$$\frac{d}{dt} \mathbb{E}(M_t | \mathcal{F}_\tau) = 0, \quad t \geq \tau,$$

is fulfilled if the values $l_{s,i}$ satisfy the difference equations:

$$\beta f_s \left(l_{s-1,i+1} - l_{s,i}\right) + \mu g_{s+i} \left(l_{s,i-1} - l_{s,i}\right) = g_{s+i}\theta h_{s+i}\, l_{s,i}, \quad (s,i) \in \mathcal{S}^*, \qquad [9.58]$$

where $\mathcal{S}^*$ is the set $\mathcal{S}$ restricted to the constraint $\{i \geq 1\}$.

Let us put forward for $l_{s,i}$ an expression of the form:

$$l_{s,i} = l_s^{(1)}\, l_{s+i}^{(2)}, \quad (s,i) \in \mathcal{S}^*, \qquad [9.59]$$

for some univariate functions $l^{(1)}$ and $l^{(2)}$. Dividing the relations [9.58] by $l_{s,i}$ (assumed $\neq 0$), we then obtain:

$$\beta f_s \left(\frac{l_{s-1}^{(1)}}{l_s^{(1)}} - 1\right) + \mu g_{s+i} \left(\frac{l_{s+i-1}^{(2)}}{l_{s+i}^{(2)}} - 1\right) - g_{s+i}\theta h_{s+i} = 0, \quad (s,i) \in \mathcal{S}^*. \qquad [9.60]$$

To verify [9.60], it suffices to take an arbitrary constant λ such that:

$$\beta f_s \left(\frac{l_{s-1}^{(1)}}{l_s^{(1)}} - 1\right) = \lambda, \quad \text{and} \quad g_{s+i} \left(\mu \frac{l_{s+i-1}^{(2)}}{l_{s+i}^{(2)}} - \mu - \theta h_{s+i}\right) = -\lambda,$$

for $(s,i) \in \mathcal{S}^*$, which is equivalent to:

$$l_s^{(1)} = l_{s-1}^{(1)} \frac{\beta f_s}{\lambda + \beta f_s}, \qquad [9.61]$$

$$l_{s+i}^{(2)} = l_{s+i-1}^{(2)} \frac{\mu g_{s+i}}{(\mu + \theta h_{s+i}) g_{s+i} - \lambda}. \qquad [9.62]$$

Thus, by inserting [9.61] and [9.62] into [9.59] then into [9.57], we find a family of martingales $M_t(\lambda)$ that depend on λ to choose.

In the present situation, we take for λ a pole of the function $l^{(1)}$ given by [9.61], hence:

$$\lambda_k = -\beta f_k, \quad k = 0, \ldots, n. \qquad [9.63]$$

Other choices would be of course possible (see below). From now on, we thus add an index k in the notation of $l_s^{(1)}, l_{s+i}^{(2)}$, which will be written as $l_{k,s}^{(1)}, l_{k,s+i}^{(2)}$. Then, we define the first $l_{k,s}^{(1)}$ by:

$$l_{k,0}^{(1)} = \ldots = l_{k,k-1}^{(1)} = 0, \qquad [9.64]$$

and for the others, we obtain from [9.61] with [9.63]:

$$l_{k,s}^{(1)} = k! \prod_{j=1}^{s-k} \frac{f_{k+j}}{f_{k+j} - f_k} \equiv \prod_{j=1}^{s-k} a_{k,j}, \quad k \le s \le n, \tag{9.65}$$

after using the notation [9.55]. As for the $l_{k,s+i}^{(2)}$, required for $s + i = k, \ldots,$ $n + m,\ i \ge 1$ by [9.63], we get from [9.62] with [9.63]:

$$\begin{aligned} l_{k,s+i}^{(2)} &= \prod_{j=1}^{s+i-k} \frac{\mu g_{k+j}}{(\mu + \theta h_{k+j}) g_{k+j} + \beta f_k} \\ &\equiv \prod_{j=1}^{s+i-k} b_{k,j}, \quad k \le s + i \le n + m,\ i \ge 1, \end{aligned} \tag{9.66}$$

using again the notation [9.55]. It remains to insert [9.64], [9.65] and [9.66] into [9.59] and then [9.57] to deduce the martingale [9.56]. □

At the end T, that is, when there are no more infectives present, the optional theorem applied to [9.56] leads to the $(n + 1)$ relations [9.14], namely:

$$\mathbb{E}\big[S_{T,[k]} \big(\prod_{j=1}^{S_T - k} u_{k,j} \big) Y_T\big] = n_{[k]} \big(\prod_{j=1}^{n-k} a_{k,j} \prod_{j=1}^{n+m-k} b_{k,j} \big),$$

where $a_{k,j}$ and $b_{k,j}$ are given by [9.55], and $u_{k,j}$ is the product $a_{k,j}\, b_{k,j}$. The methodology described in section 9.2 is then applicable to [9.14] as indicated previously.

REMARK.– Another family of martingales can be constructed by taking for λ a pole of the function $l^{(2)}$, that is by [9.63]:

$$\lambda_k = (\mu + \theta h_k) g(k), \quad k = 0, \ldots, n. \tag{9.67}$$

These martingales are however not mentioned here because they are less practical to use.

More generally, it is possible to consider the family of martingales $M_t(\lambda)$ indexed by λ. Then, the relation $\mathbb{E}[M_T(\lambda)] = M_0(\lambda)$ will translate an identity between rational fractions in λ. Assuming that the poles [9.63] and [9.67] are all distinct, these fractions can be decomposed into simple elements, and the identication pole per pole will then provide the same results as the two families above.

Let us now derive a Wald-type identity for the epidemic model.

COROLLARY 9.3.–

$$\mathbb{E}\Big[\Big(\prod_{j=1}^{S_T}\frac{\mu}{\mu+\theta h_j}\Big)\, e^{-\theta\int_0^T h(S_\tau+I_\tau)I_\tau d\tau}\Big] = \prod_{j=1}^{n+m}\frac{\mu}{\mu+\theta h_j}. \qquad [9.68]$$

Proof. This time, we work with the general martingale $M_t(\lambda)$ (discussed in the remark) and for which we take here $\lambda = 0$. Then, [9.61] and [9.62] simplify to:

$$l_s^{(1)} = l_{s-1}^{(1)}, \quad \text{and} \quad l_{s+i}^{(2)} = l_{s+i-1}^{(2)}\frac{\mu}{\mu+\theta h_{s+i}},$$

so that $M_t(0)$ is given by:

$$M_t(0) = \Big(\prod_{j=1}^{S_t+I_t}\frac{\mu}{\mu+\theta h_j}\Big)Y_t.$$

Thus, $\mathbb{E}[M_T(0)] = M_0(0)$ yields the identity [9.68]. □

Differentiating [9.68] with respect to θ and putting $\theta = 0$ leads to:

$$\mathbb{E}\Big(\int_0^T h(S_\tau+I_\tau)I_\tau d\tau\Big) = \frac{1}{\mu}\mathbb{E}\Big(\sum_{j=S_T+1}^{n+m} h_j\Big).$$

So, when h is a constant, this reduces to:

$$\mathbb{E}\Big(\int_0^T I_\tau d\tau\Big) = \frac{1}{\mu}\big(n+m-\mathbb{E}(S_T)\big), \qquad [9.69]$$

that is, on average, the final severity of the infection is equal to the total number of infectives multiplied by their expected infectivity duration. Note that the result [9.69] is valid for different processes of epidemic type.

9.6. Time-dependent bivariate death process

Consider the bivariate Markovian death process $\{(X_t^{(1)}, X_t^{(2)}),\ t \in \mathbb{R}_+\}$ that starts in state (n_1, n_2) and whose infinitesimal transition probabilities are:

$$\begin{cases} \mathbb{P}(X_{t+dt}^{(1)} = i_1 - 1, X_{t+dt}^{(2)} = i_2 | X_t^{(1)} = i_1, X_t^{(2)} = i_2) \\ \quad = \mu_{i_1,i_2}^{(1)}\, dt + o(dt),\ i_1 \geq 1, i_2 \geq 0, \\ \mathbb{P}(X_{t+dt}^{(1)} = i_1, X_{t+dt}^{(2)} = i_2 - 1 | X_t^{(1)} = i_1, X_t^{(2)} = i_2) \\ \quad = \mu_{i_1,i_2}^{(2)}\, dt + o(dt),\ i_1 \geq 0, i_2 \geq 1. \end{cases} \qquad [9.70]$$

All mortality rates involved are assumed to be positive, but $\mu_{0,i_2}^{(1)} = 0 = \mu_{i_1,0}^{(2)}$ so that ultimate absorption in state $(0,0)$ is guaranteed.

Our goal is to obtain the exact distribution of the population state at any time t and to highlight the hidden algebraic nature of this solution. For this, we will again use a martingale technique and the theory of A–G pseudopolynomials.

9.6.1. *Family of martingales*

Let $\mathcal{F}_t$ be the σ–fields generated by $\{(X_\tau^{(1)}, X_\tau^{(2)}),\ 0 \leq \tau \leq t\}$, and $\mathcal{B}$ be the space of states $\{(i_1, i_2) : i_1 = 0, \ldots, n_1,\ i_2 = 0, \ldots, n_2\}$.

LEMMA 9.4.– *Consider any real bivariate function $h(i_1, i_2)$ defined on $\mathcal{B}$. Next, we denote*

$$\beta_{i_1,i_2} = h(i_1, i_2) + \mu_{i_1,i_2}^{(1)} + \mu_{i_1,i_2}^{(2)}, \qquad [9.71]$$

supposed to be distinct, and define a family of parameters $a_{k_1,k_2}(i_1, i_2)$, for $(k_1, k_2) \in \mathcal{B}$, as obtained in [9.77] and (9.78). Then, the process:

$$M_{t,k_1,k_2} \equiv a_{k_1,k_2}(X_t^{(1)}, X_t^{(2)})\, e^{-\int_0^t h(X_\tau^{(1)}, X_\tau^{(2)})d\tau + \beta_{k_1,k_2} t}, \quad t \in \mathbb{R}_+, \qquad [9.72]$$

is a martingale with respect to $\mathcal{F}_t$.

Proof. We are looking for a martingale of the form:

$$a(X_t^{(1)}, X_t^{(2)})\, e^{-\int_0^t l(X_\tau^{(1)}, X_\tau^{(2)})d\tau}, \qquad [9.73]$$

where the real function $l(i_1, i_2)$ on $\mathcal{B}$ is supposed to be known. From the usual martingale constraint (see [9.37]), we find as a sufficient condition that the values $a(i_1, i_2)$ satisfy the difference relations:

$$[l(i_1, i_2) + \mu^{(1)}_{i_1,i_2} + \mu^{(2)}_{i_1,i_2}]\, a(i_1, i_2)$$
$$= \mu^{(1)}_{i_1,i_2}\, a(i_1 - 1, i_2) + \mu^{(2)}_{i_1,i_2}\, a(i_1, i_2 - 1), \quad (i_1, i_2) \in \mathcal{B}, \qquad [9.74]$$

where $a(-1, i_2)$ and $a(i_1, -1)$ take arbitrary values.

Now, let us choose (k_1, k_2) in $\mathcal{B}$ and require of the function l that the coefficient $[\ldots]$ in [9.74] vanishes for $(i_1, i_2) = (k_1, k_2)$, that is, $l(k_1, k_2) = -\mu^{(1)}_{k_1,k_2} - \mu^{(2)}_{k_1,k_2}$. For this, we introduce an arbitrary real function $h(i_1, i_2)$ on $\mathcal{B}$ and rewrite l as:

$$l(i_1, i_2) = h(i_1, i_2) - h(k_1, k_2) - \mu^{(1)}_{k_1,k_2} - \mu^{(2)}_{k_1,k_2}, \quad (i_1, i_2) \in \mathcal{B}. \quad [9.75]$$

Denoting β_{i_1,i_2} by [9.71], the l.h.s. of [9.74] becomes:

$$[h(i_1, i_2) - h(k_1, k_2) - \mu^{(1)}_{k_1,k_2} - \mu^{(2)}_{k_1,k_2} + \mu^{(1)}_{i_1,i_2} + \mu^{(2)}_{i_1,i_2}]\, a(i_1, i_2)$$
$$= (\beta_{i_1,i_2} - \beta_{k_1,k_2}) a(i_1, i_2). \qquad [9.76]$$

So, the $a(i_1, i_2)$ will depend on (k_1, k_2), hence the new notation $a_{k_1,k_2}(i_1, i_2)$.

Using [9.74] and [9.76], we see that $a_{k_1,k_2}(i_1, i_2)$ can be easily computed by recursion. In fact, this yields:

$$a_{k_1,k_2}(i_1, i_2) = \begin{cases} 0 \ \text{for} \ i_1 < k_1 \ \text{or} \ i_2 < k_2, \\ 1 \ \text{for} \ (i_1, i_2) = (k_1, k_2), \end{cases} \qquad [9.77]$$

and for $(i_1, i_2) > (k_1, k_2)$ (which means $i_1 \geq k_1, i_2 \geq k_2, i_1 + i_2 > k_1 = k_2$):

$$a_{k_1,k_2}(i_1, i_2) = \frac{\mu^{(1)}_{i_1,i_2}\, a_{k_1,k_2}(i_1 - 1, i_2) + \mu^{(2)}_{i_1,i_2}\, a_{k_1,k_2}(i_1, i_2 - 1)}{\beta_{i_1,i_2} - \beta_{k_1,k_2}}, \qquad [9.78]$$

provided that $\beta_{i_1,i_2} \neq \beta_{k_1,k_2}$. □

A slightly more explicit expression of $a_{k_1,k_2}(i_1,i_2)$ in [9.78] can be derived. We denote $\mathcal{P}_{k_1,k_2}(i_1,i_2)$ the set of all possible paths ω going from (i_1,i_2) to (k_1,k_2) in steps of size -1 in any of the two coordinates. Such a path ω counts i_1-k_1 steps of the form $(j_1,j_2)\to(j_1-1,j_2)$ and i_2-k_2 steps of the form $(j_1,j_2)\to(j_1,j_2-1)$, hence the following representation.

PROPERTY 9.1.– *For $(i_1,i_2)>(k_1,k_2)$:*

$$a_{k_1,k_2}(i_1,i_2)=\sum_{\omega\in\mathcal{P}_{k_1,k_2}(i_1,i_2)}\Pi_\omega, \qquad [9.79]$$

where Π_ω is the product along the path ω of i_1-k_1 factors:

$$\mu^{(1)}_{j_1,j_2}/(\beta_{j_1,j_2}-\beta_{k_1,k_2}),\quad \textit{for a step}\ (j_1,j_2)\to(j_1-1,j_2), \qquad [9.80]$$

and of i_2-k_2 factors:

$$\mu^{(2)}_{j_1,j_2}/(\beta_{j_1,j_2}-\beta_{k_1,k_2}),\quad \textit{for a step}\ (j_1,j_2)\to(j_1,j_2-1). \qquad [9.81]$$

Applying the stopping theorem to the martingales [9.72], we obtain a set of $(n_1+1)(n_2+1)$ relations for the population distribution at some stopping time of interest. In particular, given any fixed time t, this yields the following relations that will then be exploited to derive the transient population distribution.

PROPOSITION 9.4.– *For $0\le k_1\le n_1$ and $0\le k_2\le n_2$:*

$$\mathbb{E}\Big[a_{k_1,k_2}(X^{(1)}_t,X^{(2)}_t)\,e^{-\int_0^t h(X^{(1)}_\tau,X^{(2)}_\tau)d\tau}\Big]=a_{k_1,k_2}(n_1,n_2)\,e^{-\beta_{k_1,k_2}t}. \qquad [9.82]$$

9.6.2. *Bi-index A-G pseudopolynomials*

To determine those involved here, we adapt the first method of calculation proposed in the univariate case in section 3.3 and which was based on the decomposition of $\bar{G}_n(\ |U)$ with respect to $\mathcal{E}$ (see [3.44]). More precisely, we take an arbitrary $\mathcal{D}$ and two families $\mathcal{E}$ and U in order to expand a bi-index A–G pseudopolynomial of degree (n_1,n_2) by:

$$\bar{G}_{n_1,n_2}(z|\mathcal{E},U) = \sum_{(i_1,i_2)=(0,0)}^{(n_1,n_2)} g_{i_1,i_2}(n_1,n_2)\, e_{i_1,i_2}(z), \qquad [9.83]$$

the coefficients $g_{i_1,i_2}(n_1,n_2)$ having the advantage of being independent of z. From [5.2], we know that these satisfy the relations:

$$\sum_{(i_1,i_2)=(0,0)}^{(n_1,n_2)} g_{i_1,i_2}(n_1,n_2)\, e_{i_1-k_1,i_2-k_2}(u_{k_1,k_2})$$

$$= \delta_{n_1,k_1}\delta_{n_2,k_2}, \quad k_1 \le n_1,\ k_2 \le n_2. \qquad [9.84]$$

Thus, by varying (k_1,k_2), [9.84] provides recursively $g_{n_1,n_2}(n_1,n_2)$, $g_{n_1-1,n_2}(n_1,n_2)$, $g_{n_1,n_2-1}(n_1,n_2)$, ..., $g_{0,0}(n_1,n_2)$. Note that the $g_{i_1,i_2}(n_1,n_2)$ depend on $\mathcal{E}$ and U only via the terms $e_{i_1-k_1,i_2-k_2}(u_{k_1,k_2})$.

In fact, similarly to section 9.2, we choose precisely $\mathcal{E}$ and U to satisfy the identities:

$$e_{i_1-k_1,i_2-k_2}(u_{k_1,k_2}) = a_{k_1,k_2}(i_1,i_2), \quad \text{for } i_1 \ge k_1,\ i_2 \ge k_2, \qquad [9.85]$$

where the $a_{k_1,k_2}(i_1,i_2)$ are the quantities given by [9.79] with [9.80] and [9.81]. We will check later whether such a choice is actually possible. For now, we focus on obtaining more explicitly the $g_{i_1,i_2}(n_1,n_2)$ in [9.83] using the choice [9.85]. It is clear from [9.84] that $g_{n_1,n_2}(n_1,n_2) = 1$; for the others, we get the following representation.

PROPOSITION 9.5.– *For* $(i_1,i_2) < (n_1,n_2)$*:*

$$g_{i_1,i_2}(n_1,n_2) = \sum_{\omega\in\mathcal{P}_{n_1,n_2}(i_1,i_2)} \tilde{\Pi}_\omega, \qquad [9.86]$$

where $\tilde{\Pi}_\omega$ *is the product along the path* ω *of* $n_1 - i_1$ *factors:*

$$\mu^{(1)}_{j_1,j_2}/(\beta_{j_1-1,j_2} - \beta_{n_1,n_2}), \quad \textit{for a step}\ (j_1,j_2) \to (j_1-1,j_2), \qquad [9.87]$$

and of $n_2 - i_2$ *factors:*

$$\mu^{(2)}_{j_1,j_2}/(\beta_{j_1,j_2-1} - \beta_{n_1,n_2}), \quad \textit{for a step}\ (j_1,j_2) \to (j_1,j_2-1). \qquad [9.88]$$

Proof. Let us replace the $g_{i_1,i_2}(n_1,n_2)$ in the l.h.s. of [9.84] by [9.86] with [9.87] and [9.88], and take [9.85] for the $e_{i_1-k_1,i_2-k_2}(u_{k_1,k_2})$. We will show that the value thus obtained is 0, as required by the r.h.s. of [9.84].

Indeed, by means of these two substitutions, the l.h.s. of [9.84] becomes:

$$\sum_{i_1=k_1}^{n_1}\sum_{i_2=k_2}^{n_2}\Big(\sum_{\omega_1\in\mathcal{P}_{n_1,n_2}(i_1,i_2)}\tilde{\Pi}_{\omega_1}\Big)\Big(\sum_{\omega_2\in\mathcal{P}_{i_1,i_2}(k_1,k_2)}\Pi_{\omega_2}\Big). \qquad [9.89]$$

Of course, every path $\omega_1 \in \mathcal{P}_{n_1,n_2}(i_1,i_2)$ followed by a path $\omega_2 \in \mathcal{P}_{i_1,i_2}(k_1,k_2)$ generates a path $\hat{\omega} \in \mathcal{P}_{n_1,n_2}(k_1,k_2)$. Inversely, every path $\hat{\omega} \in \mathcal{P}_{n_1,n_2}(k_1,k_2)$ can be decomposed into a first path like ω_1 followed by a second path like ω_2, the breaking point (i_1,i_2) having $n_1-k_1+n_2-k_2+1$ possible positions along $\hat{\omega}$. Therefore, [9.89] can be rewritten as:

$$\sum_{\hat{\omega}\in\mathcal{P}_{n_1,n_2}(k_1,k_2)}\Big(\sum_{\omega_1\cup\omega_2=\hat{\omega}}\tilde{\Pi}_{\omega_1}\Pi_{\omega_2}\Big), \qquad [9.90]$$

where $\omega_1 \in \mathcal{P}_{n_1,n_2}(i_1,i_2)$ and $\omega_2 \in \mathcal{P}_{i_1,i_2}(k_1,k_2)$ for different points (i_1,i_2).

We will now prove that for each path $\hat{\omega}$, the inner sum over ω_1,ω_2 in [9.90] is null, which will imply the desired result. For this, we start by observing that any $\hat{\omega}$ corresponds to a succession of two-dimensional states (j_1,j_2) occurring in a specific order such that the number j_1+j_2 decreases by 1 at each step. Thus, $\hat{\omega}$ can be represented equivalently as a sequence of one-dimensional states j_1+j_2 occurring from n_1+n_2 to k_1+k_2 according to the previous order. This bijection can be extended to the numbers β_{j_1,j_2} under the form $\beta_{j_1+j_2}$ and the $n_1+n_2=n$, $k_1+k_2=k$ and $i_1+i_2=i$, we can express the inner sum in [9.90] by:

$$\sum_{i=k}^{n}\Big(\prod_{j=i+1}^{n}\frac{\mu_j}{\beta_{j-1}-\beta_n}\prod_{j=k+1}^{i}\frac{\mu_j}{\beta_j-\beta_k}\Big). \qquad [9.91]$$

After putting the term $i=n$ in front and some simplifications, [9.91] becomes:

$$\prod_{j=k+1}^{n}\frac{\mu_j}{\beta_j-\beta_k}+\sum_{i=k}^{n-1}\Big(\prod_{j=i+1}^{n}\frac{\mu_j}{\beta_{j-1}-\beta_n}\prod_{j=k+1}^{i}\frac{\mu_j}{\beta_j-\beta_k}\Big)$$

$$= \prod_{j=k+1}^{n} \frac{\mu_j}{\beta_j - \beta_k} + \mu_{k+1} \cdots \mu_n \sum_{i=k}^{n-1} \Big(\prod_{j=i+1}^{n} \frac{1}{\beta_{j-1} - \beta_n} \prod_{j=k+1}^{i} \frac{1}{\beta_j - \beta_k} \Big)$$

$$= \prod_{j=k+1}^{n} \frac{\mu_j}{\beta_j - \beta_k} + \mu_{k+1} \cdots \mu_n \sum_{i=k}^{n-1} \Big(\prod_{j=k+1}^{i} \frac{1}{\beta_j - \beta_k} \prod_{j=i}^{n-1} \frac{1}{\beta_j - \beta_n} \Big). \qquad [9.92]$$

Finally, we resort to the preliminary result [3.94] proved in section 3.5, case 3. With γ_i changed to μ_i, it asserts that when $k \neq n$:

$$\prod_{j=k+1}^{n} \frac{\mu_j}{\beta_j - \beta_k} + \sum_{i=k}^{n-1} \frac{\mu_n}{\beta_i - \beta_n} \Big(\prod_{j=k+1}^{i} \frac{\mu_j}{\beta_j - \beta_k} \prod_{j=i+1}^{n-1} \frac{\mu_j}{\beta_j - \beta_n} \Big) = 0. \qquad [9.93]$$

The l.h.s. of [9.93] is identical to the last expression [9.92] of [9.91], which is therefore equal to 0 as announced. □

As expected, a downward recurrence relation can be derived for $g_{i_1,i_2}(n_1, n_2)$ (similarly to the upward recurrence [9.78] for $a_{k_1,k_2}(i_1, i_2)$). Indeed, let us distinguish in [9.86] the paths $\omega \in \mathcal{P}_{n_1,n_2}(i_1, i_2)$ according to whether they pass through the state $(i_1 + 1, i_2)$ or through the state $(i_1, i_2 + 1)$. Since the last step is specified by [9.87] or [9.88], we obtain $g_{n_1,n_2}(n_1, n_2) = 1$ and for $(i_1, i_2) < (n_1, n_2)$:

$$g_{i_1,i_2}(n_1, n_2) = \frac{\mu^{(1)}_{i_1+1,i_2}\, g_{i_1+1,i_2}(n_1, n_2) + \mu^{(2)}_{i_1,i_2+1}\, g_{i_1,i_2+1}(n_1, n_2)}{\beta_{i_1,i_2} - \beta_{n_1,n_2}}, \qquad [9.94]$$

with $g_{n_1+1,n_2}(n_1, n_2) = g_{n_1,n_2+1}(n_1, n_2) = 0$.

9.6.3. *Time-dependent analysis*

From the A–G expansion [5.3] of $e_{i_1,i_2}(z)$, we have after inserting [9.85]:

$$e_{i_1,i_2}(z) = \sum_{(k_1,k_2)=(0,0)}^{(i_1,i_2)} a_{k_1,k_2}(i_1, i_2)\, \bar{G}_{k_1,k_2}(z|\mathcal{E}, U), \quad (i_1, i_2) \in \mathbb{N}_0^2.$$

Note that the sum above can be taken until (n_1, n_2) since by [9.77], $a_{k_1,k_2}(i_1, i_2) = 0$ when $k_1 > i_1$ or $k_2 > i_2$. Substituting $(X_t^{(1)}, X_t^{(2)})$ for

(i_1, i_2), we then obtain:

$$e_{X_t^{(1)}, X_t^{(2)}}(z) = \sum_{(k_1,k_2)=(0,0)}^{(n_1,n_2)} a_{k_1,k_2}(X_t^{(1)}, X_t^{(2)}) \, \bar{G}_{k_1,k_2}(z|\mathcal{E}, U). \quad [9.95]$$

Let us multiply both sides of [9.95] by:

$$Y_t = e^{-\int_0^t h(X_\tau^{(1)}, X_\tau^{(2)}) d\tau},$$

and take the expectation. Thanks to the martingale relations [9.82], we deduce that:

$$\mathbb{E}\Big[e_{X_t^{(1)}, X_t^{(2)}}(z)\, Y_t\Big] = \sum_{(k_1,k_2)=(0,0)}^{(n_1,n_2)} a_{k_1,k_2}(n_1, n_2)\, e^{-\beta_{k_1,k_2} t} \, \bar{G}_{k_1,k_2}(z|\mathcal{E}, U). \quad [9.96]$$

The question now is how to choose $\mathcal{D}$ and $\mathcal{E}$ in order to make the l.h.s. of [9.96] simple enough to facilitate the study of the process at time t. Recall that U plays an indirect role via [9.85] and could therefore be quite complicated.

To solve this problem, we choose for $e_{i_1,i_2}(z)$ a given function, say $\tilde{m}_{i_1,i_2}(z)$, for all z in at least one large subset of $\mathcal{D}$, denoted $\mathcal{D}_1$. More precisely, let us take:

$$\mathcal{D} = \mathbb{R}^2 \,\text{x}\, \{0, 1\}, \quad [9.97]$$

any z of $\mathcal{D}$ thus having the form $z = (z_1, z_2, \epsilon)$, with $(z_1, z_2) \in \mathbb{R}^2$ and $\epsilon = 0$ or 1. The subset $\mathcal{D}_1$ is then defined as:

$$\mathcal{D}_1 = \{z \in \mathcal{D} \text{ where } \epsilon = 1\}, \quad [9.98]$$

any z of $\mathcal{D}_1$ having the form $z = (z_1, z_2, 1)$, with $(z_1, z_2) \in \mathbb{R}^2$. Next, we fix:

$$e_{i_1,i_2}(z) = \tilde{m}_{i_1,i_2}(z) \text{ on } \mathcal{D}_1, \quad [9.99]$$

where the functions $\tilde{m}_{i_1,i_2}(z)$, rewritten as $m_{i_1,i_2}(z_1, z_2)$, are known and assumed to be linearly independent. Finally, we take for U the elements of $\mathcal{D}$ given by:

$$u_{i_1,i_2} = (i_1, i_2, 0), \;\; \big(\text{i.e. } (z_1 = i_1, z_2 = i_2, \epsilon = 0)\big), \quad [9.100]$$

and for $\mathcal{E}$ the functions defined on $\mathcal{D}$ as:

$$e_{i_1,i_2}(z) = \left(m_{i_1,i_2}(z_1,z_2)\right)^{\epsilon}\left(a_{z_1,z_2}(i_1+[z_1],i_2+[z_2])\right)^{1-\epsilon}, \qquad [9.101]$$

$[z_1]$, $[z_2]$ denoting the integer parts of z_1, z_2.

From [9.100] and [9.101], we get:

$$e_{i_1,i_2}(u_{k_1,k_2}) = a_{k_1,k_2}(i_1+k_1,i_2+k_2),$$

so that the identities [9.85] are verified as claimed, and from [9.98] and [9.101]:

$$e_{i_1,i_2}(z) = m_{i_1,i_2}(z_1,z_2) \text{ on } \mathcal{D}_1,$$

that is, [9.99] is also satisfied. We can therefore apply [9.96] on the subset $\mathcal{D}_1$, hence the following result (using the expansion [9.83] and proposition 9.5).

PROPOSITION 9.6.– *For* $(z_1,z_2) \in \mathbb{R}^2$*:*

$$\mathbb{E}\left[m_{X_t^{(1)},X_t^{(2)}}(z_1,z_2)\,Y_t\right]$$

$$= \sum_{(k_1,k_2)=(0,0)}^{(n_1,n_2)} a_{k_1,k_2}(n_1,n_2)\,e^{-\beta_{k_1,k_2}t}\,\bar{G}_{k_1,k_2}(z_1,z_2|\mathcal{E},U), \qquad [9.102]$$

where the bivariate A–G pseudopolynomials can be expanded as:

$$\bar{G}_{k_1,k_2}(z_1,z_2|\mathcal{E},U) = \sum_{(i_1,i_2)=(0,0)}^{(k_1,k_2)} g_{i_1,i_2}(k_1,k_2)\,m_{i_1,i_2}(z_1,z_2),$$

the coefficients $g_{i_1,i_2}(k_1,k_2)$ *being given by [9.86] with [9.87] and [9.88].*

Let us illustrate this proposition with some special functions $m_{i_1,i_2}(z_1,z_2)$.

EXAMPLE 9.2.– Choosing $m_{i_1,i_2}(z_1,z_2) = z_1^{i_1} z_2^{i_2}$, [9.102] gives us $\mathbb{E}(z_1^{X_t^{(1)}} z_2^{X_t^{(2)}} Y_t)$, in particular the generating function of $(X_t^{(1)}, X_t^{(2)})$ when $h = 0$.

For $m_{i_1,i_2}(z_1,z_2) = i_1^{z_1} i_2^{z_2}$, we obtain $\mathbb{E}[(X_t^{(1)})^{z_1}(X_t^{(2)})^{Z_2} Y_t]$, including the joint moments of $(X_t^{(1)}, X_t^{(2)})$ of any order.

When $m_{i_1,i_2}(z_1, z_2) = \mathbf{1}_{\{i_1=z_1, i_2=z_2\}}$ where here, $(z_1, z_2) \equiv (j_1, j_2) \in \mathcal{B}$, [9.102] leads to:

$$\mathbb{E}\left[\mathbf{1}_{\{X_t^{(1)}=j_1, X_t^{(2)}=j_2\}} Y_t\right] = \sum_{(k_1,k_2)=(0,0)}^{(n_1,n_2)} a_{k_1,k_2}(n_1, n_2)\, e^{-\beta_{k_1,k_2} t}\, g_{j_1,j_2}(k_1, k_2),$$

so the joint probabilities of $(X_t^{(1)}, X_t^{(2)})$ when $h = 0$.

10

Towards a Non-homogeneous Theory

We begin by pursuing our analysis for the first meeting of a compound Poisson process, assumed to be non-stationary (time varying), with a lower boundary. We then generalize this approach by examining a first meeting problem for a compound Poisson random field.

The content of this chapter is new, the subject having never been treated in a publication. It allows us to underline the robustness and the power of a methodology based on A–G pseudopolynomials. Exact results are often complex and recursive, but current computers manage to exploit them efficiently. This original approach can certainly be pursued in various ways, as this last chapter will show.

10.1. A non-stationary compound Poisson process

Let $t_0 \geq 0$ be any fixed time and consider a non-stationary compound Poisson process $\{N_{t_0,t},\, t \geq t_0\}$. In other words, the jumps occur according a non-stationary Poisson process of parameter λ_t, and the successive jump sizes are independent discrete positive random variables, a jump size at time t being of p.m.f. $\{q_n(t), n \geq 1\}$, where $q_1(t) > 0$, and whose p.g.f. is:

$$f_t(s) \equiv \sum_{n=1}^{\infty} q_n(t)\, s^n, \quad 0 \leq s \leq 1.$$

Note that the p.g.f. of $N_{t_0,t}$ is given by:

$$L_{t_0,t}(s) \equiv \mathbb{E}(s^{N_{t_0,t}}) = \exp\left\{ -\int_{t_0}^{t} [1 - f_\tau(s)]\, \lambda_\tau d\tau \right\}. \qquad [10.1]$$

Indeed, by arguing on the interval $(t_0, t + \Delta t)$, we get:

$$\begin{aligned} L_{t_0,t+\Delta t}(s) &= \mathbb{E}[s^{N_{t_0,t}}\, s^{N_{t_0,t+\Delta t}-N_{t_0,t}}] \\ &= \mathbb{E}(s^{N_{t_0,t}})\,\mathbb{E}[1 - \lambda_t \Delta t + \lambda_t \Delta t\, s^{X_t} + o(\Delta t)] \\ &= L_{t_0,t}(s)\,\{1 - [1 - f_t(s)]\lambda_t \Delta t + o(\Delta t)\}, \end{aligned}$$

hence as $\Delta t \to 0$:

$$\frac{d}{dt} L_{t_0,t}(s) = -[1 - f_t(s)]\lambda_t\, L_{t_0,t}(s),$$

which leads to [10.1].

10.1.1. *Revisited framework*

We start by extending the framework described in section 2.1 to suit our current needs.

The set $\mathfrak{D}$: We work with $\mathfrak{D} = \mathbb{R}_+$, an element of $\mathfrak{D}$ corresponding to time t. Initially $t \geq t_0$ (t_0 fixed), but this assumption will be relaxed below.

The family $\mathcal{E}$: Let us introduce a family $\mathcal{E}_{t_0}$ of functions $e_n(t_0, \ ) : \mathcal{D} \to \mathbb{R}_+$, $n \in \mathbb{N}_0$, defined by:

$$e_n(t_0, t) = \frac{\mathbb{P}(N_{t_0,t} = n)}{\mathbb{P}(N_{t_0,t} = 0)}, \quad t \geq t_0. \tag{10.2}$$

Thus, we have:

$$e_0(t_0, t) = 1, \quad e_n(t, t) = \delta_{n,0}, \tag{10.3}$$

and for any τ in (t_0, t):

$$e_n(t_0, t) = \frac{\mathbb{P}(N_{t_0,\tau} + N_{\tau,t} = n)}{\mathbb{P}(N_{t_0,\tau} + N_{\tau,t} = 0)} = \sum_{i=0}^{n} \frac{\mathbb{P}(N_{t_0,\tau} = i)\,\mathbb{P}(N_{\tau,t} = n - i)}{\mathbb{P}(N_{t_0,\tau} = 0)\,\mathbb{P}(N_{\tau,t} = 0)},$$

since the process has independent increments, so that by the definition [10.2]:

$$e_n(t_0, t) = \sum_{i=0}^{n} e_i(t_0, \tau)\, e_{n-i}(\tau, t). \tag{10.4}$$

Denote the p.g.f. of the $e_n(t_0, t)$ as:

$$F_{t_0,t}(s) \equiv \sum_{n=0}^{\infty} e_n(t_0,t)\, s^n = \frac{\mathbb{E}(s^{N_{t_0,t}})}{\mathbb{P}(N_{t_0,t}=0)}$$

$$= \frac{L_{t_0,t}(s)}{L_{t_0,t}(0)} = \exp\left\{ \int_{t_0}^{t} f_\tau(s)\, \lambda_\tau\, d\tau \right\}, \qquad [10.5]$$

by virtue of [10.1]. Using [10.4], we have for $t_0 < \tau < t$:

$$F_{t_0,t}(s) = F_{t_0,\tau}(s)\, F_{\tau,t}(s), \qquad [10.6]$$

and therefore:

$$F_{\tau,t}(s) = \frac{F_{t_0,t}(s)}{F_{t_0,\tau}(s)}. \qquad [10.7]$$

In fact, the r.h.s. of [10.7] has a probabilistic meaning as soon as $t_0 \leq \min\{\tau, t\}$. It also does not change if t_0 is replaced by $t_0' < t_0$. We may thus use formula [10.7] for arbitrary values of τ and t provided that $t_0 \leq \min\{\tau, t\}$. Once this precaution is taken, let τ, t, t' any times with $t_0 \leq \min\{\tau, t, t'\}$. From [10.7], we see that:

$$F_{\tau,t}(s)\, F_{t,t'}(s) = \frac{F_{t_0,t}(s)}{F_{t_0,\tau}(s)} \frac{F_{t_0,t'}(s)}{F_{t_0,t}(s)}$$

$$= \frac{F_{t_0,t'}(s)}{F_{t_0,\tau}(s)} = F_{\tau,t'}(s), \qquad [10.8]$$

which shows the validity of [10.6] and [10.7]. The generalized $e_n(t_0, t)$ with $t < t_0$ can be negative and so loose their interpretation. For example, if $t < t_0$, we still have $e_0(t_0, t) = 1$ since for $t' < t < t_0$:

$$e_0(t', t) = e_0(t', t_0)\, e_0(t_0, t), \;\text{ i.e. } 1 = 1\, e_0(t_0, t),$$

while $e_1(t_0, t) = -e_1(t, t_0)$ and is thus negative since for $t' < t < t_0$:

$$e_1(t', t) = e_0(t', t_0)\, e_1(t_0, t) + e_1(t', t_0)\, e_0(t_0, t) = e_1(t_0, t) + e_1(t', t_0),$$

thus $e_1(t_0, t) = e_1(t', t) - e_1(t', t_0)$, but by [10.4], $e_1(t', t_0) = e_1(t', t) + e_1(t, t_0)$ hence $e_1(t_0, t) < 0$. However, this loss of meaning has no practical importance as long as we are dealing with algebraic operations.

In the following, a key assumption is that the functions $e_n(t_0, t)$, $n \geq 0$, are linearly independent. It is not easy to guarantee this condition in general, and we limit ourselves to convenient sufficient conditions.

1) So, if only the Poisson parameter λ_t depends on time, we have from [10.2]:

$$e_n(t_0, t) = \sum_{k=0}^{n} \frac{\Lambda_t^k}{k!} q_n^{*k}, \qquad [10.9]$$

where $\Lambda_t = \int_{t_0}^{t} \lambda_\tau d\tau$. Using [10.9], we can write, for any $j \in \mathbb{N}_0$:

$$\sum_{n=0}^{j} \alpha_n \, e_n(t_0, t) = \sum_{k=0}^{j} \frac{\Lambda_t^k}{k!} \sum_{n=k}^{j} \alpha_n \, q_n^{*k}. \qquad [10.10]$$

If the l.h.s. of [10.10] is null for all t, letting $t \to t_0$ will then yield:

$$\sum_{n=0}^{j} \alpha_n \, q_n^{*0} = 0, \quad j \geq 0,$$

hence $\alpha_0 = 0$ for $j = 0$. In addition, this allows us to return to [10.10] by starting the summation from $k = 1$ only, which will give $\alpha_1 = 0$ for $j = 1$ (because $q_1 > 0$), and then $\alpha_2 = \ldots = \alpha_j = 0$. The family $\mathcal{E}_{t_0}$ is therefore linearly independent.

2) When the jump sizes depend also on time, we obtain from [10.5] that, for example:

$$\begin{aligned}
e_1(t_0, t) &= \int_{t_0}^{t} q_1(\tau) \, \lambda_\tau d\tau, \\
2e_2(t_0, t) &= \Big[\int_{t_0}^{t} q_1(\tau) \, \lambda_\tau d\tau \Big]^2 + \int_{t_0}^{t} 2q_2(\tau) \, \lambda_\tau d\tau, \qquad [10.11] \\
3!e_3(t_0, t) &= \Big[\int_{t_0}^{t} q_1(\tau) \, \lambda_\tau d\tau \Big]^3 + 3\Big[\int_{t_0}^{t} q_1(\tau) \, \lambda_\tau d\tau \Big] \\
&\quad \times \Big[\int_{t_0}^{t} 2q_2(\tau) \, \lambda_\tau d\tau \Big] + \int_{t_0}^{t} 3!q_3(\tau) \, \lambda_\tau d\tau.
\end{aligned}$$

Under certain conditions (not given here), we can deduce from [10.11] that:

$$n!e_n(t_0,t) \sim \Big[\int_{t_0}^{t} q_1(\tau)\,\lambda_\tau d\tau\Big]^n, \quad \text{as } t \to t_0. \tag{10.12}$$

In such a case, the relation $\sum_{n=0}^{j} \alpha_n\, e_n(t_0,t) = 0$ for all t would lead again to $\alpha_j = \ldots = \alpha_0 = 0$, hence $\mathcal{E}_{t_0}$ is linearly independent.

The space $\mathfrak{F}$: define by $\mathfrak{F}_{t_0}$ the vector space generated by $\mathcal{E}_{t_0}$. Similarly, let $\mathfrak{F}_{t_1}$ be the vector space generated by $\mathcal{E}_{t_1}$. We will prove that $\mathcal{E}_{t_0}$ and $\mathcal{E}_{t_1}$ generate a unique space that we will denote by $\mathfrak{F}$.

Indeed, consider any pseudopolynomial f_n of degree n in $\mathfrak{F}_{t_0}$. By definition, its representation on the basis $\mathcal{E}_{t_0}$ is:

$$f_n(t) = \sum_{i=0}^{n} a_i(t_0)\, e_i(t_0,t). \tag{10.13}$$

with $a_n(t_0) \neq 0$. Because of [10.4], f_n can be reexpressed as:

$$\begin{aligned} f_n(t) &= \sum_{i=0}^{n} a_i(t_0)\,\Big(\sum_{r=0}^{i} e_{i-r}(t_0,t_1)\, e_r(t_1,t)\Big) \\ &= \sum_{r=0}^{n}\Big(\sum_{i=r}^{n} a_i(t_0)\, e_{i-r}(t_0,t_1)\Big)\, e_r(t_1,t), \end{aligned}$$

and therefore:

$$f_n(t) = \sum_{r=0}^{n} a_r(t_1)\, e_r(t_1,t), \tag{10.14}$$

after putting:

$$a_r(t_1) = \sum_{i=r}^{n} a_i(t_0)\, e_{i-r}(t_0,t_1), \quad 0 \le r \le n. \tag{10.15}$$

Formula [10.14] shows that f_n belongs to $\mathfrak{F}_{t_1}$, thus $\mathfrak{F}_{t_0} \subseteq \mathfrak{F}_{t_1}$. The converse being also true, we obtain $\mathfrak{F}_{t_0} = \mathfrak{F}_{t_1} \equiv \mathfrak{F}$. Note that [10.15] provides the link when we go from $\mathcal{E}_{t_0}$ to $\mathcal{E}_{t_1}$. As $a_n(t_1) = a_n(t_0) \neq 0$, the degree of f_n is the same in the two bases.

The operator Δ: with the basis $\mathcal{E}_{t_0}$, we associate the linear operator Δ_{t_0} by defining:

$$\Delta_{t_0} e_{n+1}(t_0, \ ) = e_n(t_0, \ ), \ n \in \mathbb{N}_0, \quad \text{and} \quad \Delta_{t_0} e_0(t_0, \ ) = 0, \qquad [10.16]$$

which we extend to $\mathfrak{F}_{t_0} = \mathfrak{F}$ in a natural way. Similarly, let Δ_{t_1} be the operator corresponding to the basis $\mathcal{E}_{t_1}$. Here too, we will prove that Δ_{t_0} and Δ_{t_1} operate identically on $\mathfrak{F}$, hence a common notation Δ.

Indeed, using [10.4] (with the discussion after) and [10.16] we have:

$$\Delta_{t_1} e_{n+1}(t_1, \ ) = e_n(t_1, \ ) = \sum_{r=0}^{n} e_{n-r}(t_1, t_0)\, e_r(t_0, \ )$$

$$= \sum_{r=0}^{n} e_{n-r}(t_1, t_0)\, \Delta_{t_0} e_{r+1}(t_0, \ ) = \Delta_{t_0} \Big(\sum_{r=0}^{n} e_{n-r}(t_1, t_0)\, e_{r+1}(t_0, \ ) \Big)$$

$$= \Delta_{t_0} \Big(e_{n+1}(t_1, \ ) - e_{n+1}(t_1, t_0) e_0(t_0, \ ) \Big) = \Delta_{t_0} e_{n+1}(t_1, \ ). \qquad [10.17]$$

10.1.2. *A-G pseudopolynomials*

A pseudopolynomial f_n can be represented in the bases $\mathcal{E}_{t_0}$ and $\mathcal{E}_{t_1}$ as:

$$f_n(\ ) = \sum_{i=0}^{n} a_i(t_0)\, e_i(t_0, \ ) = \sum_{r=0}^{n} a_r(t_1)\, e_r(t_1, \ ).$$

From [10.15], we see that:

$$\sum_{r=1}^{n} a_r(t_1)\, e_{r-1}(t_1, \ ) = \sum_{r=1}^{n} \Big(\sum_{i=r}^{n} a_i(t_0)\, e_{i-r}(t_0, t_1) \Big)\, e_{r-1}(t_1, \ )$$

$$= \sum_{i=1}^{n} a_i(t_0) \Big(\sum_{r=1}^{i} e_{i-r}(t_0, t_1)\, e_{r-1}(t_1, \ ) \Big) = \sum_{i=1}^{n} a_i(t_0)\, e_{i-1}(t_0, \ ).$$

Thus, we can define Δf_n unambiguously by:

$$\Delta f_n(\ ) = \sum_{i=1}^{n} a_i(t_0)\, e_{i-1}(t_0, \ ) = \sum_{r=1}^{n} a_r(t_1)\, e_{r-1}(t_1, \ ), \qquad [10.18]$$

hence by recurrence $\Delta^j f_n$, $1 \leq j \leq n$.

For example, in the basis $\mathcal{E}_{t_0}$, [10.18] yields:

$$\Delta^j f_n(\ ) = \sum_{i=j}^{n} a_i(t_0)\, e_{i-j}(t_0,\ ),$$

which also shows that $\Delta^j f_n(t_0) = a_j(t_0)$. Thus, a Maclaurin type formula applies as follows.

COROLLARY 10.1.–

$$f_n(\ ) = \sum_{i=0}^{n} \Delta^i f_n(t_0)\, e_i(t_0,\ ). \quad [10.19]$$

Now, consider a family of dates $U = \{u_i,\ i \in \mathbb{N}_0\}$. We define an associated family $\mathcal{G}(U)$ of A–G pseudopolynomials, denoted $\bar{G}_n(\ |U)$, by requiring the constraints [2.31] as before.

DEFINITION 10.1.– *For all* $n \geq 0$*:*

$$\Delta^m \bar{G}_n(u_m|U) = \delta_{m,n}, \quad 0 \leq m \leq n. \quad [10.20]$$

Let us check that such a family exists and is unique.

Proof. In the basis $\mathcal{E}_{t_1}$, t_1 arbitrary, $\bar{G}_n(\ |U)$ is of the form:

$$\bar{G}_n(\ |U) = \sum_{r=0}^{n} a_{n,r}(t_1)\, e_r(t_1,\ ),$$

for some coefficients $a_{n,r}(t_1)$ (adding an index n here is necessary). Using [10.16], the condition [10.20] means that:

$$\sum_{r=m}^{n} a_{n,r}(t_1)\, e_{r-m}(t_1, u_m) = \delta_{m,n}, \quad 0 \leq m \leq n. \quad [10.21]$$

Let us choose $t_1 = u_m$. Then, [10.21] becomes from [10.3]:

$$a_{n,m}(u_m) = \delta_{m,n}, \quad 0 \leq m \leq n. \quad [10.22]$$

Applying formula [10.15], we may consider a time t_0, say, to rewrite [10.22] as:

$$\sum_{i=m}^{n} a_{n,i}(t_0)\, e_{i-m}(t_0, u_m) = \delta_{m,n}, \quad 0 \leq m \leq n. \quad [10.23]$$

Finally, it suffices to observe that [10.23] determines in a unique manner the terms $a_{n,i}(t_0)$ obtained by recursion for $i = n, \ldots, 0$, hence $\bar{G}_n(\ |U)$ in the basis $\mathcal{E}_{t_0}$. □

REMARK.– When the u_i are all equal to u,

$$\bar{G}_n(\ |U = \{u, u, \ldots\}) = e_n(u,\ ), \quad n \geq 0.$$

Indeed, we know that $\Delta^m e_n(u,t) = e_{n-m}(u,t)$, thus for $t = u$:

$$\Delta^m e_n(u,u) = \delta_{m,n}, \quad 0 \leq m \leq n,$$

as required by [10.20] for $\bar{G}_n(\ |\{u, u, \ldots\})$.

As expected, the following two central properties hold here too (compare with propositions 2.8 and 2.11).

PROPOSITION 10.1.–

$$\Delta^m \bar{G}_n(\ |U) = \bar{G}_{n-m}(\ |E^m U), \quad 0 \leq m \leq n. \qquad [10.24]$$

Proof. From [10.18], we know that $\Delta^m \bar{G}_n(\ |U)$, $0 \leq m \leq n$, is a pseudopolynomial of degree $n - m$, which we denote h_{n-m}. We then take Δ^r of h_{n-m} and obtain from [10.20]:

$$\Delta^r h_{n-m}(u_{m+r}) = \Delta^{m+r} \bar{G}_n(u_{m+r}|U) = \delta_{m+r,n} = \delta_{n-m,r}, \quad 0 \leq r \leq n-m.$$

Returning to [10.20], we deduce that h_{n-m} corresponds to $\bar{G}_{n-m}(\ |E^m U)$. □

PROPOSITION 10.2.– *Any pseudopolynomial f_n admits an A–G type expansion:*

$$f_n(\ ) = \sum_{i=0}^{n} \Delta^i f_n(u_i)\, \bar{G}_i(\ |U). \qquad [10.25]$$

Proof. Since $\mathcal{G}(U)$ constitutes a new basis of $\mathcal{F}$, f_n can be expressed as:

$$f_n(\ ) = \sum_{i=0}^{n} c_i\, \bar{G}_i(\ |U). \qquad [10.26]$$

This implies that:

$$\Delta^r f_n(u_r) = \sum_{i=r}^{n} c_i \, \Delta^r \bar{G}_i(u_r|U) = \sum_{i=r}^{n} c_i \, \delta_{r,i} = c_r,$$

hence [10.25] after substitution in [10.26]. □

Along with the A–G class, the current framework is well suited to define a family of Appell pseudopolynomials that are useful in particular in the theory of insurance ruin.

10.1.3. *First meeting problem*

We seek the distribution of the first meeting level of a non-stationary compound Poisson process $\{N_{x,t},\, t \geq x\}$ with a lower boundary specified by the set $F = \{(v_i, i),\, i \in \mathbb{N}_0\}$ where $v_0 \geq x$. To this end, we start by modifying the previous definitions a bit and introduce a backward version of $\mathfrak{F}$, Δ and $\mathcal{G}(U)$.

Backward formulation: given any fixed time t_1, we consider a backward family of linearly independent functions:

$$\ddot{\mathcal{E}}_{t_1} = \{e_n(\ ,t_1),\, n \in \mathbb{N}_0\}. \qquad [10.27]$$

It can be shown that all the families of this type generate a unique vectorial space denoted as $\tilde{\mathfrak{F}}$.

Then, we define a backward linear operator $\tilde{\Delta}$ by:

$$\tilde{\Delta} e_{n+1}(\ ,t_1) = e_n(\ ,t_1), \quad n \in \mathbb{N}_0, \quad \text{and} \quad \tilde{\Delta} e_0(t_0,\ ) = 0, \qquad [10.28]$$

which again does not depend on t_1.

This leads us to construct families $\tilde{\mathcal{G}}(U)$ of A–G pseudopolynomials associated with $\tilde{\Delta}$ as they are associated with Δ. They should not be confused, of course. So, from [3.50], we have for any $\bar{G}_n(\ |U) \in \bar{\mathfrak{F}}$:

$$\bar{G}_n(\ |U) = e_n(t_0,\ ) - \sum_{i=0}^{n-1} e_{n-i}(t_0, u_i)\, \bar{G}_i(\ |U),$$

while for any $\tilde{\tilde{G}}_n(\ |U) \in \tilde{\mathfrak{F}}$:

$$\tilde{\tilde{G}}_n(\ |U) = e_n(\ , t_1) - \sum_{i=0}^{n-1} e_{n-i}(u_i, t_1)\, \tilde{\tilde{G}}_i(\ |U).$$

First level distribution: let K be the level of first encounter of the process $\{N_{x,t},\, t \geq x\}$ with the lower boundary F. Its m.p.f. is given by the formula [10.29] which generalizes the result [7.14] obtained in the stationary case.

PROPOSITION 10.3.–

$$\mathbb{P}(K = n|x, V) = e^{-\int_x^{v_n} \lambda_\tau d\tau}\, \tilde{\tilde{G}}_n(x|V), \quad n \geq 0. \qquad [10.29]$$

Proof. We directly see that:

$$\mathbb{P}(K = n|x, V) = \sum_{r=0}^{n} \mathbb{P}(N_{x,v_0} = r)\, \mathbb{P}(K = n - r|v_0, E^r V). \quad [10.30]$$

Setting:

$$\mathbb{P}(K = n|x, V) = \mathbb{P}(N_{x,v_n} = 0)\, R_n(x|V), \quad n \geq 0, \qquad [10.31]$$

we observe that:

$$R_0(x|V) = 1, \quad R_n(v_n|V) = \delta_{n,0}, \qquad [10.32]$$

and by inserting [10.31] in [10.30]:

$$\mathbb{P}(N_{x,v_n} = 0)\, R_n(x|V) = \sum_{r=0}^{n} \mathbb{P}(N_{x,v_0} = r)\, \mathbb{P}(N_{v_0,v_n} = 0) \times R_{n-r}(v_0|E^r V), \quad n \geq 1. \qquad [10.33]$$

This identity [10.33] can be rewritten as:

$$\begin{aligned} R_n(x|V) &= \sum_{r=0}^{n} R_{n-r}(v_0|E^r V)\, P(N_{x,v_0} = r)\, \frac{\mathbb{P}(N_{v_0,v_n} = 0)}{\mathbb{P}(N_{x,v_n} = 0)} \\ &= \sum_{r=0}^{n} R_{n-r}(v_0|E^r V)\, e_r(x, v_0), \quad n \geq 1, \end{aligned} \qquad [10.34]$$

using the definition [10.2] of $e_r(x, v_0)$ since $\mathbb{P}(N_{x,v_n} = 0) = \mathbb{P}(N_{x,v_0} = 0)\mathbb{P}(N_{v_0,v_n} = 0)$.

By [10.34], we deduce that each $R_n(\ |V)$ is a pseudopolynomial of degree n in the space $\tilde{\mathfrak{F}}$. Moreover, it is identified with $\tilde{\tilde{G}}_n(\ |V)$ since from [10.19] and [10.24]:

$$\tilde{\tilde{G}}_n(\ |V) = \sum_{r=0}^{n} \tilde{\Delta}^r \tilde{\tilde{G}}_n(v_0|V)\, e_r(\ , v_0) = \sum_{r=0}^{n} \tilde{\tilde{G}}_{n-r}(v_0|V)\, e_r(\ , v_0),$$

which is to be compared to [10.34]. □

10.2. A compound Poisson random field

The previous approach has the merit of being simple and efficient, at least at first sight. Nevertheless, one would prefer to deal with a more general process than the non-stationary compound process. Moreover, how could one study a first encounter problem with disjoint intervals placed in any order in $\mathbb{R}_+$, and what would happen for disjoint Borel sets of $\mathbb{R}^{\mathbb{N}}$? After some attempts (including the previous one), we managed to develop such a general theory for a compound Poisson random field.

DEFINITION 10.2.– *Let $(C, \mathcal{C})$ be a measurable space (for example, $(\mathbb{R}^n, \mathcal{B}_{\mathbb{R}^n})$). Consider a σ-finite measure μ on $(C, \mathcal{C})$, taken non-atomic. An associated random field of compound Poisson type is defined as follows.*

To any set $D \in \mathcal{C}$ is attached a random variable N_D with values in $\mathbb{N}_0$ and whose p.g.f. is of the form:

$$L_D(s) \equiv \mathbb{E}(s^{N_D}) = \exp\Big\{ -\int_D [1 - f_\tau(s)]\, \mu(d\tau)\Big\}, \qquad [10.35]$$

where $f_\tau(s)$ is a p.g.f. given by:

$$f_\tau(s) = \sum_{n=1}^{\infty} q_n(\tau)\, s^n, \quad 0 \le s \le 1, \qquad [10.36]$$

with $q_1(\tau) > 0$. Moreover, for disjoint sets $D_i \in \mathcal{C}$, $i \ge 1$, the random variables N_{D_i} are assumed to be stochastically independent.

EXAMPLE 10.1..– A typical case is the non-stationary compound Poisson process of section 10.1. Here, the space $(C, \mathcal{C})$ is $(\mathbb{R}_+, \mathcal{B}_{\mathbb{R}_+})$, the number of jumps up to time t is a Poisson random variable $\mathcal{P}_t$ with parameter $\Lambda_t = \int_0^t \lambda_\tau d\tau$, and the jump sizes are i.i.d. arithmetic random variables with p.m.f. $\{q_n(t), n \geq 1\}$. Obviously, the current model is much more general.

Note from [10.35] that:

$$\mathbb{P}(N_D = 0) = \mathbb{E}(0^{N_D}) = e^{-\int_D \mu(d\tau)} = e^{-\mu(D)}. \quad [10.37]$$

After inserting [10.36], the p.g.f. of N_D is such that:

$$\frac{L_D(s)}{\mathbb{P}(N_D = 0)} = \exp\Big\{\int_D f_\tau(s)\,\mu(d\tau)\Big\}$$
$$= \exp\Big\{\sum_{n=1}^{\infty} \Lambda_n(D)\, s^n\Big\}, \quad [10.38]$$

where $\Lambda_n(D)$ are the measures:

$$\Lambda_n(D) = \int_D q_n(\tau)\,\mu(d\tau), \quad [10.39]$$

which are absolutely continuous with respect to μ and of sum $\mu(D)$. Thus, the distribution of N_D is determined by the set $\Lambda(D)$ of non-negative reals:

$$\Lambda(D) = \{\Lambda_n(D), n \geq 1\}. \quad [10.40]$$

10.2.1. *New framework*

We will consider a completely different framework than the one of section 10.1.1.

The set $\mathfrak{D}$: this time, we choose to work with $\mathfrak{D} = \mathbb{R}^{\mathbb{N}}$, so an element $\mathbf{x}$ of $\mathfrak{D}$ is a sequence of reals. An addition operator $\mathbf{x} + \mathbf{y}$ is defined as usual by the addition of each component of the two sequences of reals.

The family $\mathcal{E}$: in several of the previous cases, we used the family $\mathcal{E} = \{e_n(\mathbf{x}) = x_1 \ldots x_n/n!, n \in \mathbb{N}_0\}$ on $\mathbb{R}^{\mathbb{N}}$. In the present case, we introduce a family $\mathcal{E}$ of functions $e_n(\mathbf{x}) : \mathfrak{D} \to \mathbb{R}$, $n \in \mathbb{N}_0$, through their generating function $F_{\mathbf{x}}(s)$ given below.

DEFINITION 10.3.– *For all* $\mathbf{x} \in \mathfrak{D}$*:*

$$F_{\mathbf{x}}(s) \equiv \sum_{n=0}^{\infty} e_n(\mathbf{x})\, s^n = \prod_{j\geq 1} e^{x_j\, s^j} = e^{\sum_{j\geq 1} x_j\, s^j}. \quad [10.41]$$

Explicitly, we get from the definition [10.41]:

$$e_n(\mathbf{x}) = \sum_{\substack{i_1, i_2, \ldots, i_n \in \{0,\ldots,n\},\\ i_1+2i_2+\ldots+ni_n=n}} \frac{x_1^{i_1}}{i_1!}\frac{x_2^{i_2}}{i_2!} \cdots \frac{x_n^{i_n}}{i_n!}, \quad n \geq 0, \quad [10.42]$$

which is a variant of the Bell polynomials. In terms of convolutions, since $(\sum_{j\geq 1} x_j\, s^j)^k = \sum_{n\geq k} x_n^{*k}\, s^n$, we also obtain from [10.41]:

$$e_n(\mathbf{x}) = \sum_{k=0}^{n} \frac{x_n^{*k}}{k!}, \quad n \geq 0. \quad [10.43]$$

We observe that each e_n depends only on $x_1, \ldots, x_n$, hence $e_n(\mathbf{x}) = e_n(x_1, \ldots, x_n)$. Moreover, it is a polynomial in x_1 of degree n, in x_2 of degree $[n/2]$, …, and in x_n of degree 1. Of course:

$$e_0(\mathbf{x}) = 1, \quad e_n(\mathbf{0}) = \delta_{n,0}, \quad [10.44]$$

that is, the family $\mathcal{E}$ is standard at $\mathbf{0}$.

PROPOSITION 10.4.– *Here are some additional simple properties.*

a) $\mathcal{E}$ *is linearly independent.*

b) $\mathcal{E}$ *satisfies the binomial identity, that is:*

$$e_n(\mathbf{x}+\mathbf{y}) = \sum_{i=0}^{n} e_i(\mathbf{x})\, e_{n-i}(\mathbf{y}), \quad \mathbf{x}, \mathbf{y} \in \mathfrak{D}. \quad [10.45]$$

c) The operators Δ^j *correspond to the partial derivatives* $\partial/\partial x_j$*, that is:*

$$\frac{\partial}{\partial x_j} e_n(\mathbf{x}) = e_{n-j}(\mathbf{x}), \quad j = 1, \ldots, n. \quad [10.46]$$

d) Any $e_n(\mathbf{x})$*,* $n \geq 1$*, admits the expansions:*

$$e_n(x_1, x_2, \ldots, x_n) = \sum_{j=0}^{n} \frac{x_1^j}{j!} e_{n-j}(0, x_2, \ldots, x_n),$$

$$e_n(x_1, \ldots, x_{n-1}, x_n) = x_n + e_n(x_1, \ldots, x_{n-1}, 0). \quad [10.47]$$

Proof. The assertion (a) follows from the particular polynomial form of the e_n. As for (b), it directly results from [10.41] that yields the relation:

$$F_{\mathbf{x}+\mathbf{y}}(s) = F_{\mathbf{x}}(s)\, F_{\mathbf{y}}(s).$$

Furthermore, (c) comes from

$$\frac{\partial}{\partial x_j} F_{\mathbf{x}}(s) = F_{\mathbf{x}}(s)\, s^j,$$

giving the identity [10.46]. Finally, we have:

$$\begin{aligned} F_{(x_1,x_2,\ldots,x_n)}(s) &= e^{x_1 s}\left[F_{(x_1,x_2,\ldots,x_n)}(s)\right]_{x_1=0} \\ &= \Big(\sum_{j=0}^{\infty} \frac{x_1^j}{j!}\, s^j\Big)\Big(\sum_{r=0}^{\infty} e_r(0,x_2,\ldots,x_n)\, s^r\Big), \end{aligned}$$

hence the first expansion in [10.47]. Similarly:

$$\begin{aligned} F_{(x_1,\ldots,x_{n-1},x_n)} &= e^{x_n s^n}\left[F_{(x_1,\ldots,x_{n-1},x_n)}(s)\right]_{x_n=0} \\ &= \Big(\sum_{j=0}^{\infty} \frac{x_n^j}{j!}\, s^{nj}\Big)\Big(\sum_{r=0}^{\infty} e_r(x_1,\ldots,x_{n-1},0)\, s^r\Big), \end{aligned}$$

hence the second expansion. □

The space $\mathfrak{F}$: by the property (**a**), the family $\mathcal{E}$ can serve as a basis for a vector space $\mathfrak{F}$. Let us emphasize that unlike in section 10.1.1, this time we have a linearly independent family that allows us to always generate $\mathfrak{F}$.

The operator Δ: the property (**c**) states that Δ is the partial derivative $\partial/\partial x_1$ and, more generally, Δ^j for $1 \le j \le n$ corresponds to $\partial/\partial x_j$.

Given any $\mathbf{a} \in \mathfrak{D}$, a shift operator $S^{\mathbf{a}}$ is defined as usual by writing:

$$S^{\mathbf{a}} e_n(\mathbf{x}) = e_n(\mathbf{x} + \mathbf{a}), \quad n \in \mathbb{N}_0.$$

This is indeed an element of $\mathfrak{F}$. To verify it, simply apply the multivariate Taylor theorem to $e_n(\mathbf{x}+\mathbf{a})$, and then use [10.46] that gives $\partial e_n(\mathbf{x})/\partial x_j = e_{n-j}(\mathbf{x})$, $\partial^2 e_n(\mathbf{x})/\partial x_j \partial x_k = e_{n-j-k}(\mathbf{x})$, ..., so that $e_n(\mathbf{x}+\mathbf{a})$ is a linear combination of $e_r(\mathbf{x})$ for $r = 0, \ldots, n$, as desired.

Moreover, Δ and $S^{\mathbf{a}}$ commute since:

$$\begin{aligned} \Delta S^{\mathbf{a}} F_{\mathbf{x}}(s) &= \Delta F_{\mathbf{x}+\mathbf{a}}(s) = s F_{\mathbf{x}+\mathbf{a}}(s) \\ &= S^{\mathbf{a}}\, s F_{\mathbf{x}}(s) = S^{\mathbf{a}}\, \Delta F_{\mathbf{x}}(s). \end{aligned}$$

10.2.2. A-G pseudopolynomials

Recall that $\Lambda(D)$, for any $D \in \mathcal{C}$, is the set of measures defined by [10.40]; it corresponds to a particular element x of $\mathfrak{D}$. We will consider the family of polynomials $\mathcal{E}$ that is generated through the function $F_{\Lambda(D)}(s)$ given by [10.41]. Thus:

$$F_{\Lambda(D)}(s) = \sum_{n=0}^{\infty} e_n[\Lambda(D)]\, s^n = e^{\sum_{j\geq 1} \Lambda_j(D)\, s^j}, \tag{10.48}$$

so that each $e_n[\Lambda(D)]$ is given by [10.42] or [10.43], and:

$$\mathcal{E} = \{e_n[\Lambda(D)] \equiv e_n[\Lambda_1(D), \ldots, \Lambda_n(D)],\, n \in \mathbb{N}_0\}. \tag{10.49}$$

The associated vectorial space is denoted $\mathfrak{F}$.

By arguing as in section 10.2, we can define A–G pseudopolynomials and obtain A–G and Taylor-type expansions. We begin by adapting the definitions [2.20] and [2.14].

DEFINITION 10.4.– *Any countable (or finite) family $W = \{\Lambda(D_i),\, i \in \mathbb{N}_0\}$ of elements of $\mathfrak{D}$ is associated with the family $\mathcal{G}(W)$ of A–G pseudopolynomials defined by:*

$$\begin{cases} \bar{G}_0(\ |W) = e_0, \\ \bar{G}_n(\ |W) = I_{\Lambda(D_0)}\, I_{\Lambda(D_1)} \cdots I_{\Lambda(D_{n-1})}\, e_0(\ ), \quad n \geq 1, \end{cases} \tag{10.50}$$

where $I_{\Lambda(D)} : \mathcal{E} \to \mathfrak{F}$ is the integral operator given by:

$$I_{\Lambda(D)} e_n(\ ) = e_{n+1}(\ ) - e_{n+1}[\Lambda(D)]\, e_0(\ ), \quad n \in \mathbb{N}_0. \tag{10.51}$$

Proposition 2.8 and formula [2.31], inter alia, remain valid; the proofs are similar and omitted.

PROPOSITION 10.5.– *For all $n \geq 0$:*

$$\Delta^m \bar{G}_n(\ |W) = \bar{G}_{n-m}(\ |E^m W), \quad 0 \leq m \leq n, \tag{10.52}$$

where $E^m W = \{\Lambda(D_{m+i}),\ i \in \mathbb{N}_0\}$, *and:*

$$\Delta^m \bar{G}_n[\Lambda(D_m)|W] = \delta_{m,n}, \quad 0 \le m \le n, \tag{10.53}$$

this property characterizing the family $\mathcal{G}(W)$.

The A–G type expansion [2.29] is also easily generalized. This gives, in particular, a recursive method to calculate $\bar{G}_n(\ |W),\ n \ge 1$, from $\bar{G}_0(\ |W) = 1$.

PROPOSITION 10.6.– *Any pseudopolynomial* $f_n \in \mathfrak{F}$ *can be expressed as:*

$$f_n(\ ) = \sum_{i=0}^{n} \Delta^i f_n[\Lambda(D_i)]\, \bar{G}_i(\ |W). \tag{10.54}$$

So, taking $f_n = e_n$ *in [10.54] gives the recursions:*

$$\bar{G}_n(\ |W) = e_n(\ ) - \sum_{i=0}^{n-1} e_{n-i}[\Lambda(D_i)]\, \bar{G}_i(\ |W). \tag{10.55}$$

Moreover, a Taylor-type formula applies to any pseudopolynomial f_n as follows.

PROPOSITION 10.7.– *Given any two disjoint sets* D_0 *and* D_1*:*

$$\begin{aligned} f_n[\Lambda(D_0 \cup D_1)] &= f_n[\Lambda(D_0) + \Lambda(D_1)] \\ &= \sum_{i=0}^{n} \Delta^i f_n[\Lambda(D_0)]\, e_i[\Lambda(D_1)]. \end{aligned} \tag{10.56}$$

Taking $D_0 = \emptyset$ *in [10.56] provides a Maclaurin-type expansion with* $\Lambda(\emptyset) = \mathbf{0}$.

10.2.3. *Revised first meeting concept*

Let us return to the compound Poisson random field under study. We proved before that the p.g.f. $L_D(s)$ of N_D is provided by [10.38]. Using the family $\mathcal{E}$ and formula [10.48], we can rewrite [10.38] as:

$$\frac{L_D(s)}{\mathbb{P}(N_D = 0)} = \sum_{n=0}^{\infty} e_n[\Lambda(D)]\, s^n. \tag{10.57}$$

From [10.57], the p.m.f. of N_D is thus given by:

$$\mathbb{P}(N_D = n) = \mathbb{P}(N_D = 0)\, e_n[\Lambda(D)], \quad n \geq 0. \qquad [10.58]$$

For clarity, we now prefer to denote the variable N_D instead by $N(D)$. The first meeting of the random field $\{N(D),\ D \in \mathcal{C}\}$ with a lower boundary will be reformulated as follows.

DEFINITION 10.5.– *Let $\{D_i,\ i \in \mathbb{N}_0\}$ be any fixed sequence of pairwise disjoint measurable sets of $\mathcal{C}$. The first meeting level with these sets is the random variable K satisfying:*

$$K \textit{ is the first index } n \textit{ such that } N\big(\cup_{i=0}^{n} D_i\big) = n. \qquad [10.59]$$

We first check that the variable K is well defined by [10.59]. Indeed, $N(D_0) \geq 0$, thus:

– if $N(D_0) = 0$, then $K = 0$;

– if $N\big(\cup_{i=0}^{n} D_i\big) > n$ for all $n \geq 0$, then $K = \infty$;

– if there exists $n \geq 1$ such that $N\big(\cup_{i=0}^{j} D_i\big) > j$ for $j = 0, \ldots, n-1$ and $N\big(\cup_{i=0}^{n} D_i\big) \leq n$, then we have:

$$n > N\big(\cup_{i=0}^{n-1} D_i\big) + N(D_n) > n - 1 + N(D_n),$$

which implies $N(D_n) < 1$, that is, is equal to 0, so:

$$N\big(\cup_{i=0}^{n} D_i\big) = N\big(\cup_{i=0}^{n-1} D_i\big) = n, \text{ hence } K = n.$$

In other words, [10.59] states that:

$$\{K = n\} \text{ means } \Big\{N\big(\cup_{i=0}^{n} D_i\big) = n\Big\} \bigcap_{j=0}^{n-1} \Big\{N\big(\cup_{i=0}^{j} D_i\big) > j\Big\}. \qquad [10.60]$$

We emphasize that the D_i can be arranged in any given order. In particular, if $C = \mathbb{R}$, the order has nothing to do with the natural order on $\mathbb{R}$. Of course, the choice of this natural order is also allowed, which leads, for example, to the intervals:

$$D_0 = (x, v_0], D_1 = (v_0, v_1], D_2 = (v_1, v_2], \ldots, \quad \text{for } x \leq v_0 \leq v_1 \leq v_2, \ldots,$$

thus giving a lower boundary function as before. This time, however, the Borel sets D_i are no longer necessarily intervals, they can be taken far from each other and especially put in any order. Their only constraint is that they must be two by two disjoint.

Let $\mathcal{E}$ be the family [10.49] of multivariate polynomials $e_n(\ )$, $n \geq 0$, given by [10.42] or [10.43]. Consider the family $\mathcal{G}(W)$ of A–G pseudopolynomials associated with a family W of elements in $\mathfrak{D}$ by the integrals [10.50] with [10.51]. The p.m.f. of the first meeting level K is provided by the formula [10.61] in which W is defined by [10.62].

PROPOSITION 10.8.–

$$\mathbb{P}\big(K = n\,|\{D_i,\, i \in \mathbb{N}_0\}\big) = \mathrm{e}^{-\mu(\cup_{i=0}^{n} D_i)}\, \bar{G}_n\big[\Lambda(D_0)\,|W\big], \quad n \geq 0, \qquad [10.61]$$

where the family W is defined as:

$$W = \Big\{ -\sum_{j=1}^{i} \Lambda(D_j),\, i \in \mathbb{N}_0 \Big\}. \qquad [10.62]$$

Proof. We can write:

$$\mathbb{P}(K = n) = \sum_{k=0}^{n} \mathbb{P}[N(D_0) = k]\, \mathbb{P}[K = n|N(D_0) = k],$$

and of course:

$$\mathbb{P}[N(\cup_{i=0}^{n} D_i) = 0] = \mathbb{P}[N(D_0) = 0]\, \mathbb{P}[N(\cup_{i=1}^{n} D_i) = 0],$$

so that after division and using [10.58]:

$$\frac{\mathbb{P}(K = n)}{\mathbb{P}[N(\cup_{i=0}^{n} D_i) = 0]} = \sum_{k=0}^{n} e_k[\Lambda(D_0)]\, \frac{\mathbb{P}[K = n|N(D_0) = k]}{\mathbb{P}[N(\cup_{i=1}^{n} D_i) = 0]}. \qquad [10.63]$$

To make the similarity of the two members of [10.63] evident, we introduce a specific notation by setting, for any $k \geq 0$:

$$\hat{D}_0(k) = \cup_{i=1}^{k} D_i, \quad \text{and} \quad \hat{D}_r(k) = D_{k+r},\ r \geq 1, \qquad [10.64]$$

thus $\hat{D}_0(0) = \emptyset$ and $\hat{D}_r(0) = D_r$, $r \geq 1$, and for $j \geq k$:

$$\cup_{i=1}^{j} D_i = \hat{D}_0(k) \cup_{i=k+1}^{j} D_i = \hat{D}_0(k) \cup_{r=1}^{j-k} \hat{D}_r(k) = \cup_{r=0}^{j-k} \hat{D}_r(k). \quad [10.65]$$

Using [10.60] then [10.64] along with [10.65], the conditional probability $\mathbb{P}[K = n|N(D_0) = k]$ becomes:

$$\begin{aligned}&\mathbb{P}\big[N\big(\cup_{i=0}^{n} D_i\big) = n,\ N\big(\cup_{i=0}^{j} D_i\big)\\&> j \text{ for } j = 0, \ldots, n-1|N(D_0) = k\big]\\&= \mathbb{P}\big[N\big(\cup_{i=0}^{n-k} \hat{D}_i(k)\big) = n-k,\ N\big(\cup_{i=0}^{j} \hat{D}_i(k)\big)\\&> j \text{ for } j = 0, \ldots, n-k-1\big].\end{aligned} \quad [10.66]$$

Denote the l.h.s. of [10.63] by $B_n(D_0, D_1, \ldots)$, that is:

$$B_n\big(\{D_i,\ i \in \mathbb{N}_0\}\big) = \frac{\mathbb{P}(K = n)}{\mathbb{P}[N(\cup_{i=0}^{n} D_i) = 0]}, \quad n \geq 0. \quad [10.67]$$

Because of [10.66], we then see that [10.63] takes the simple form:

$$B_n\big(\{D_i,\ i \in \mathbb{N}_0\}\big) = \sum_{k=0}^{n} e_k[\Lambda(D_0)]\, B_{n-k}\big(\{\hat{D}_i(k),\ i \in \mathbb{N}_0\}\big), \quad n \geq 1. \quad [10.68]$$

Moreover, from the definition [10.67]:

$$B_0(\{D_i,\ i \in \mathbb{N}_0\}) = \mathbb{P}(K = 0)/\mathbb{P}[N(D_0) = 0] = 1, \quad [10.69]$$

so that [10.68] with [10.69] determines recursively B_n, $n \geq 0$.

We will identify B_n with the A–G pseudopolynomial $\bar{G}_n$ defined in [10.61] and [10.62]. Indeed, the Taylor-type formula [10.56] yields:

$$\bar{G}_n[\Lambda(D_0)\,|W] = \sum_{k=0}^{n} \bar{G}_{n-k}[\Lambda(\emptyset)\,|E^k W]\, e_k[\Lambda(D_0)], \quad [10.70]$$

where $\Lambda(\emptyset) = \mathbf{0}$. By the definition [10.62] of W, each $\bar{G}_{n-k}$ on the r.h.s. of [10.70] is written explicitly as:

$$\bar{G}_{n-k}\big[\mathbf{0}\,|E^k W\big] = \bar{G}_{n-k}\Big[\mathbf{0}\,|\{-\sum_{j=1}^{k+i} \Lambda(D_j),\ i \geq 0\}\Big],$$

and using the formula [2.50] and the change of notation [10.64], we obtain:

$$\bar{G}_{n-k}\big[\mathbf{0}\,|E^kW\big] = \bar{G}_{n-k}\Big[\sum_{j=1}^{k}\Lambda(D_j)\,|\{-\sum_{j=k+1}^{k+i}\Lambda(D_j),\, i\geq 0\}\Big]$$

$$= \bar{G}_{n-k}\Big[\Lambda\big(\hat{D}_0(k)\big)\,|\{-\sum_{j=1}^{i}\Lambda(\hat{D}_j(k)),\, i\geq 0\}\Big]$$

$$=\equiv \bar{G}_{n-k}\big[\Lambda\big(\hat{D}_0(k)\big)\,|\hat{W}(k)\big], \qquad [10.71]$$

where $\hat{W}(k)$ corresponds to W after substituting $\Lambda\big(\hat{D}_i(k)\big)$ for $\Lambda(D_i)$. Therefore, the expansion [10.70] becomes:

$$\bar{G}_n[\Lambda(D_0)\,|W] = \sum_{k=0}^{n}\bar{G}_{n-k}\big[\Lambda\big(\hat{D}_0(k)\big)\,|\hat{W}(k)\big]\, e_k[\Lambda(D_0)]. \qquad [10.72]$$

Observe that the structure of [10.72] is exactly the same as that of [10.68]. So, we can perform the announced identification, and [10.67] finally leads to the result [10.61]. □

COROLLARY 10.1.– Let $(C, \mathcal{C})$ be $(\mathbb{R}_+, \mathcal{B}_{\mathbb{R}_+})$, and consider the intervals that follow each other (without overlapping) $D_0 = (x, v_0], D_1 = (v_0, v_1]$ and $D_2 = (v_1, v_2], \ldots$ where $x \leq v_0 \leq v_1 \leq v_2, \ldots$.

Note that $\cup_{j=1}^{i} D_j$ corresponds to the interval $(v_0, v_i]$. Thus, the results [10.61] and [10.62] give here, for all $n \geq 0$:

$$\mathbb{P}\big(K = n\,|\{D_i,\, i\geq 0\}\big) = e^{-\mu(\cup_{j=0}^{n} D_j)}$$
$$\times\, \bar{G}_n\big[\Lambda\big(D_0\big)\,|\{-\Lambda\big(\cup_{j=1}^{i} D_j)\big),\, i\geq 1\}\big]. \qquad [10.73]$$

Homogeneous case: Suppose that $\mu(D) = \lambda V(D)$ (where $V(D)$ means the volume of D) and $q_n(\tau) = q_n$, $n \geq 1$, independently of τ. Therefore, $\Lambda_n(D) = \mu(D)\, q_n = \lambda V(D)\, q_n$, and:

$$\Lambda(D) = \{\Lambda_n(D) = \lambda V(D)\, q_n,\, n \geq 1\}.$$

The polynomial family $\mathcal{E}$ generated by $F_{\Lambda(D)}(s)$ (see [10.48]) then simplifies as:

$$F_{\Lambda(D)}(s) = \sum_{n=0}^{\infty} e_n[\Lambda(D)]\, s^n = \exp\Big\{\sum_{j=1}^{\infty} \Lambda_j(D) s^j\Big\}$$

$$= \sum_{k=0}^{\infty} \frac{[\lambda V(D)]^k}{k!} \Big(\sum_{j=1}^{\infty} q_j s^j\Big)^k = \sum_{k=0}^{\infty} \frac{[\lambda V(D)]^k}{k!} \Big(\sum_{n=k}^{\infty} q_n^{*k} s^n\Big)$$

$$= \sum_{n=0}^{\infty} s^n \Big(\sum_{k=0}^{n} \frac{[\lambda V(D)]^k}{k!}\, q_n^{*k}\Big),$$

which implies that:

$$e_n[\Lambda(D)] = \sum_{k=0}^{n} \frac{[\lambda V(D)]^k}{k!}\, q_n^{*k}. \qquad [10.74]$$

Therefore, $e_n[\Lambda(D)]$ has a form similar to that encountered in section 7.2 for a compound Poisson process.

When $C = \mathbb{R}_+$ and D is an interval $(0, t]$, then $\lambda V(D) = \lambda t$ and [10.74] reduces to $e_n(t)$ given by [7.17]. Consequently, the $\bar{G}_n$ appearing in [10.73] are identical to the A–G pseudopolynomials derived in section 7.2.

References

Monographs by Gontcharoff

The two monographs by Gontcharoff on the (so-called) A–G polynomials constitute the starting point of our research on the more general family of A–G pseudopolynomials.

Gontcharoff, W. (1930). Recherches sur les dérivées successives des fonctions analytiques. Généralisation de la série d'Abel. *Annales Scientifiques de l'E.N.S.*, 47, 1–78.

Gontcharoff, W. (1937). *Détermination des fonctions entières par interpolation*. Hermann et Cie, Paris.

Papers on A-G polynomials

We first present below a set of works in which we used the A-G polynomial families for the study of stochastic models in applied probability. Let us recall that several articles which motivated our interest in the A–G polynomials for this field of application are cited in the footnotes of the monograph. Furthermore, several articles using A–G polynomials have been written by other (co)authors but are not cited below.

Denuit, M., Lefèvre, C., Picard, P. (2003). Polynomial structures in order statistics distributions. *Journal of Statistical Planning and Inference*, 113, 151–178.

Lefèvre, C. and Picard, P. (1990). A non-standard family of polynomials and the final size distribution of Reed-Frost epidemic processes. *Advances in Applied Probability*, 22, 25–48.

Lefèvre, C. and Picard, P. (1996). On the first-crossing of a Poisson process in a lower boundary. In *Athens Conference on Applied Probability and Time Series Analysis*. Springer, New York.

Lefèvre, C. and Picard, P. (2015). Risk models in insurance and epidemics: A bridge through randomized polynomials. *Probability in the Engineering and Informational Sciences*, 29, 399–420.

Lefèvre, C. and Picard, P. (2016). Polynomials, random walks and risk processes: A multivariate framework. *Stochastics: An International Journal of Probability and Stochastic Processes,* 88, 1147–1172.

Lefèvre, C. and Picard, P. (2017). On the outcome of epidemics with detections. *Journal of Applied Probability,* 54, 890–904.

Lefèvre, C. and Picard, P. (2023). Abel-Gontcharoff polynomials, parking trajectories and ruin probabilities. *Dependence Modeling,* 11, 20230107.

Picard, P. and Lefèvre, C. (1990). A unified analysis of the final state and severity distribution in collective Reed-Frost epidemic processes. *Advances in Applied Probability,* 22, 269–294.

Picard, P. and Lefèvre, C. (1993). Distribution of the final state and severity of epidemics with fatal risk. *Stochastic Processes and their Applications*, 48, 277–294.

Papers on A–G pseudopolynomials

We now list most of the works in which we introduced the A–G pseudopolynomial families and used them for the study of more complex stochastic models. It should be emphasized that the methodology adopted in the current monograph is more elaborate and precise than in our previous articles.

Lefèvre, C. and Picard, P. (1996). Abelian-type expansions and non-linear death processes (II). *Advances in Applied Probability*, 28, 877–894.

Lefèvre, C. and Picard, P. (1999). Abel-Gontcharoff pseudopolynomials and the final outcome of SIR epidemic models (III). *Advances in Applied Probability*, 31, 532–549.

Lefèvre, C. and Picard, P. (2001). Busy periods for some queues with deterministic interarrival or service times. In *Probability and Statistical Models with Applications* (Volume dedicated to T. Cacoullos), Charalambides, C., Koutras, M.V., Balakrishnan, N. (eds). Chapman and Hall, Boca Raton.

Lefèvre, C. and Picard, P. (2006). A nonhomogeneous risk model for insurance. *Computers and Mathematics with Applications*, 51, 325–334.

Lefèvre, C. and Picard, P. (2011). Polynomial structures in rank statistics distributions. *Journal of Statistical Planning and Inference*, 141, 1380–1393.

Lefèvre, C. and Picard, P. (2014). Appell pseudopolynomials and Erlang-type risk models. *Stochastics: An International Journal of Probability and Stochastic Processes*, 86, 676–695.

Picard, P. and Lefèvre, C. (1996). First crossing of basic counting processes with lower non-linear boundaries: A unified approach through pseudopolynomials (I). *Advances in Applied Probability*, 28, 853–876.

Picard, P. and Lefèvre, C. (1999). On the algebraic structure in Markovian processes of death and epidemic types. *Advances in Applied Probability*, 31, 742–757.

Picard, P. and Lefèvre, C. (2003). On the first meeting or crossing of two independent trajectories for some counting processes. *Stochastic Processes and their Applications*, 104, 217–242.

Index

A

B

C

D, E

F, G

I, L

Due to the specific nature of this monograph, it was very difficult to follow the traditional form of an index. Only a few keywords are mentioned in the index list. We advise the reader to refer to the table of contents.

M

N

P

R

S, T

Other titles from

iSTE

in

Mathematics and Statistics

2025

DUPUY Jean-François
Generalized Linear Models: Problems with Censored, Missing, and Zero-inflated Data

GAIVORONSKI Alexei A., KNOPOV Pavel S., NORKIN Vladimir I., ZASLAVSKYI Volodymyr A.
Stochastic Modeling and Optimization Methods for Critical Infrastructure Protection 1: Stochastic Modeling
Stochastic Modeling and Optimization Methods for Critical Infrastructure Protection 2: Methods and Tools

2024

BOURLÈS Henri
System Theory – A Modern Approach 1: Linear Ordinary and Functional Differential Equations
(New Mathematical Methods, Systems and Applications Set – Volume 1)

MAKAROV Volodymyr, MAYKO Nataliya
Traditional Functional-Discrete Methods for the Problems of Mathematical Physics: New Aspects

2023

AIT BEN HASSI El Mustapha
The Theory of Distributions: Introduction

KOROLIOUK Dmitri, SAMOILENKO Igor
Asymptotic and Analytic Methods in Stochastic Evolutionary Systems

PARROCHIA Daniel
Graphs, Orders, Infinites and Philosophy

2022

CHAKRAVARTHY Srinivas R.
Introduction to Matrix-Analytic Methods in Queues 1: Analytical and Simulation Approach – Basics
Introduction to Matrix-Analytic Methods in Queues 2: Analytical and Simulation Approach – Queues and Simulation

LESFARI Ahmed
Integrable Systems

RADCHENKI Vadym M.
General Stochastic Measures: Integration, Path Properties and Equations

DE SAPORTA Benoîte, ZILI Mounir
Martingales and Financial Mathematics in Discrete Time

SIMON Jacques
Distributions
(Analysis for PDEs Set – Volume 3)

2021

KOROLIOUK Dmitri, SAMOILENKO Igor
Random Evolutionary Systems: Asymptotic Properties and Large Deviations

MOKLYACHUK Mikhail
Convex Optimization: Introductory Course

POGORUI Anatoliy, SWISHCHUK Anatoliy, RODRÍGUEZ-DAGNINO Ramón M.
Random Motions in Markov and Semi-Markov Random Environments 1: Homogeneous Random Motions and their Applications
Random Motions in Markov and Semi-Markov Random Environments 2: High-dimensional Random Motions and Financial Applications

PROVENZI Edoardo
From Euclidean to Hilbert Spaces: Introduction to Functional Analysis and its Applications

2020

BARBU Vlad Stefan, VERGNE Nicolas
Statistical Topics and Stochastic Models for Dependent Data with Applications

CHABANYUK Yaroslav, NIKITIN Anatolii, KHIMKA Uliana
Asymptotic Analyses for Complex Evolutionary Systems with Markov and Semi-Markov Switching Using Approximation Schemes

KOROLIOUK Dmitri
Dynamics of Statistical Experiments

MANOU-ABI Solym Mawaki, DABO-NIANG Sophie, SALONE Jean-Jacques
Mathematical Modeling of Random and Deterministic Phenomena

SIMON Jacques
Continuous Functions
(Analysis for PDEs Set – Volume 23)

2019

BANNA Oksana, MISHURA Yuliya, RALCHENKO Kostiantyn, SHKLYAR Sergiy
Fractional Brownian Motion: Approximations and Projections

GANA Kamel, BROC Guillaume
Structural Equation Modeling with lavaan

KUKUSH Alexander
Gaussian Measures in Hilbert Space: Construction and Properties

LUZ Maksym, MOKLYACHUK Mikhail
Estimation of Stochastic Processes with Stationary Increments and Cointegrated Sequences

MICHELITSCH Thomas, PÉREZ RIASCOS Alejandro, COLLET Bernard, NOWAKOWSKI Andrzej, NICOLLEAU Franck
Fractional Dynamics on Networks and Lattices

VOTSI Irene, LIMNIOS Nikolaos, PAPADIMITRIOU Eleftheria, TSAKLIDIS George
Earthquake Statistical Analysis through Multi-state Modeling
(Statistical Methods for Earthquakes Set – Volume 2)

2018

AZAÏS Romain, BOUGUET Florian
Statistical Inference for Piecewise-deterministic Markov Processes

IBRAHIMI Mohammed
Mergers & Acquisitions: Theory, Strategy, Finance

PARROCHIA Daniel
Mathematics and Philosophy

2017

CARONI Chysseis
First Hitting Time Regression Models: Lifetime Data Analysis Based on Underlying Stochastic Processes
(Mathematical Models and Methods in Reliability Set – Volume 4)

CELANT Giorgio, BRONIATOWSKI Michel
Interpolation and Extrapolation Optimal Designs 2: Finite Dimensional General Models

CONSOLE Rodolfo, MURRU Maura, FALCONE Giuseppe
Earthquake Occurrence: Short- and Long-term Models and their Validation
(Statistical Methods for Earthquakes Set – Volume 1)

D'AMICO Guglielmo, DI BIASE Giuseppe, JANSSEN Jacques, MANCA Raimondo
Semi-Markov Migration Models for Credit Risk
(Stochastic Models for Insurance Set – Volume 1)

GONZÁLEZ VELASCO Miguel, del PUERTO GARCÍA Inés, YANEV George P.
Controlled Branching Processes
(Branching Processes, Branching Random Walks and Branching Particle Fields Set – Volume 2)

HARLAMOV Boris
Stochastic Analysis of Risk and Management
(Stochastic Models in Survival Analysis and Reliability Set – Volume 2)

KERSTING Götz, VATUTIN Vladimir
Discrete Time Branching Processes in Random Environment
(Branching Processes, Branching Random Walks and Branching Particle Fields Set – Volume 1)

MISHURA Yuliya, SHEVCHENKO Georgiy
Theory and Statistical Applications of Stochastic Processes

NIKULIN Mikhail, CHIMITOVA Ekaterina
Chi-squared Goodness-of-fit Tests for Censored Data
(Stochastic Models in Survival Analysis and Reliability Set – Volume 3)

SIMON Jacques
Banach, Fréchet, Hilbert and Neumann Spaces
(Analysis for PDEs Set – Volume 1)

2016

CELANT Giorgio, BRONIATOWSKI Michel
Interpolation and Extrapolation Optimal Designs 1: Polynomial Regression and Approximation Theory

CHIASSERINI Carla Fabiana, GRIBAUDO Marco, MANINI Daniele
Analytical Modeling of Wireless Communication Systems
(Stochastic Models in Computer Science and Telecommunication Networks Set – Volume 1)

GOUDON Thierry
Mathematics for Modeling and Scientific Computing

KAHLE Waltraud, MERCIER Sophie, PAROISSIN Christian
Degradation Processes in Reliability
(Mathematial Models and Methods in Reliability Set – Volume 3)

KERN Michel
Numerical Methods for Inverse Problems

RYKOV Vladimir
Reliability of Engineering Systems and Technological Risks
(Stochastic Models in Survival Analysis and Reliability Set – Volume 1)

2015

DEVOLDER Pierre, JANSSEN Jacques, MANCA Raimondo
Basic Stochastic Processes

LE GAT Yves
Recurrent Event Modeling Based on the Yule Process
(Mathematical Models and Methods in Reliability Set – Volume 2)

DE SAPORTA Benoîte, DUFOUR François, ZHANG Huilong
Numerical Methods for Simulation and Optimization of Piecewise Deterministic Markov Processes

2014

COOKE Roger M., NIEBOER Daan, MISIEWICZ Jolanta
Fat-tailed Distributions: Data, Diagnostics and Dependence
(Mathematical Models and Methods in Reliability Set – Volume 1)

MACKEVIČIUS Vigirdas
Integral and Measure: From Rather Simple to Rather Complex

PASCHOS Vangelis Th
Combinatorial Optimization – 3-volume series – 2nd edition
Concepts of Combinatorial Optimization / Concepts and Fundamentals – Volume 1
Paradigms of Combinatorial Optimization – Volume 2
Applications of Combinatorial Optimization – Volume 3

2013

COUALLIER Vincent, GERVILLE-RÉACHE Léo, HUBER Catherine, LIMNIOS Nikolaos, MESBAH Mounir
Statistical Models and Methods for Reliability and Survival Analysis

JANSSEN Jacques, MANCA Oronzio, MANCA Raimondo
Applied Diffusion Processes from Engineering to Finance

SERICOLA Bruno
Markov Chains: Theory, Algorithms and Applications

2012

BOSQ Denis
Mathematical Statistics and Stochastic Processes

CHRISTENSEN Karl Bang, KREINER Svend, MESBAH Mounir
Rasch Models in Health

DECREUSEFOND Laurent, MOYAL Pascal
Stochastic Modeling and Analysis of Telecom Networks

DEVOLDER Pierre, JANSSEN Jacques, MANCA Raimondo
Stochastic Methods for Pension Funds

2011

MACKEVIČIUS Vigirdas
Introduction to Stochastic Analysis: Integrals and Differential Equations

MAHJOUB Ridha
Recent Progress in Combinatorial Optimization – ISCO2010

RAYNAUD Hervé, ARROW Kenneth
Managerial Logic

2010

BAGDONAVIČIUS Vilijandas, KRUOPIS Julius, NIKULIN Mikhail
Nonparametric Tests for Censored Data

IOSIFESCU Marius, LIMNIOS Nikoloas, OPRISAN Gheorghe
Introduction to Stochastic Models

VASSILIOU P.C.G.
Discrete-time Asset Pricing Models in Applied Stochastic Finance

2008

ANISIMOV Vladimir
Switching Processes in Queuing Models

FICHE Georges, HÉBUTERNE Gérard
Mathematics for Engineers

HUBER Catherine, LIMNIOS Nikolaos, MESBAH Mounir, NIKULIN Mikhail
Mathematical Methods in Survival Analysis, Reliability and Quality of Life

JANSSEN Jacques, MANCA Raimondo, VOLPE Ernesto
Mathematical Finance

2007

HARLAMOV Boris
Continuous Semi-Markov Processes

2006

CLERC Maurice
Particle Swarm Optimization

Printed and bound by CPI Group (UK) Ltd, Croydon, CR0 4YY

24/08/2026

14947835-0004